Methods and Applications of LINEAR PROGRAMMING

LEON COOPER

Associate Dean and Director of the Graduate Division, Southern Methodist University Institute of Technology, Dallas, Texas

DAVID STEINBERG

Associate Professor, Department of Mathematical Studies, Southern Illinois University at Edwardsville, Edwardsville, Illinois

1974

W. B. SAUNDERS COMPANY Philadelphia • London • Toronto

W. B. Saunders Company: West Washington Square
Philadelphia, PA 19105

12 Dyott Street
London, WC1A 1DB

833 Oxford Street
Toronto, Ontario M8Z 5T9, Canada

Methods and Applications of Linear Programming ISBN 0-7216-2694-7

 Library of Congress catalog card number 73-89935.

Last digit is the print number: 9 8 7 6 5 4 3 2 1

Preface

Linear programming has become one of the most important and effective tools in use today in modern management analysis. Hence, a thorough study of the techniques and applications of linear programming should be an integral part of any curriculum in operations research, management science, or systems analysis.

This text deals in a rigorous but readily comprehensible way with the theory, computational procedures, and selected applications of linear programming. It may be used for either a one or two semester course at the junior, senior, or first year graduate student level, and should be of value for self-study as well. The text assumes a knowledge of basic linear algebra.

Chapters 1 and 3 contain an introduction to and a review of the necessary mathematical background. Chapter 2 presents a variety of applications of linear programming. A detailed development of the simplex method is given in Chapters 4 and 5. Duality theory and related algorithms are discussed in Chapters 8 and 9. Chapter 7 contains a general discussion of the subject of computer implementation of linear programming; and Chapters 6 and 13 provide a more detailed investigation of some of the computational techniques currently being employed in most commercially written computer codes for solving linear programming problems, such as the revised simplex method, the composite simplex method, decomposition, upper bounding and generalized upper bounding techniques. The remaining chapters provide a wide variety of special topics, including the transportation problem (Chapter 11), post-optimal analysis (Chapter 12), and network flow problems (Chapter 14).

An introduction to integer programming is included, in Chapters 15 and 16. While integer programming problems are not, strictly speaking, linear, the great importance of such problems, and the extensive use of linear programming methods in their solution, warrant this inclusion.

A standard one semester introductory course in linear programming should minimally consist of Chapters 1 through 5, 8, and 10. Selections from Chapters 6, 7, 9, and 12 may also be included, at the discretion of the instructor.

*   *   *   *   *   *

We wish to express our utmost appreciation of and admiration for the professional skills (and patience) of Mrs. Sharon Schaefer and Miss Marilyn Vaughan, who did such an outstanding job of typing the manuscript.

LEON COOPER

DAVID I. STEINBERG

Contents

1

Introduction

1-1 OPTIMIZATION PROBLEMS AND LINEAR PROGRAMMING PROBLEMS

By the term "optimization problem" we mean the problem of finding the greatest possible numerical value (maximization) or least possible value (minimization) of some numerical or symbolic mathematical function of any number of independent variables. In addition, there may be constraints on the variables, in the form of mathematical equations or inequalities, e.g., algebraic inequalities or differential equations. Hence, in an optimization problem we seek values of the aforementioned variables which do not violate the several constraints imposed on them, but which lead to an optimal (maximal or minimal) value of the function which is to be optimized.

The earliest optimization methods, usually called "classical optimization methods," led to the use of differential calculus and later to the development and use of the calculus of variations. These have been used particularly in the physical sciences and in the various fields of engineering and applied science, and also more recently in economic theory. For an overview of classical optimization methods the reader is referred to Cooper and Steinberg,[1] and Wilde and Beightler.[2] For an introductory treatment of the calculus of variations one can consult Hildebrand.[3]

In this book we shall restrict our attention to what has become in the last 25 years, the most widely used optimization method for modeling physical, economic, engineering, and business problems of all kinds. The general problem type is invariably called a *linear programming problem.* Mathematically, it is relatively easy to describe. It is as follows:

Maximize $f(x) = c_1x_1 + c_2x_2 + \dots + c_px_p$

subject to:

$$\begin{aligned} a_{11}x_1 + a_{12}x_2 + \dots + a_{1p}x_p &\{\le, =, \ge\} b_1 \\ a_{21}x_1 + a_{22}x_2 + \dots + a_{2p}x_p &\{\le, =, \ge\} b_2 \\ &\vdots \\ a_{m1}x_1 + a_{m2}x_2 + \dots + a_{mp}x_p &\{\le, =, \ge\} b_m \\ x_j \ge 0, \quad j &= 1, 2, \dots, p \end{aligned} \tag{1-1}$$

It is understood in the above mathematical statement that for each of the constraints

$$\sum_{j=1}^{p} a_{ij}x_j\{\leq, =, \geq\}b_i, \qquad i = 1, 2, \ldots, m \tag{1-2}$$

only *one* of the relations $\{\leq, =, \geq\}$ can obtain at any one time. The statement in that form indicates that one of those relations must hold for the original statement of a linear programming problem. The function $f(x) = \sum_{j=1}^{p} c_jx_j$ is called the *objective function* and it will be noted that this function is linear in the variables x_j, $j = 1, 2, \ldots, p$. Similarly, each of the constraints of (1–1), i.e., $\sum_{j=1}^{p} a_{ij}\,x_j\{\leq, =, \geq\}b_i, i = 1, 2, \ldots, m$ and also $x_j \geq 0$, $j = 1, 2, \ldots, p$ are linear expressions with respect to the variables x_j, $j = 1, 2, \ldots, p$.

The importance of the problem format expressed by the statement given in (1–1) is that a great variety of economic, scientific, industrial, business, and societal problems can be meaningfully expressed in terms of the linear model that (1–1) represents. Generally speaking, these problems deal with the optimal allocation or deployment of resources of all kinds that are limited in total amount (the constraints). In most such problems there are many possible allocations that satisfy the constraints of the problem. What is sought, however, is the one or more out of the total set of these allocations that will lead to the greatest or least value of some objective function. Usually, the objective function is some kind of expression of "profit" or "cost." For example, we might be concerned with the determination of the least cost diet to be fed to chickens, where the variables are the amounts of each possible foodstuff that could make up a standard total quantity of a chicken's daily requirement. The constraints relate to total minimal requirements of nutritive components, which will be present in varying amounts in each of the available foodstuffs. We seek the precise mixture for a given situation which will result in the least cost formulation. Numerous detailed examples of many different kinds of problem areas in which linear programming formulations have been made are given in Chapter 2.

The term "programming" in the name of our subject "linear programming" should be considered a synonym (even though it is not) for "optimization." It is often confused with the use of the word "programming" for digital computers. There is no real connection, however. The use of this word arose in the context of the formulation of logistics problems for the U.S. Air Force, where the "programming" of activities is in common parlance. Hence, when we speak of linear programming problems, we mean linear optimization problems.

The reader who is interested in a detailed historical treatment of the origins of linear programming may consult reference 4. However, some brief mention of the origin of our subject is in order.

In 1941 Hitchcock[5] formulated a problem now known as the transportation problem, which is a linear programming problem with a fixed structure in terms of the coefficients a_{ij} in (1–1). There were earlier intimations relating to linear programming and the transportation problem by Kantorovitch.[6,7] It was not until the work of Dantzig[8] that the general linear programming problem was formulated. Even more important, however, was that he developed an extremely effective numerical solution technique, the simplex method, for its solution.

1–2 AN EXAMPLE OF A LINEAR PROGRAMMING PROBLEM

A supplier of feed for poultry raisers wishes to formulate a least cost feed. Suppose, for simplicity, we assume that each chicken requires at least 250, 150, and 400 units of certain nutritive elements each day (e.g., vitamins, protein, etc.). Let us call these elements 1, 2, 3. Suppose further that there are four possible natural feed materials he can use to feed the chickens and that we know how many units of each of these nutritive elements is contained in each pound of feed material. In addition, we know the cost of a pound of each of the feed materials. Suppose the actual numbers are as given in Table 1–1.

TABLE 1–1 FEED BLENDING DATA

Nutritive Elements	Feed Materials *A*	*B*	*C*	*D*	Minimum Daily Requirement
1	20	15	10	30	250
2	25	15	20	25	150
3	25	20	20	30	400
Unit Cost	3	1	2	4	

The feed supplier would like to know how much of each of the four feed materials he should blend together in order to minimize his costs and still satisfy the minimum daily nutritional requirements. Even for a problem of this exceedingly small size (4 feed materials and 3 nutritive elements) it is not at all obvious how to determine a solution. The degree of interdependence between the variables makes a trial and error procedure either laborious or hazardous or both. Let us formulate this problem in terms of a linear programming model.

Let x_j, $j = 1, 2, 3, 4$ represent the amounts (pounds) of feed materials to be mixed together. These are the quantities we wish to determine. Furthermore, we wish to determine these x_j in such a way that the total cost of the mixture we put together is a minimum. Hence, we wish to minimize (referring to the last line of Table 1–1)

$$z = 3x_1 + x_2 + 2x_3 + 4x_4 \tag{1-3}$$

However, there are certain conditions that the x_j must satisfy. These come about because we have specified that each pound of the mixed feed material must contain a minimal amount of each of three nutritive elements. For example, each pound of mixed feed must contain 250 units of nutritive element 1. For each pound of feed material A that we use, we supply 20 units. For each pound of feed material B that we use, we supply 15 units and so forth. Hence the values of $x_j, j = 1, 2, 3, 4$ must satisfy a constraint (inequality) of the form

$$20x_1 + 15x_2 + 10x_3 + 30x_4 \geq 250 \tag{1-4}$$

We write (1–4) as an inequality because under certain circumstances it may be desirable in terms of cost minimization to supply more than the minimum amount required of one particular nutritive element. We allow for this possibility by using an inequality statement. Similarly, we have two other constraints, one for each of the requirements for nutritive elements 2 and 3. These are

$$\begin{aligned} 25x_1 + 15x_2 + 20x_3 + 25x_4 &\geq 150 \\ 25x_1 + 20x_2 + 20x_3 + 30x_4 &\geq 400 \end{aligned} \tag{1-5}$$

We have one further restriction on the variables x_j, namely that they not be allowed to be negative numbers, i.e., they must be either zero or positive. Therefore, we have

$$x_j \geq 0, \quad j = 1, 2, 3, 4 \tag{1-6}$$

If we now put together (1–3), (1–4), (1–5), and (1–6) we have the following linear programming model of our feed supplier's problem†

$$\text{Min } z = 3x_1 + x_2 + 2x_3 + 4x_4$$

subject to:

$$\begin{aligned} 20x_1 + 15x_2 + 10x_3 + 30x_4 &\geq 250 \\ 25x_1 + 15x_2 + 20x_3 + 25x_4 &\geq 150 \\ 25x_1 + 20x_2 + 20x_3 + 30x_4 &\geq 400 \\ x_j \geq 0, \quad j = 1, 2, 3, 4 & \end{aligned} \tag{1-7}$$

† The statement of (1–7) is to be interpreted as: Find values of x_1, x_2, x_3, x_4 which satisfy all the constraints and make z as small as possible.

The solution to (1–7) will tell the feed supplier how much of each of the feed materials to mix together to satisfy the nutritional requirements at least cost to him.

The problem of the feed supplier will serve to illustrate the essential features of assuming a linear model. The two mathematical properties that characterize linearity are the *additivity* property and the *proportionality* or *multiplicative* property. By the additivity property we mean that if 10 pounds of feed material A are used and 20 pounds of feed material B are mixed with it, we will then have $25(10) + 15(20) = 550$ units of nutritive element 2, i.e., the amounts obtained from the two materials are additive. There are physical situations where additivity fails. For example, volumes of liquids of greatly different chemical structure are not necessarily additive.

The proportionality property would say, for example, that if 10 pounds of feed material D cost \$40 then 100 pounds would cost \$400. This is strict proportionality and could conceivably not be true in practice. Economies of scale, quantity discounts, and similar production and marketing factors often violate this kind of strict proportionality. It is important to recognize that for any given physical situation where either or both of these assumptions does not hold, a linear programming formulation is inappropriate. Nonlinear programming formulations[1] are possible for such cases although the computational art of nonlinear programming is not nearly as satisfactory as that which is available for linear programming.

1–3 GEOMETRIC INTERPRETATION OF LINEAR PROGRAMMING

It is instructive to consider a geometric representation of a linear programming problem. Such a representation leads to some valuable insights concerning the nature of solutions to linear programming problems, which we shall verify and prove in Chapter 4. While it is a simple matter to solve linear programming problems in two variables geometrically, this approach is useless for problems involving greater numbers of variables.

Let us consider the following linear programming problem

$$\text{Max } z = 8x_1 + 11x_2$$

subject to:

$$\begin{aligned} 5x_1 + 4x_2 &\leq 40 \\ -x_1 + 3x_2 &\leq 12 \\ x_1, x_2 &\geq 0 \end{aligned} \tag{1–8}$$

As in all linear programming problems, the variables are constrained to be non-negative.† Hence, we need only examine the non-negative quadrant of two dimensional Euclidean space in our geometric analysis. This is what is shown in Figure 1–1. The two arrows, one pointing upward from the x_1 axis and one to the right of the x_2 axis, indicate that we are confined to the non-negative quadrant of the x_1–x_2 plane.

Let us now examine the other constraints of the problem. Consider the constraint

$$5x_1 + 4x_2 \leq 40 \tag{1-9}$$

The straight line $5x_1 + 4x_2 = 40$ is plotted in Figure 1–1. The inequality expressed in (1–9) allows points (x_1, x_2) to be on this straight line or on one side of it. A simple test using the point $(x_1, x_2) = (0, 0)$ clearly shows that $5(0) + 4(0) < 40$. Hence, we have shown an arrow pointing toward the origin. This is the allowable region for (x_1, x_2) with respect to the constraint (1–9). A similar analysis of the constraint

$$-x_1 + 3x_2 \leq 12 \tag{1-10}$$

shows the allowable region to be in a direction towards the x_1 axis. Accordingly, an arrow points from the constraint in this direction. If *all* the constraints are to be satisfied simultaneously, we must then consider the

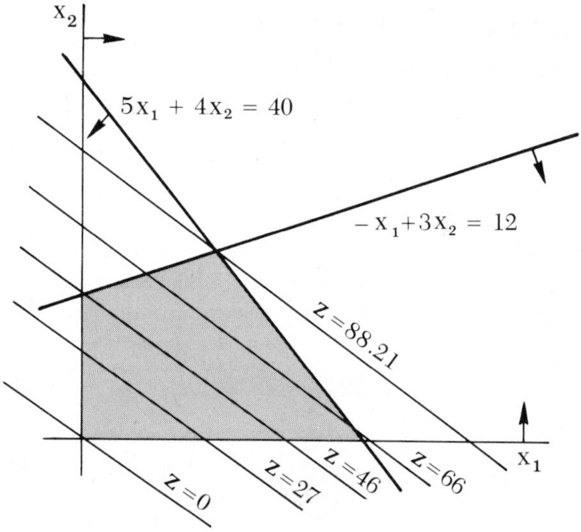

Figure 1–1

† This is not a limitation in practice. If some problem requires negative variables, an equivalent linear programming problem with non-negative variables may be solved. This is discussed in a later section.

intersection of the two straight lines

$$5x_1 + 4x_2 = 40$$
$$-x_1 + 3x_2 = 12$$

with the co-ordinate axes $x_1 = 0$ and $x_2 = 0$. This intersection gives rise to the shaded polygon shown in Figure 1–1. Any point inside this polygon or on any of its bounding lines is a *feasible* (allowable) set of values (x_1, x_2), which satisfies the constraints of the problem.

The shaded polygon of Figure 1–1 has an infinite number of feasible solutions. We must now consider how to find the one or more of these solutions which are optimal, i.e., having the greatest value of the objective function. Our objective function is

$$z = 8x_1 + 11x_2$$

For some particular value of z we can plot a contour line, i.e., a line along which all the combinations of values of (x_1, x_2) give the same value of the objective function, namely the one we have selected. A series of such constant z lines is shown in Figure 1–1. It can be seen that the direction of increasing z is away from the origin. It can also be readily seen that the largest value of z is taken on by the objective function at a single point. This is the point generated by the intersection of two of the constraint boundaries, i.e., by the intersection of the straight lines

$$5x_1 + 4x_2 = 40$$
$$-x_1 + 3x_2 = 12$$

At this point $x_1 = 3.789$, $x_2 = 5.263$, and $z = 88.21$. If we try to move into the polygon, we will decrease z. If we try to increase z, we will no longer be feasible, i.e., the constraints will not be satisfied. Hence, the point $(x_1, x_2) = (3.789, 5.263)$ is the optimal solution and $z = 88.21$ is the optimal value of the objective function.

It is always tempting to try to generalize from a simple example such as (1–8) to any linear programming problem. It is a remarkable fact that such a generalization is possible. All the significant aspects relating to the nature of the feasible region and the location of the optimal solution in our simple example also hold for problems with larger numbers of variables and constraints. Let us state what these aspects of the problem solution are that we may have noted.

1. The intersection of the non-negative quadrant boundaries and the other constraint boundaries generated a convex polygon. This will also be true in the n-dimensional case. In Chapter 3 it will be shown

that the intersection of any number of half-spaces and/or hyperplanes generates a convex polyhedron in a higher dimensional space ($n \geq 2$).

2. A solution to our linear programming problem example was at an extreme point or "corner" of the convex polygon of feasible solutions. There were no optimal solutions in the interior of the polygon.† This will also be true for the general case of a convex polyhedron in an n-dimensional space. An optimal solution will occur at an extreme point of the convex polyhedron. Hence, only extreme points need be examined as candidates for the optimal solution.

What we have noted in the foregoing is stated in the language of geometry. In Chapter 4 we shall prove these general assertions in the context of an algebraic formulation of the problem. However, we will also prove and point out the correspondence of our algebraic formulation and results with their geometric counterparts.

1–4 FURTHER GEOMETRIC EXAMPLES

It is convenient to use the geometric interpretation of linear programming problems to illustrate points about the nature of the possible solutions to linear programming problems and also to exhibit some anomalous situations which may arise.

In the example (1–8) which was shown in Figure 1–1, there was a unique optimum solution. However, this is not always the case. Consider the following problem

$$\text{Max } z = 6x_1 + 2.5x_2$$

subject to

$$\begin{aligned} 7x_1 + 9x_2 &\leq 63 \\ 12x_1 + 5x_2 &\leq 60 \\ x_1, x_2 &\geq 0 \end{aligned} \qquad (1\text{–}11)$$

This problem is illustrated in Figure 1–2. Following the same kind of argument we used with the problem (1–8) we can see that the optimal value of the objective function is $z = 30$. It can also be seen that either of the extreme points ("corners") which are designated A and B are optimal solutions. Hence, in this case, two extreme points yield the same optimal value of the objective function. However, it is also apparent that *any* point on the line segment $\overline{AB}$ is also an optimal solution, since it lies on the line $z = 30$. Hence, there are

†In the example, there was only one optimal solution. In problems with more optimal solutions than one, they will be boundary solutions.

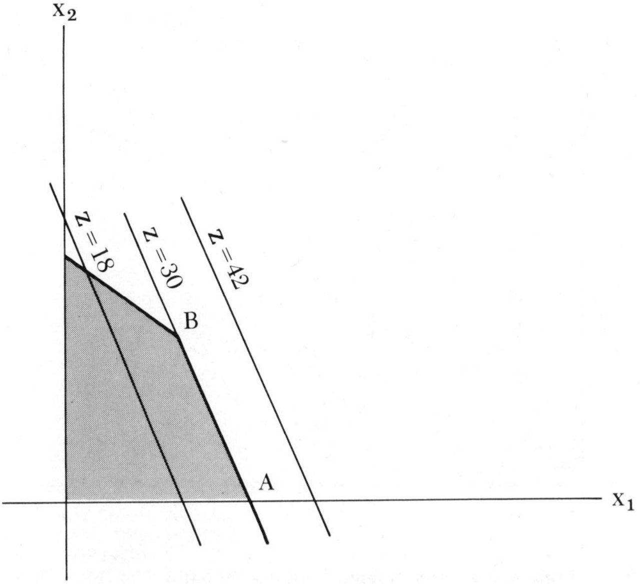

Figure 1–2 Non-unique optimal solution—equations (1–11)

an infinite number of optimal solutions to the problem given by (1–11). In linear programming, when such a situation exists, we say that there are alternative optimal solutions or simply, alternative optima. This is a common occurrence in linear programming problems. The simplex method, which we shall develop in Chapter 4, will always find one of the solutions of the type designated A and B, i.e., extreme point solutions. It will never find any of the others on the line segment $\overline{AB}$. Furthermore, the simplex method can be used to find all extreme point solutions, if one desires to do so.

In order to illustrate another situation that may arise, let us consider the following problem

$$\text{Max } z = 4x_1 + 3x_2$$

subject to

$$\begin{aligned} -3x_1 + 2x_2 &\leq 6 \\ -x_1 + 3x_2 &\leq 18 \\ x_1, x_2 &\geq 0 \end{aligned} \qquad (1\text{–}12)$$

This problem is shown in Figure 1–3. It is apparent that there is no largest value of z, i.e., it is always possible to find values of (x_1, x_2) that will make z larger than any preassigned value. In short, z can be made arbitrarily large. When such a situation exists in a linear programming problem, the problem is said to have an *unbounded* solution.

In the example of Figure 1–3, z becomes arbitrarily large as both x_1 and

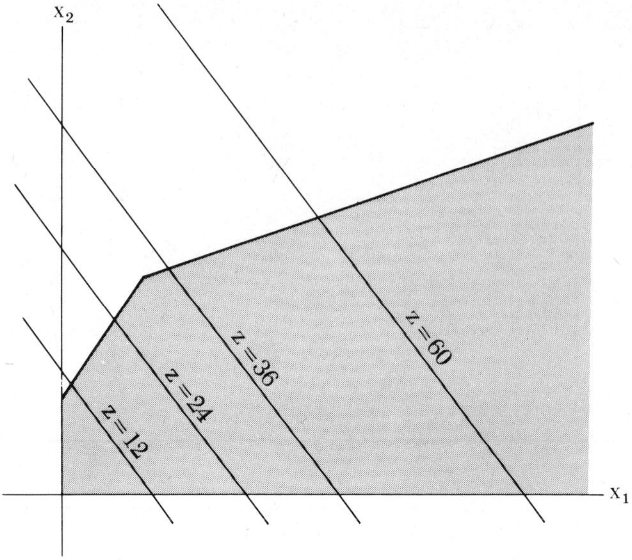

Figure 1–3 Unbounded solution—equations (1–12)

x_2 are allowed to assume increasingly large values. However, it is possible for a linear programming problem to have an unbounded solution even if only one variable becomes arbitrarily large. Let us consider the following example:

$$\text{Max } z = 3x_1 - 4x_2$$

subject to:

$$\begin{aligned} x_1 - x_2 &\geq 0 \\ x_2 &\leq 6 \\ x_1, x_2 &\geq 0 \end{aligned} \tag{1–13}$$

The problem given in (1–13) is depicted in Figure 1–4. It is clear that this problem has an unbounded solution. We see that as x_1 becomes arbitrarily large, so does the objective function. However, x_2 remains finite. No matter how large x_1 becomes, x_2 is never greater than 6.

A very different kind of unboundedness is illustrated by the following problem:

$$\text{Max } z = -4x_1 + 10x_2$$

subject to:

$$\begin{aligned} -3x_1 + 2x_2 &\leq 3 \\ -2x_1 + 5x_2 &\leq 20 \\ x_1, x_2 &\geq 0 \end{aligned} \tag{1–14}$$

which is shown in graphical form in Figure 1–5.

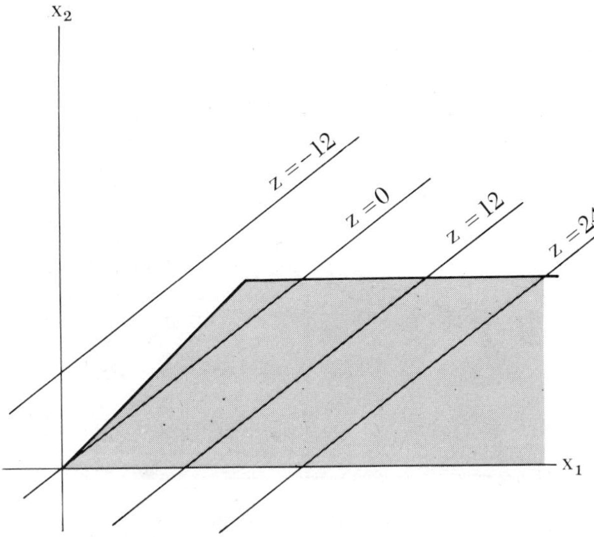

Figure 1–4 Unbounded solution—equations (1–13)

In this problem there is a finite maximum value of the objective function. It is clear that this maximum value is $z = 40$. As we have noted before, there is an optimal extreme point solution which occurs at the intersection of

$$-3x_1 + 2x_2 = 3$$
$$-2x_1 + 5x_2 = 20$$

This point is $(x_1, x_2) = (\frac{25}{11}, \frac{54}{11})$. However, any point along the upper boundary ($z = 40$) of the convex polygon is also an optimal solution. It is clear from Figure 1–5, that even though z cannot exceed 40, values of (x_1, x_2) which will yield a value of $z = 40$ can become arbitrarily large.

These last three examples, which are concerned with unboundedness either of the optimal value of the objective function or of the values of the variables which correspond to the optimal value of the objective function or to both such situations, are considered to be pathological in a practical sense. What we mean by this is that such cases should not be encountered in any well-formulated problem whose variables purport to represent physical or economic entities of some kind. Nevertheless, such situations may be encountered because of inadvertent errors which occur during the formulation of a linear programming problem or during purely mechanical manipulation of large masses of data. It is important to be able to recognize that a problem has an unbounded solution, since this is an indication of some sort of error. The simplex method which is described in Chapter 4 and, indeed, all other methods described in later chapters, possess simple tests to determine the existence of an unbounded objective function.

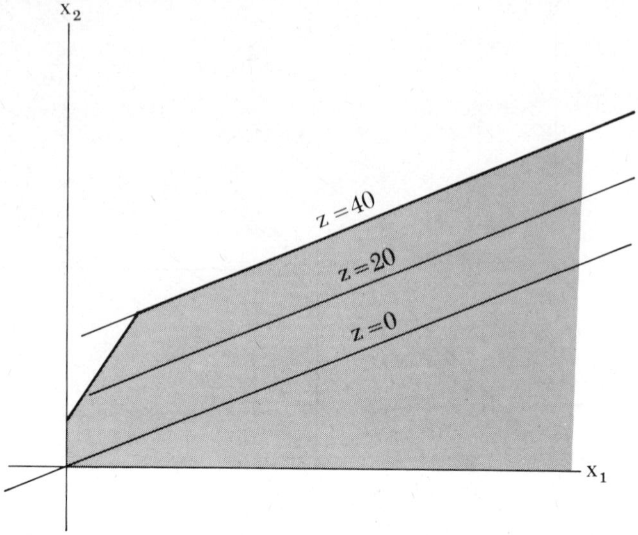

Figure 1–5 Unboundedness in variables—equations (1–14)

Another pathological situation which may arise because of errors of problem formulation is that of the nonexistence of any feasible solution. This can occur because there do not exist a set of x_j, $j = 1, 2, \ldots, p$ which simultaneously satisfy the constraints

$$\sum_{j=1}^{p} a_{ij}x_j\{\leq, =, \geq\}b_i, \qquad i = 1, 2, \ldots, m$$

or the restrictions

$$x_j \geq 0 \qquad j = 1, 2, \ldots, p$$

or both of these sets of conditions. For example, consider the following problem:

$$\text{Max } z = 3x_1 + 5x_2$$

subject to:

$$\begin{aligned} 5x_1 + 5x_2 &\leq 25 \\ 9x_1 + 13x_2 &\geq 117 \\ x_1, x_2 &\geq 0 \end{aligned} \tag{1–15}$$

This problem is represented in Figure 1–6. It is readily seen that there are no solutions $x_1, x_2 \geq 0$ which simultaneously satisfy both of the constraints

$$5x_1 + 5x_2 \leq 25$$
$$9x_1 + 13x_2 \geq 117$$

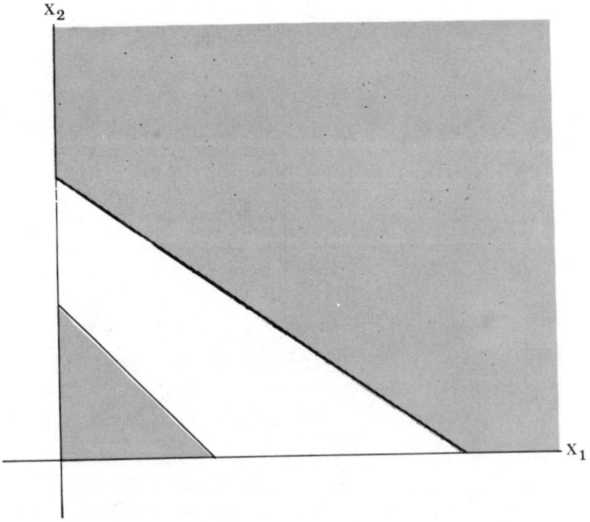

Figure 1–6 No feasible solution—equations (1–15)

It is obviously absurd to make such contradictory demands in any realistic problem. However, it is possible to do so unintentionally when writing mathematical expressions.

A second kind of infeasibility is of the kind in which we can simultaneously satisfy the structural constraints of the problem but not the nonnegativity restrictions. This is illustrated by the following problem:

$$\text{Max } z = 3x_1 + 2x_2$$

subject to:

$$\begin{aligned} x_1 + 2x_2 &\leq 2 \\ 2x_1 + x_2 &\geq 6 \\ x_1, x_2 &\geq 0 \end{aligned} \qquad (1\text{–}16)$$

which is represented in Figure 1–7. It is seen that even though there exist values of (x_1, x_2) that satisfy both

$$\begin{aligned} x_1 + 2x_2 &\leq 2 \\ 2x_1 + x_2 &\geq 6 \end{aligned}$$

there are no such values which also satisfy the conditions that $x_1 \geq 0$, $x_2 \geq 0$. Hence, in this case also, there are no feasible solutions to the problem (1–16).

1–5 INTRODUCTION TO PART I

We have presented in this chapter a simple example of a linear programming problem to illustrate its nature. Subsequently, we have given geometric

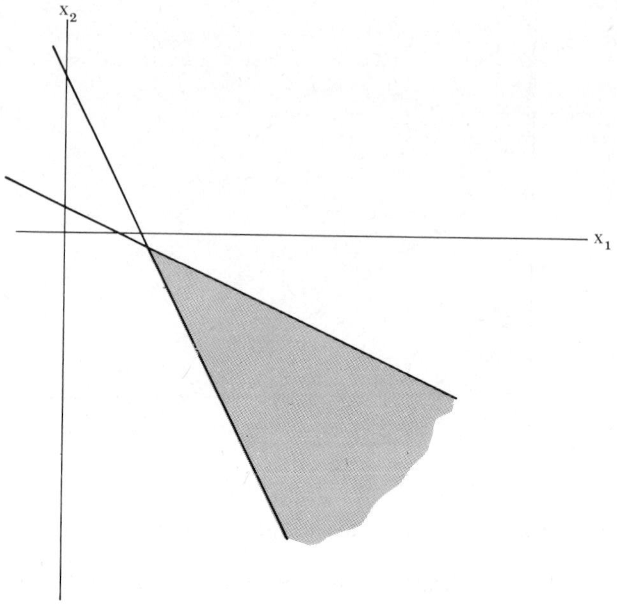

Figure 1–7 No feasible solution—equations (1–16)

examples to gain some tentative insight, which turns out to be correct, into the characteristics of optimal solutions to linear programming problems. In Chapter 2 we will present a host of different applications of linear programming, to demonstrate the almost universal applicability of this type of model formulation. In Chapter 3, a brief summary of the necessary mathematical background will be presented for reference. In Chapters 4, 5, 6, and 7, a thorough development of all necessary results relating to the theory and use of the primal simplex method will be made.

A few words on the simplex method are in order. We noted in our geometric examples that an optimal solution always occurred at an extreme point of the convex set of feasible solutions. The simplex method is an iterative computational method. It starts at some extreme point solution. If this solution is not optimal, we determine an adjacent extreme point which, in a maximization problem, will yield a greater value of the objective function. This process is continued until we have an indication that no further improvement is possible. Hence, the optimal solution is at hand. The derivation and proofs of the basic simplex method are presented in Chapter 4, where the process we have described in geometric language is put in an algebraic framework, so that the iterative calculations may be simply carried out for problems with large numbers of variables and constraints.

REFERENCES

1. Cooper, L., and D. I. Steinberg: *Introduction to Methods of Optimization.* Philadelphia, W. B. Saunders Company, 1970.
2. Wilde, D. J., and C. S. Beightler: *Foundations of Optimization.* Englewood Cliffs, N.J., Prentice-Hall Inc., 1967.
3. Hildebrand, F. B.: *Methods of Applied Mathematics.* New York, Prentice-Hall, 1952.
4. Dantzig, G. B.: *Linear Programming and Extensions.* Princeton, N.J., Princeton University Press, 1963.
5. Hitchcock, F. L.: The Distribution of a Product from Several Sources to Numerous Localities. *J. Math. Phys.*, *20*:224–230, 1941.
6. Kantorovitch, L. V.: *Calcul Economique et Utilisation des Resources.* Paris, Dunod, 1963. Translation of original Russian edition.
7. Kantorovitch, L. V.: On the Translocation of Masses, Comptes rendus (Doklady) de l'academie des sciences de l'USSR, **37**, No. 7–8, 199–201, 1942.
8. Dantzig, G. B.: Maximization of a Linear Function of Variables Subject to Linear Inequalities. In T. C. Koopmans (ed.): *Activity Analysis of Production and Allocation*, New York, John Wiley and Sons, 1951.

2

Applications of Linear Programming

In the previous chapter we defined what is meant by a linear programming problem and discussed the assumptions and restrictions imposed on any model so formulated. During the last several decades linear programming has risen to become one of the most important—and effective—tools available for dealing with certain types of large scale problems. Linear programming models have been formulated to help solve problems in most large industries, and in many governmental situations as well.

For example, linear programming has frequently been used to determine the allocation of land resources to the raising of different types of crops, in the scheduling and routing of commercial aircraft, in the design and use of communications networks, in the development of minimum cost shipping and production schedules, and in descriptions of large scale chemical plant systems. Many of these problems involve hundreds of constraints and thousands of variables.

Linear programming has also been employed in the development of economic models, to describe the economic interaction of many industries,[3,4] and to determine economic policy for underdeveloped nations.[2,5]

Perhaps the best way to illustrate the wide variety of applications of linear programming is by presenting a number of examples.

In this chapter, then, we shall be concerned with the formulation of various models as linear programming problems. Although the reader may consider some of the models described below to be oversimplified (and thus unrealistic) it should be borne in mind that these models are presented for illustration purposes only, and they can, in most cases, be made more realistic (and more complex) within the linear programming framework if desired.

MODEL 1

A small metal parts manufacturing company, most of whose business comes from special orders, has decided to make use of the idle time of its equipment, if this should prove to be profitable. The company has determined that two particular parts, A and B, could readily be sold at per unit profits of \$3 and \$2, respectively, in quantities up to twelve dozen of each part per day. In order to make each unit of part A, 5 minutes are required on a lathe, 7 minutes on a mill, and 4 minutes on a grinder. To make each unit of part B, 3, 9, and 7 minutes, respectively, are required on a lathe, mill, and grinder. The company has available one machine of each type. These machines are currently idle as follows:

Machine	Lathe	Mill	Grinder
Minutes per day idle	65	100	90

How many parts of type A and B, if any, should the company produce in order to make maximum use (in terms of profit) of the idle time of its equipment?

Solution

Let x_A = no. of parts of type A to be produced

x_B = no. of parts of type B to be produced

Then, the total profit is

$$z = 3x_A + 2x_B \tag{2-1}$$

Because of sales expectations, the company does not wish to produce more than 12 dozen of each part:

$$\begin{aligned} x_A &\le 144 \\ x_B &\le 144 \end{aligned} \tag{2-2}$$

The total number of minutes required on the lathe to produce x_A units of part A and x_B units of part B is $(5x_A + 3x_B)$ and this quantity cannot exceed the available lathe time, 65 minutes:

$$5x_A + 3x_B \le 65 \tag{2-3}$$

Similarly, the restrictions on available time for the mill and grinder yield, respectively,

$$7x_A + 9x_B \le 100 \tag{2-4}$$

$$4x_A + 7x_B \le 90 \tag{2-5}$$

Since x_A and x_B cannot be negative, we must also require

$$x_A \geq 0 \tag{2–6}$$

$$x_B \geq 0 \tag{2–7}$$

Thus, the linear programming formulation of Model 1 is described by

$$\text{maximize } z = 3x_A + 2x_B$$

subject to the constraints (2–2) through (2–7).

Exercise 1: Suppose the company in Model 1 has two lathes available, for 45 and 35 minutes per day, respectively. If all other conditions are unchanged, formulate the new model as a linear programming problem.

MODEL 2

An automobile manufacturing company has five assembly plants and twelve dealerships located throughout the country. The company would like to determine which plants should assemble the automobiles for each dealer, in order to minimize the shipping costs. The following information is available:

Dealership No.	1	2	3	4	5	6	7	8	9	10	11	12
No. Cars Ordered	35	42	28	52	17	33	62	61	43	37	28	42

Assembly Plant	*A*	*B*	*C*	*D*	*E*
Capacity (cars)	100	90	120	80	90

Shipping Costs (per car)

		Dealers											
From \ *To*		1	2	3	4	5	6	7	8	9	10	11	12
Plants	*A*	25	36	31	20	27	33	27	21	24	35	21	24
	B	37	23	28	35	19	31	23	25	28	27	19	31
	C	26	19	17	28	27	36	31	24	21	27	17	30
	D	42	25	15	33	23	28	29	33	26	29	20	27
	E	19	31	45	37	31	34	27	31	32	30	22	25

Thus, each dealership requires a certain number of cars, and each assembly plant can provide cars up to its capacity; each dealership can receive cars from more than one assembly plant, if this procedure would reduce the overall costs.

Solution

Let x_{ij} = no. of cars shipped from plant i to dealer j

$$i = A, B, C, D, E; \qquad j = 1, 2, \ldots, 12$$

The total shipping cost is given by

$$Z = (25x_{A1} + 36x_{A2} + \ldots + 24x_{A,12}) + (37x_{B1} + 23x_{B2} + \ldots + 31x_{B,12}) + \ldots + (19x_{E1} + 31x_{E2} + \ldots + 25x_{E,12}) \quad (2\text{–}8)$$

The orders of each dealership must be satisfied. Thus, the total number of cars shipped to dealer 1 must be 35 cars:

$$x_{A1} + x_{B1} + x_{C1} + x_{D1} + x_{E1} = 35 \quad (2\text{–}9)$$

Similarly, for dealers 2 through 12 we obtain

$$\begin{aligned} x_{A2} + x_{B2} + x_{C2} + x_{D2} + x_{E2} &= 42 \\ x_{A3} + x_{B3} + x_{C3} + x_{D3} + x_{E3} &= 28 \\ &\vdots \\ x_{A,12} + x_{B,12} + x_{C,12} + x_{D,12} + x_{E,12} &= 42 \end{aligned} \quad (2\text{–}10)$$

Moreover, the total number of cars shipped from each assembly plant cannot exceed its capacity. For plant A, the total number of cars shipped is $(x_{A1} + x_{A2} + \ldots + x_{A,12})$. Therefore,

$$x_{A1} + x_{A2} + \ldots + x_{A,12} \leq 100 \quad (2\text{–}11)$$

Similarly, for plants B, C, D, and E:

$$\begin{aligned} x_{B1} + x_{B2} + \ldots + x_{B,12} &\leq 90 \\ x_{C1} + x_{C2} + \ldots + x_{C,12} &\leq 120 \\ x_{D1} + x_{D2} + \ldots + x_{D,12} &\leq 80 \\ x_{E1} + x_{E2} + \ldots + x_{E,12} &\leq 90 \end{aligned} \quad (2\text{–}12)$$

And, of course, we cannot ship negative quantities:

$$x_{ij} \geq 0, \quad i = A, B, C, D, E; \qquad j = 1, 2, \ldots, 12 \quad (2\text{–}13)$$

The linear programming problem describing Model 2 is, therefore, to minimize the cost (2–8) subject to the constraints (2–9) through (2–13).

COMMENT: The linear programming problem which resulted from Model 2 occurs so frequently that it has been given a special name, the *transportation problem.* In general, suppose there are m shipping origins (e.g., assembly plants, warehouses) and n destinations (e.g., dealerships, retail outlets). Let a_i be the capacity (or available supply) of the ith origin, $i = 1, 2, \ldots, m$, let b_j denote the number of units ordered (or required) by the jth destination, and let c_{ij} be the shipping cost per unit from origin i to destination j. If

$$x_{ij} = \text{no. of units shipped from origin } i \text{ to destination } j$$
$$i = 1, 2, \ldots, m \quad \text{and} \quad j = 1, 2, \ldots, n$$

then the general transportation problem may be described thusly:

$$\text{Minimize } z = \sum_{j=1}^{n} \sum_{i=1}^{m} c_{ij} x_{ij} \tag{2–14}$$

subject to

$$\sum_{j=1}^{n} x_{ij} \leq a_i, \qquad i = 1, 2, \ldots, m \tag{2–15}$$

$$\sum_{i=1}^{m} x_{ij} \geq b_j, \qquad j = 1, 2, \ldots, n \tag{2–16}$$

$$x_{ij} \geq 0 \qquad i = 1, 2, \ldots, m \quad \text{and} \quad j = 1, 2, \ldots, n \tag{2–17}$$

The inequalities (2–15) and/or the inequalities (2–16) are sometimes replaced by equality constraints, as was the case in Model 2. The special structure of the constraints for the transportation problem enables us to develop special methods for solving such problems which are computationally superior to those for solving general linear programming problems. These methods will be discussed in Chapters 10 and 14.

Exercise 2: Suppose that, in the situation described in Model 2, it is decided that assembly plant A will not be used unless it will assemble at least 30 cars. Can this additional restriction be incorporated into a linear programming formulation? Discuss this situation in detail. [Hint: Can the above model be formulated in terms of two linear programming problems?]

Exercise 3: Formulate as a linear programming problem:

A Hollywood producer is currently developing a TV special, to be televised in eight weeks. The producer has decided that the special shall consist of four fourteen-minute musical skits. The producer's staff of seventeen includes seven writers, four choreographers, and six musical comedy specialists. However, each of these artistes is capable of performing the functions of the others, so it is not necessary to have, for example, at least one writer working on each skit. But, the producer does feel that there should be at least three artistes working on each skit. Based on their past performance and on the nature of the four skits, the producer has devised a rating system which indicates the relative abilities of each of the artistes (the lower the number, the better the rating) as follows:

	Skit			
Type of Artiste	1	2	3	4
Writer	2	7	4	2
Choreographer	5	6	3	3
M-C specialist	3	4	7	5

Moreover, for each artiste who is idle (i.e., not working on one of the four skits) the producer assigns a rating number of 6. Because of time restrictions, each artiste can only work on one skit. How should the producer assign the artistes to skits so that the total rating (sum of the ratings) is a minimum?

Can this problem be formulated as a transportation problem (as defined by equation (2–14) and inequalities (2–15) through (2–17))?

MODEL 3

The investment counselor of a bank has been given \$1.5 million dollars to invest for a period of three years. The counselor has determined that there are three reasonably attractive investment opportunities available currently: Investment *A* yields an annual rate of return of 7%, investment *B* yields 5% the first year and 8% per annum thereafter; investment *C* yields a flat 20% at the end of the third year, and may not be invested in after this year. The counselor has also learned that at the beginning of the second year (only) a new investment opportunity, investment *D*, will become available, yielding 16% at the end of the third year. The investment counselor wishes to determine how much money to invest—and when—in each of the possible investments, so that the amount of money he will have at the beginning of the fourth year is a maximum.

Solution

Let A_i = amount of money (in dollars) to be invested at the beginning of year i in investment A; $i = 1, 2, 3$

B_i = amount of money to be invested at the beginning of year i in investment B; $i = 1, 2, 3$.

C_1 = amount of money to be invested at the beginning of year 1 in investment C

D_2 = amount of money to be invested at the beginning of year 2 in investment D.

For each of the three years, let us calculate how much money is available to be invested during that year. At the beginning of year 1, \$1,500,000 is available for investment. For years 2 and 3, the amount available for investment is \$1,500,000 minus all previous investments plus income from previous investments. Thus, for year 1,

$$A_1 + B_1 + C_1 \leq \$1{,}500{,}000 \tag{2–18}$$

For year 2, the sum of previous investments is $(A_1 + B_1 + C_1)$ and the income from previous investments is $[(A_1 + 0.07A_1) + (B_1 + 0.05B_1)]$. There is no income from investment C_1 until after year 3. Thus, the amount to be invested in year 2, $(A_2 + B_2 + D_2)$, is constrained by

$$A_2 + B_2 + D_2 \leq 1{,}500{,}000 - (A_1 + B_1 + C_1) + (1.07A_1 + 1.05B_1) \tag{2–19}$$

For year 3, the sum of previous investments is $(A_1 + B_1 + C_1 + A_2 + B_2 + D_2)$; the income from previous investments is $[(A_1 + 0.07A_1) + (A_2 + 0.07A_1) + (B_1 + 0.05B_1) + (B_2 + 0.08B_2)]$. Thus, as before,

$$\begin{aligned} A_3 + B_3 \leq 1{,}500{,}000 &- (A_1 + B_1 + C_1 + A_2 + B_2 + D_2) \\ &+ (1.07A_1 + 1.07A_2 + 1.05B_1 + 1.08B_2) \end{aligned} \tag{2–20}$$

The amount of money on hand at the beginning of the fourth year, Z, is

$$\begin{aligned} Z = 1{,}500{,}000 &- (A_1 + B_1 + C_1 + A_2 + B_2 + D_2 + A_3 + B_3) \\ &+ (1.07A_1 + 1.07A_2 + 1.07A_3 + 1.05B_1 + 1.08B_2 \\ &+ 1.08B_3 + 1.20C_1 + 1.16D_2) \end{aligned} \tag{2–21}$$

Thus, the resulting linear programming problem is to minimize Z, as given by equation (2–21), subject to the constraints (2–18), (2–19), (2–20), and the non-negativity restrictions

$$A_i \geq 0 \qquad i = 1, 2, 3$$

$$B_i \geq 0 \qquad i = 1, 2, 3$$

$$C_1 \geq 0$$

$$D_2 \geq 0$$

[Note: The constant 1,500,000 in the objective function (2–21) can be deleted, since its contribution to Z is unchanged for any values of the variables. That is, the optimal values of the variables for the above problem and for the same problem with the objective function

$$\text{Maximize } \hat{Z} = Z - 1{,}500{,}000 \qquad (2\text{–}21)$$

will be the same.]

MODEL 4

A nut company sells three different assortments of mixed nuts. Each assortment contains varying amounts of almonds, pecans, cashews, and walnuts. In order to preserve the company's reputation for quality, certain maximum or minimum percentages of the various nuts are required for each type of assortment:

Assortment Name	Requirements	Selling Price per Pound
Regular	Not more than 20% cashews Not less than 40% walnuts Not more than 25% pecans	59¢
Deluxe	Not more than 35% cashews Not less than 25% almonds	69¢
Blue Ribbon	Between 30–50% cashews Not less than 30% almonds	85¢

The company would like to determine the exact amounts of almonds, pecans, cashews, and walnuts which should go into each assortment, in order to maximize its profit. In order to calculate the profit per pound, for each assortment, we must know how much the ingredients cost:

Type of Nut	Cost per Pound	Maximum Quantity Available per week (pounds)
Almonds	25¢	2000
Pecans	35¢	4000
Cashews	50¢	5000
Walnuts	30¢	3000

As indicated in the above table, the company's nut farm supplies a limited quantity of each type of nut.

Thus, the quantity to be maximized is total net profit per week.

Solution

Let R_i = no. of pounds of nuts of kind i in the Regular assortment; $i = 1, 2, 3, 4$

D_i = no. of pounds of nuts of kind i in the Deluxe assortment; $i = 1, 2, 3, 4$

B_i = no. of pounds of nuts of kind i in the Blue Ribbon assortment; $i = 1, 2, 3, 4$

where:

Kind of Nut	Almonds	Pecans	Cashews	Walnuts
i	1	2	3	4

The total number of pounds of each kind of nut will then be

i	Kind of Nut	Number of Pounds Used Per Week
1	Almonds	$R_1 + D_1 + B_1$
2	Pecans	$R_2 + D_2 + B_2$
3	Cashews	$R_3 + D_3 + B_3$
4	Walnuts	$R_4 + D_4 + B_4$

From the maximum quantities available, we obtain the four constraints:

$$\begin{aligned} R_1 + D_1 + B_1 &\leq 2000 \\ R_2 + D_2 + B_2 &\leq 4000 \\ R_3 + D_3 + B_3 &\leq 5000 \\ R_4 + D_4 + B_4 &\leq 3000 \end{aligned} \tag{2-22}$$

Next, we must ensure that the requirements for each assortment are satisfied. Since these requirements are given in terms of per cents, rather than in pounds, we shall calculate the following.

Assortment	Total Pounds Produced Per Week
Regular	$R_1 + R_2 + R_3 + R_4$
Deluxe	$D_1 + D_2 + D_3 + D_4$
Blue Ribbon	$B_1 + B_2 + B_3 + B_4$

Thus, the per cent of cashews in the Regular assortment is $100R_3/(R_1 + R_2 + R_3 + R_4)$. This quantity cannot exceed 20%:

$$\frac{100R_3}{R_1 + R_2 + R_3 + R_4} \leq 20$$

Since $(R_1 + R_2 + R_3 + R_4) > 0$ (we shall assume it is desirable to produce some of each type of assortment), we can multiply both sides of the above inequality by $(R_1 + R_2 + R_3 + R_4)$, yielding

$$100R_3 \leq 20(R_1 + R_2 + R_3 + R_4)$$

or,

$$20R_1 + 20R_2 - 80R_3 + 20R_4 \geq 0 \tag{2–23}$$

In a similar manner, the other two requirements for the Regular assortment are described thusly:

$$\frac{100R_4}{R_1 + R_2 + R_3 + R_4} \geq 40$$

$$\frac{100R_2}{R_1 + R_2 + R_3 + R_4} \leq 25$$

These two inequalities can be rewritten as

$$\begin{aligned} 40R_1 + 40R_2 + 40R_3 - 60R_4 &\leq 0 \\ 25R_1 - 75R_2 + 25R_3 + 25R_4 &\geq 0 \end{aligned} \tag{2–24}$$

Following the same procedure for the Deluxe and Blue Ribbon assortments yields

$$\frac{100D_3}{D_1 + D_2 + D_3 + D_4} \leq 35$$

$$\frac{100D_1}{D_1 + D_2 + D_3 + D_4} \geq 25$$

$$30 \leq \frac{100B_3}{B_1 + B_2 + B_3 + B_4} \leq 50$$

$$\frac{100B_1}{B_1 + B_2 + B_3 + B_4} \geq 30$$

Again, these inequalities may be expressed as follows:

$$\begin{aligned} 35D_1 + 35D_2 - 65D_3 + 35D_4 &\geq 0 \\ -75D_1 + 25D_2 + 25D_3 + 25D_4 &\leq 0 \\ 30B_1 + 30B_2 - 70B_3 + 30B_4 &\leq 0 \\ 50B_1 + 50B_2 - 50B_3 + 50B_4 &\geq 0 \\ -70B_1 + 30B_2 + 30B_3 + 30B_4 &\leq 0 \end{aligned} \tag{2–25}$$

The constraints (2–23), (2–24), (2–25), along with the usual non-negativity restrictions

$$R_i, D_i, B_i \geq 0 \qquad i = 1, 2, 3, 4 \tag{2–26}$$

ensure that the requirements for each assortment will be satisfied.

Now, let us calculate the net profit:

$$\text{Net Profit} = \text{Total Sales} - \text{Total Costs}$$

The total sales are:

$$59(R_1 + R_2 + R_3 + R_4) + 69(D_1 + D_2 + D_3 + D_4) + 85(B_1 + B_2 + B_3 + B_4).$$

The total costs are:

$$25(R_1 + D_1 + B_1) + 35(R_2 + D_2 + B_2) + 50(R_3 + D_3 + B_3) + 30(R_4 + D_4 + B_4).$$

Therefore, we wish to maximize, subject to the constraints (2–22) through (2–26), the net profit Z:

$$\begin{aligned} Z &= 59(R_1 + R_2 + R_3 + R_4) + 69(D_1 + D_2 + D_3 + D_4) \\ &\quad + 85(B_1 + B_2 + B_3 + B_4) - 25(R_1 + D_1 + B_1) \\ &\quad -35(R_2 + D_2 + B_2) - 50(R_3 + D_3 + B_3) - 30(R_4 + D_4 + B_4) \\ &= 34R_1 + 24R_2 + 9R_3 + 29R_4 \\ &\quad + 44D_1 + 34D_2 + 19D_3 + 39D_4 \\ &\quad + 60B_1 + 50B_2 + 35B_3 + 55B_4 \end{aligned} \tag{2–27}$$

COMMENT: Model 4 is an example of a whole class of linear programming problems called "blend problems." Another example of a blend problem occurs in the petroleum industry (optimal blends of different grades of gasoline and of oil). However, blend problems in general have no special structure, as does the transportation problem.

Exercise 4: Formulate as a linear programming problem:

A popular brand of cattle feed is a mixture of corn grain, hay, and wheat germ. The feed is mixed to order, so that each farmer specifies the mixture he wants. Suppose a farmer wants his mixture to contain at least ten pounds of protein, five pounds of fat, and four pounds of vitamin and mineral substances. The following information is available to the farmer:

	Contents (per 100 pounds)			
Ingredient	*Protein* (*lbs*)	*Fat* (*lbs*)	*Vitamins & Minerals* (*lbs*)	**Cost (per 100 pounds)**
Corn grain	2	7	2	$1.25
Hay	5	4	2	$1.40
Wheat germ	8	2	3	$1.75

Determine how many pounds of corn grain, hay, and wheat germ the farmer should order for his mixture, so that the cost is a minimum.

Exercise 5: The dietitian for a large hospital must see to it that each patient receives certain minimum amounts of various vitamins and minerals. The dietitian feels that, with careful planning of meals, this can be accomplished without having to incur the additional costs of giving each patient vitamin pills. However, in order to maintain a reasonable degree of variety in the daily meals, the dietitian must formulate a different problem for each day's menu. For simplicity, let us assume that the dietitian is planning for one meal, and has available roast beef, potatoes, and green beans, and that the meal must provide at least the following:

Protein	Hydrocarbons	Thiamine	Riboflavin	Niacinamide	Vitamin C
4 oz	7 oz	1 mg	1 mg	10 mg	30 mg

The dietitian knows that roast beef costs $1.39 per pound, potatoes $0.25 a pound, and green beans $0.79 per 48 ounce can. Use the following "table of contents" to determine the minimum cost meal (amounts of roast beef, potatoes, and green beans) which meets the minimum requirements:

	1 lb Roast Beef	8 oz Potatoes	8 oz Green Beans
Protein (oz)	5	1	1
Hydrocarbons (oz)	7	5	4
Thiamine (mg)	1.25	0.40	0.55
Riboflavin (mg)	1.15	0.40	0.45
Niacinamide (mg)	15	4	4
Vitamin C (mg)	40	10	15

[Note: This exercise is an example of a "diet problem." This is not the same as the "menu-planning problem," in which a menu must be planned for a long period of time (e.g., one month, in a hospital) so that food purchases can be made in large quantities. Moreover, in formulating a menu-planning problem, one must be careful to include constraints relating to variety and well-balanced meals, as well as those relating to dietary requirements. The interested reader might consider how such problems might be formulated.]

MODEL 5

A certain telephone company would like to determine the maximum number of long distance calls from Jonesville to Smithsboro that it can handle at any one time. The company has cables linking these cities via several intermediary cities as follows:

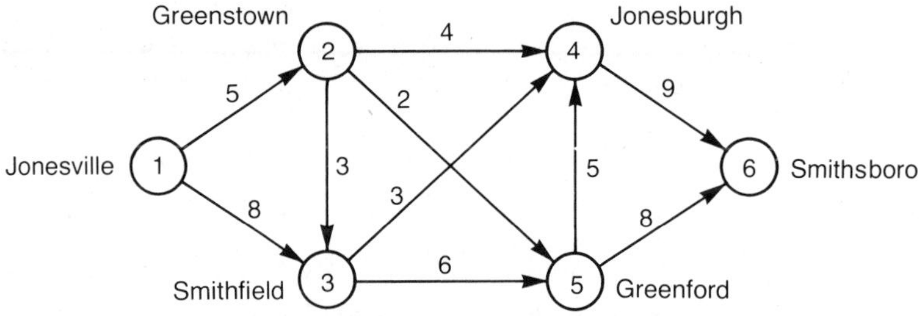

Each cable can handle a maximum number of calls simultaneously, as indicated above. For example, the number of calls routed from Greenstown to Jonesburgh cannot exceed 4 at any one time. A call from Jonesville to Smithsboro can be routed through any of the other cities, as long as there is a cable available which is not currently being used to its capacity.

In addition to determining the maximum number of calls from Jonesville to Smithsboro, the company would, of course, like to know the optimal routing of these calls. Assume that calls can be routed only in the directions indicated by the arrows.

Solution

For convenience, we shall refer to the cities by number, as shown in the figure. Then, let us define

$$x_{ij} = \text{no. of calls routed from city } i \text{ to city } j.$$

Then if we denote by M_{ij} the maximum number of simultaneous calls from city i to city j (e.g., $M_{12} = 5$, $M_{24} = 4$, etc.),

$$x_{ij} \leq M_{ij}, \quad \text{all} \quad i, j \tag{2–28}$$

For each of the intermediary cities 2, 3, 4, and 5, it must be true that the number of calls coming into that city equals the number of calls leaving that city (these cables are used only for routing calls all the way to Smithsboro). Thus, for city 2 (Greenstown),

$$x_{12} = x_{23} + x_{24} + x_{25} \tag{2–29}$$

Similarly, for cities 3, 4, and 5, respectively, we obtain

$$\begin{aligned} x_{13} &= x_{34} + x_{35} \\ x_{24} + x_{34} + x_{54} &= x_{46} \\ x_{25} + x_{35} &= x_{54} + x_{56} \end{aligned} \tag{2–30}$$

Moreover, the number of calls leaving Jonesville will equal the number of calls arriving at Smithsboro:

$$x_{12} + x_{13} = x_{46} + x_{56} \tag{2–31}$$

In fact, either side of equation (2–31) is exactly the quantity we wish to maximize:

$$\text{maximize } z = x_{12} + x_{13}$$

subject to the constraints (2–28) through (2–30) and the non-negativity restrictions

$$x_{ij} \geq 0, \qquad \text{all } i \text{ and } j.$$

COMMENT: Model 5 is an example of a special type of problem known as a *network flow problem*. Special techniques for solving such problems are available, and these will be discussed in Chapter 14. Network flow problems arise in a wide variety of situations, such as traffic flow, fluid flows through a pipe network, etc.

MODEL 6

Consider the following card game, for two players: Player X is given three cards, a red ace, a black ace, and a red deuce. Player Y is given a red ace, a black ace, and a black deuce. On each move of the game, each player simultaneously turns one of his cards face up. If both cards are red or both black, Player X wins the face value of his card (one dollar, if an ace, two dollars if the deuce); If the two cards are of opposite color, Player Y wins the face value of his card. If both players turn up deuces, then neither player wins.

Let us formulate a linear programming problem which will tell Player X what his optimal strategy should be.

Solution

The first step is to prepare a table showing the winnings of Player X for each possible outcome of the game. Labeling the strategies

1. Play red ace
2. Play black ace
3. Play deuce

we then obtain the following:

Player X wins		If Player Y plays Strategy 1	2	3
If Player X plays Strategy	1	1	−1	−2
	2	−1	1	1
	3	2	−1	0

For example, if Player X plays Strategy 1 (red ace) and Player Y plays Strategy 2 (black ace), then Player X loses one dollar (wins minus one dollar).

Since the amount Player X wins or loses on each play (and thus his total winnings) depends upon Player Y's selection of strategies, the best we can do is to maximize Player X's *expected* winnings.

Suppose we denote by x_1, x_2, x_3 the fraction of the time Player X plays each of his three strategies, respectively. Then, if Player Y plays strategy 1, Player X's expected winnings will be $1x_1 + (-1)x_2 + 2x_3$. Similarly, if Player Y plays strategy 2, Player X's expected winnings will be $(-1)x_1 + 1x_2 + (-1)x_3$, and if Player Y plays strategy 3, Player X's expected winnings will be $(-2)x_1 + 1x_2 + 0x_3$. Although we don't know which strategy Player Y will play (and indeed, he will most likely mix them up), we can require that each of the three above expressions for expected winnings *exceeds* some quantity, say w. This last statement can be expressed by the three inequalities,

$$\begin{aligned} 1x_1 + (-1)x_2 + 2x_3 &\geq w \\ (-1)x_1 + 1x_2 + (-1)x_3 &\geq w \\ (-2)x_1 + 1x_2 + 0x_3 &\geq w \end{aligned} \tag{2–32}$$

Moreover, since x_1, x_2, x_3 represent fractions of (the whole) time,

$$\begin{aligned} x_1 + x_2 + x_3 &= 1 \\ x_1, x_2, x_3 &\geq 0 \end{aligned} \tag{2–33}$$

The quantity we wish to maximize is w, since the inequalities (2–32) guarantee that Player X's expected winnings will be at least w. Thus, the desired linear programming problem is

$$\text{Maximize } z = w$$

subject to:

$$\begin{aligned} 1x_1 + (-1)x_2 + 2x_3 - w &\geq 0 \\ (-1)x_1 + 1x_2 + (-1)x_3 - w &\geq 0 \\ (-2)x_1 + 1x_2 + 0x_3 - w &\geq 0 \\ 1x_1 + 1x_2 + 1x_3 + 0w &= 1 \\ x_1, x_2, x_3 &\geq 0 \end{aligned} \tag{2-34}$$

Note that there is no non-negativity restriction on w, since if Player X is unfortunate, his expected winnings might be negative.

COMMENT: It is instructive to formulate the linear programming problem which provides the optimal strategy for Player Y. The reader should verify that if y_1, y_2, y_3 denote Player Y's three strategies, then his optimal strategy is found by solving

$$\text{Maximize } \hat{z} = v$$

Subject to:

$$\begin{aligned} y_1 + (-1)y_2 + (-2)y_3 + v &\leq 0 \\ (-1)y_1 + 1y_2 + 1y_3 + v &\leq 0 \\ 2y_1 + (-1)y_2 + 0y_3 + v &\leq 0 \\ 1y_1 + 1y_2 + 1y_3 + 0v &= 1 \\ y_1, y_2, y_3 &\geq 0 \end{aligned} \tag{2-35}$$

The two linear programming problems (2–34) and (2–35) are related in a very special way. One is called the *dual* of the other. Duality theory in linear programming is an extremely important topic, and will be discussed in considerable detail in Chapters 8 and 9. In Chapter 8 it will be shown that if the optimal value of (2–34) is w^* and the optimal value of (2–35) is v^*, then $w^* = -v^*$. This important relation tells us that if both players play their optimal strategies, they will both achieve their optimal expected winnings. Obviously, if one player wins an amount w, the other loses the same amount w, so that maximizing expected winnings is the same as minimizing expected losses. For this reason, and for the convenience of the theory developed in Chapter 8, the objective function of (2–34) is often expressed as

$$\text{Minimize } z = -w$$

With this objective function, the objective functions of the two linear programming problems are equal at optimality.

Model 6 is an example of a competitive situation. Models of this type belong to a branch of mathematics known as the theory of games. For a detailed discussion of this topic, the reader should consult references 3, 7, and 8, at the end of this chapter.

MODEL 7

A dairy has a plant which produces and supplies various flavors of ice cream to a chain of ice cream parlors. All flavors of the ice cream cost approximately the same to produce and all are packed in three-gallon containers. The dairy knows from past experience and from the parlors' own estimates, how much ice cream will be ordered by the parlors during each of the next twelve months:

Month	Jan.	Feb.	Mar.	Apr.	May	Jun.	Jul.	Aug.	Sep.	Oct.	Nov.	Dec.
No. of containers ordered	75	80	85	95	120	140	175	190	175	130	110	90

The dairy has a large freezer-warehouse, and can therefore produce a surplus in any one month, for sale in future months, if this should prove profitable. Moreover, the dairy can go into overtime production if necessary; it is possible that by so doing, the dairy will be able to save money. For example, the increased costs of overtime might be offset by reductions in storage costs.

The ice cream production capabilities of the dairy vary from month to month, due to changing demands in the plant for other products. Below is a table summarizing the production capabilities and the associated per unit costs of production:

Month	Jan.	Feb.	Mar.	Apr.	May	Jun.	Jul.	Aug.	Sep.	Oct.	Nov.	Dec.
Regular time production capacities	100	100	100	90	90	90	90	90	80	80	80	90
Overtime production capacities	40	40	40	40	40	50	50	60	60	60	50	50
Regular time cost per 3 gallons	65¢	65¢	65¢	65¢	70¢	70¢	75¢	75¢	75¢	70¢	70¢	65¢
Overtime cost per 3 gallons	85¢	85¢	85¢	85¢	90¢	90¢	95¢	95¢	95¢	90¢	90¢	85¢

For each month the ice cream is kept in storage, there is a cost of 10¢ per three gallon container.

How many gallons should the dairy produce in each month so that its total costs (production and storage) are a minimum?

Solution

In order to be able to calculate the actual storage costs associated with our solution, we must not only determine how many units (three-gallon containers) are produced each month, but also in which month each of these units is sold to the ice cream parlors. We must also distinguish between units produced on regular time and units produced on overtime. Thus, let us define

x_{ij} = no. of units produced on regular time in month i to be sold in month j $(j \geq i)$

y_{ij} = no. of units produced on overtime in month i to be sold in month j $(j \geq i)$

$$i = 1, 2, \ldots, 12 \quad \text{and} \quad j = i, i + 1, \ldots, 12$$

where the months January, February, . . . , December have been numbered consecutively from 1 to 12.

The above relation between i and j ensures that no units are to be produced *after* they are to be sold!

For each of the twelve months, there are two constraints relating to the regular time and overtime production capacities, respectively. Thus, for January,

$$x_{11} + x_{12} + \ldots + x_{1,12} \leq 100$$
$$y_{11} + y_{12} + \ldots + y_{1,12} \leq 40$$

For February,

$$x_{22} + x_{23} + \ldots + x_{2,12} \leq 100$$
$$y_{22} + y_{23} + \ldots + x_{2,12} \leq 40$$

In general, for the ith month,

$$\begin{aligned} x_{ii} + x_{i,i+1} + \ldots + x_{i,12} &\leq P_{ri} \qquad i = 1, 2, \ldots, 12 \\ y_{ii} + y_{i,i+1} + \ldots + y_{i,12} &\leq P_{oi} \qquad i = 1, 2, \ldots, 12 \end{aligned} \tag{2-36}$$

where P_{ri} and P_{oi} are, respectively, the regular and overtime production capacities in the ith month.

Now, we must ensure that the demands of the parlors for each month j are met by production either in month j or in *prior* months:

The total number of units produced from month 1 through month j, for sale in month j, is

$$(x_{1j} + x_{2j} + \ldots + x_{jj}) + (y_{1j} + y_{2j} + \ldots + y_{jj})$$

This quantity must meet the demand for month j. If we denote this demand by D_j, then

$$(x_{1j} + x_{2j} + \ldots + x_{jj}) + (y_{1j} + y_{2j} + \ldots + y_{jj}) = D_j \qquad j = 1, 2, \ldots, n \qquad (2\text{–}37)$$

For example, the demand constraint for March is

$$(x_{13} + x_{23} + x_{33}) + (y_{13} + y_{23} + y_{33}) = 85$$

Adding the non-negativity restrictions

$$x_{ij}, y_{ij} \geq 0, \qquad i = 1, 2, \ldots, n \quad \text{and} \quad j = i, i + 1, \ldots, n \qquad (2\text{–}38)$$

completes the formulation of the constraints.

Now, let us calculate the objective function. There are three sets of costs: regular time production costs, overtime production costs, and storage costs.

The total regular time production cost is easily seen to be

$$0.65(x_{11} + x_{12} + \ldots + x_{1,12}) + 0.65(x_{22} + x_{23} + \ldots + x_{2,12}) + \ldots + 0.70(x_{11,11} + x_{11,12}) + 0.65(x_{12,12})$$

The total overtime production cost is similarly found to be

$$0.85(y_{11} + y_{12} + \ldots + y_{1,12}) + 0.85(y_{22} + y_{23} + \ldots + y_{2,12}) + \ldots + 0.90(y_{11,11} + y_{11,12}) + 0.85(y_{12,12})$$

If we denote by C_{ri} and C_{oi} the per unit regular time and overtime production costs, respectively, then we can write

$$\text{Total Regular Time Production Costs} = \sum_{i=1}^{12} C_{ri}\left(\sum_{j=i}^{12} x_{ij}\right)$$

$$\text{Total Overtime Production Costs} = \sum_{i=1}^{12} C_{oi}\left(\sum_{j=i}^{12} y_{ij}\right)$$

We must also calculate the total storage cost. Each unit produced in month i for month j will be in storage for exactly $(j - i)$ months. The cost per month, which we shall denote by C_{sj}, is $C_{sj} = \$0.10$ per unit. The total number of units produced in month i for sale in month j is $(x_{ij} + y_{ij})$. Thus, the storage cost for these items is

$$C_{sj}(j - i)(x_{ij} + y_{ij})$$

The total storage cost is then obtained by summing over the twelve months.

$$\text{Total Storage Cost} = \sum_{j=1}^{12} C_{sj} \sum_{i=1}^{j} (j - i)(x_{ij} + y_{ij})$$

Combining these three costs yields the objective function:

$$\text{Minimize } z = \sum_{i=1}^{12} C_{ri}\left(\sum_{j=i}^{12} x_{ij}\right) + \sum_{i=1}^{12} C_{oi}\left(\sum_{j=i}^{12} y_{ij}\right) + \sum_{j=1}^{12} C_{sj} \sum_{i=1}^{j} (j - i)(x_{ij} + y_{ij})$$

Exercise 6: Show that the linear programming problem formulated in Model 7 is identical with a transportation problem. [Hint: re-define the variables y_{ij} as $y_{ij} = x_{n+i,j}$, where n is the total number of months (twelve in Model 7). Assign costs of $+M$, where M is a very large positive number, to variables x_{ij} which must be 0 in a feasible solution (e.g., x_{ij} for which $i > j$, $i = 1, 2, \ldots, n$)]. Show that the resulting transportation problem has $2n$ origins and $(n + 1)$ destinations.

Exercise 7: Explain why an optimal solution to Model 7 will never use overtime production in any one month as long as there is still some regular time production capacity available.

Exercise 8: Suppose we modify Model 7 as follows: The dairy makes 10 flavors of ice cream, and each has a different production cost. In particular, the regular time cost of producing one three-gallon container of flavor k in month i is C_{rik} and the corresponding overtime cost is C_{oik}. Moreover, it is known that the parlors will demand D_{jk} three-gallon containers of flavor k in month j. Assume all other quantities in Model 7 are unchanged (i.e., production capacities and storage costs are independent of flavor).

Formulate this new model as a linear programming problem.

MODEL 8

We conclude this chapter with a model developed by researchers at the National Aeronautics and Space Administration[1] for use on Apollo spacecraft. The problem may be stated as follows:

Suppose a space vehicle has N jets. These jets are used for making alterations in the direction of flight of the vehicle, and also for changing the orientation of the spacecraft (rotating it). These changes are accomplished by firing various combinations of the jets for differing lengths of time.

For each specific desired change, there are a very large number of possible combinations of jet-firing strategies. The spacecraft commander would like to choose the particular strategy which minimizes total fuel consumption (or, equivalently, the total time the jets are fired—and thus consuming fuel).

For purposes of illustration, we shall consider only the case of changing the direction of flight of the spacecraft.

Suppose that when the jth jet is fired, a force is exerted on the spacecraft, with components F_j^x, F_j^y, and F_j^z in the directions x, y, and z, respectively (assuming the usual rectangular coordinate system, with origin at the center of gravity of the spacecraft). Moreover, suppose that the spacecraft is currently moving with a velocity whose respective coordinates are v_x^0, v_y^0, v_z^0.

From Newton's second law, it is known that

$$F = ma \tag{2–39}$$

where F, m, a, denote respectively force, mass, and acceleration. Moreover,

$$a = \frac{dv}{dt}; \tag{2–40}$$

that is, acceleration is the rate of change of the velocity v with respect to time t. Substituting equation (2–30) into (2–39) yields

$$F = m\,\frac{dv}{dt} \tag{2–41}$$

If F and m are constants, we can integrate equation (2–42) and obtain

$$v(t) = \frac{F}{m}\,t + v^0$$

In other words, the velocity at time t is equal to the constant (F/m) times t plus the initial velocity v^0.

For a spacecraft with only one jet, we have one equation of the form (2–41) for each of the three coordinate directions x, y, z. If we denote by v_x, v_y, v_z respectively the desired velocity in the directions x, y, z, then we can write

$$\frac{F^x}{m} t = v_x - v_x^0$$
$$\frac{F^y}{m} t = v_y - v_y^0 \qquad (2\text{–}42)$$
$$\frac{F^z}{m} t = v_z - v_z^0$$

For a spacecraft with N jets, equations (2–42) become

$$\frac{F_1^x}{m} t_1 + \frac{F_2^x}{m} t_2 + \ldots + \frac{F_N^x}{m} t_N = v_x - v_x^0$$
$$\frac{F_1^y}{m} t_1 + \frac{F_2^y}{m} t_2 + \ldots + \frac{F_N^y}{m} t_N = v_y - v_y^0 \qquad (2\text{–}43)$$
$$\frac{F_1^z}{m} t_1 + \frac{F_2^z}{m} t_2 + \ldots + \frac{F_N^z}{m} t_N = v_z - v_z^0$$

where t_j denotes the time that the jth jet is being fired, and m is, of course, the mass of the spacecraft.

Equations (2–43) represent a system of three simultaneous linear equations in N unknowns; $t_1, t_2, \ldots, t_N$. All other quantities are known constants.

Thus, adding the non-negativity restrictions on the times t_j completes the constraint set for our linear programming problem formulation

$$t_j \geq 0, \qquad j = 1, 2, \ldots, N \qquad (2\text{–}44)$$

Finally, we must determine the objective function. As we have indicated earlier, in order to conserve as much fuel as possible, we would like to minimize the total time the jets are fired. Thus, the objective function is simply to

$$\text{Minimize } z = t_1 + t_2 + \ldots + t_N$$

subject to the constraints (2–43) and (2–44).

Exercise 9:

Formulate the following model† as a linear programming problem:

A cement manufacturing company produces 2.5 million barrels of cement annually. Although its kilns are equipped with mechanical collectors for air pollution control, they still emit two pounds of dust for every barrel of cement produced.

† This model is taken from Kohn.[6]

The company has been instructed to reduce these emissions to a total of 0.8 million pounds. To do so, the company can replace the mechanical collectors with four-field electrostatic precipitators which would reduce emissions to 0.5 pounds of dust per barrel of cement, or with five-field electrostatic precipitators which would reduce emissions to 0.2 pounds of dust per barrel of cement. If the capital and operating costs of the four-field precipitator and five-field precipitator are 14¢ and 18¢ per barrel of cement produced, respectively, determine how many barrels should be produced using each type of pollution control method, so that the total cost is a minimum.

REFERENCES

1. Frese, R. C.: An Approach to a Generalized Jet-Select Policy Through Linear Programming. U.S. Government Memorandum 9-18-69, unpublished.
2. Frisch, R.: *Planning for India: Selected Explorations in Methodology*. Bombay, London, New York, Asia Publishing House, 1960.
3. Gale, D.: *The Theory of Linear Economic Models*. New York, McGraw-Hill, 1960.
4. Hadley, G.: *Linear Programming*, Reading, Mass., Addison-Wesley, 1962.
5. Hanssman, F.: Operations Research in National Planning of Underdeveloped Countries, *Operations Research*, 9:230–248, March-April 1961.
6. Kohn, Robert E., "A Mathematical Programming Model for Air Pollution Control," *School Science & Mathematics*, June, 1969.
7. Owen, G.: *Game Theory*. Philadelphia, W. B. Saunders Co., 1968.
8. Von Neumann, J., and O. Morgenstern: *Theory of Games and Economic Behavior*. New York, John Wiley, 1944.

3

Mathematical Background

3-1 MATRICES: NOTATION AND DEFINITIONS†

Matrix: A rectangular array of numbers which will be written as

$$A = \|a_{ij}\| = \begin{bmatrix} a_{11} & a_{12} & \cdots & a_{1n} \\ a_{21} & a_{22} & \cdots & a_{2n} \\ \cdot & & & \cdot \\ \cdot & & & \cdot \\ \cdot & & & \cdot \\ a_{m1} & a_{m2} & \cdots & a_{mn} \end{bmatrix} \tag{3-1}$$

The array so defined (called the matrix A) is said to be an "m by n" matrix or "of order m by n" written "$m \times n$." All matrices in this text will have elements a_{ij} which are real scalars.

Matrix Equality: Two matrices A, B are equal (written $A = B$) if and only if A & B are of the same order, and corresponding elements are equal, i.e., $a_{ij} = b_{ij}$ for all i, j.

Matrix Addition: If a matrix A is of order $m \times n$ and a matrix B is of order $m \times n$ then $C = A + B$ is a matrix each of whose elements is $c_{ij} = a_{ij} + b_{ij}$. Only matrices of the same order may be added.

Example

$$\begin{bmatrix} 2 & 3 & 1 \\ 1 & 2 & 4 \end{bmatrix} + \begin{bmatrix} 0 & 1 & -2 \\ 3 & -1 & 2 \end{bmatrix} = \begin{bmatrix} 2 & 4 & -1 \\ 4 & 1 & 6 \end{bmatrix}$$

† The reader should be familiar with basic linear algebra (see references 1 and 2). The above definitions are merely for reference and to establish notational conventions.

Addition of matrices clearly satisfies both the commutative and associative laws, i.e.,

$$\begin{aligned} A + B &= B + A \\ A + (B + C) &= (A + B) + C \end{aligned} \tag{3-2}$$

Scalar Multiplication: A matrix A can be multiplied by a real scalar α by multiplying each element of A by α. Hence

$$\alpha A = \|\alpha a_{ij}\| \tag{3-3}$$

Example

$$3\begin{bmatrix} 1 & 2 \\ -4 & 0 \\ 5 & 7 \end{bmatrix} = \begin{bmatrix} 3 & 6 \\ -12 & 0 \\ 15 & 21 \end{bmatrix}$$

Matrix Multiplication: The product AB of two matrices A, B is defined if and only if the number of columns of A equals the number of rows of B. Hence if A is an $m \times p$ matrix and B is a $p \times n$ matrix, the product $C = \|c_{ij}\|$ will exist and will be an $m \times n$ matrix. The elements c_{ij} of C are given by

$$c_{ij} = \sum_{k=1}^{p} a_{ik} b_{kj} \qquad \begin{aligned} i &= 1, 2, \ldots, m \\ j &= 1, 2, \ldots, n \end{aligned} \tag{3-4}$$

Example

$$A = \begin{bmatrix} 1 & 2 \\ 3 & 0 \\ 5 & 4 \end{bmatrix}, \qquad B = \begin{bmatrix} 4 & 2 \\ 1 & 3 \end{bmatrix}$$

$$AB = \begin{bmatrix} 1 & 2 \\ 3 & 0 \\ 5 & 4 \end{bmatrix} \begin{bmatrix} 4 & 2 \\ 1 & 3 \end{bmatrix} = \begin{bmatrix} 1(4) + 2(1) & 1(2) + 2(3) \\ 3(4) + 0(1) & 3(2) + 0(3) \\ 5(4) + 4(1) & 5(2) + 4(3) \end{bmatrix} = \begin{bmatrix} 6 & 8 \\ 12 & 6 \\ 24 & 22 \end{bmatrix}$$

Matrix multiplication satisfies the associative and distributive laws, i.e.,

$$\begin{aligned} (AB)C &= A(BC) \\ A(B + C) &= AB + AC \end{aligned} \tag{3-5}$$

provided that the multiplications and/or additions are defined. Note that matrix multiplication is *not* commutative, i.e., $AB \neq BA$, in general. Using A, B of the example, BA is not even defined. Even when defined the two products are generally not equal.

Null Matrix: A matrix of any order, all of whose elements are zero, is called a null matrix. It is denoted O. Provided that the operations are defined, the following are true:

$$\begin{aligned} A + O &= A \\ AO &= O \\ OA &= O \end{aligned} \tag{3-6}$$

Identity Matrix: A square matrix of any order, which has ones along the principal diagonal and zeros elsewhere, is called an identity matrix. It is designated I or I_n (assuming the order is n). Hence,

$$I = \begin{bmatrix} 1 & 0 & 0 & \dots & 0 \\ 0 & 1 & 0 & \dots & 0 \\ 0 & 0 & 1 & \dots & 0 \\ \cdot & \cdot & \cdot & & \cdot \\ \cdot & \cdot & \cdot & & \cdot \\ \cdot & \cdot & \cdot & & \cdot \\ 0 & 0 & 0 & \dots & 1 \end{bmatrix}$$

Symbolically, $I = \|\delta_{ij}\|$ where δ_{ij} is the Kronecker delta, i.e.,

$$\delta_{ij} = \begin{cases} 1, & i = j \\ 0, & i \neq j \end{cases}$$

It is understood that $i, j = 1, 2, \dots, n$, if I is an nth order matrix, i.e., $n \times n$.

Scalar Matrix: If α is a real scalar, then the square matrix $C = \|\alpha\delta_{ij}\| = \alpha I$ is called a scalar matrix.

Diagonal Matrix: If α_i are real scalars, then the square matrix $D = \|\alpha_i\delta_{ij}\|$ is called a diagonal matrix.

Matrix Transpose: If we interchange the rows and columns of a matrix A, the resulting matrix, which is denoted A', or A^T, is called the transpose of matrix A. In the process of transposition, the ith row of A becomes the ith column of A', i.e., $A' = \|a'_{ij}\|$ where $a'_{ij} = a_{ji}$.

Example

$$A = \begin{bmatrix} 1 & 2 & 4 & 2 \\ 5 & 3 & 0 & -1 \\ -1 & 2 & 3 & 6 \end{bmatrix}, \qquad A' = \begin{bmatrix} 1 & 5 & -1 \\ 2 & 3 & 2 \\ 4 & 0 & 3 \\ 2 & -1 & 6 \end{bmatrix}$$

The process of matrix transposition has the following properties:

$$\begin{aligned} (A + B)' &= A' + B' \\ (AB)' &= B'A' \\ (A')' &= A \end{aligned} \tag{3–7}$$

Symmetric Matrix: A matrix A which has the property that $A = A'$ is called a symmetric matrix.

Example

$$A = \begin{bmatrix} 2 & 1 & 4 \\ 1 & 0 & 3 \\ 4 & 3 & 7 \end{bmatrix} = A'$$

Determinants: We can associate with every square matrix A, a number which is called the determinant of A. It is usually denoted by $|A|$. The determinant of A is defined as

$$|A| = \sum (\pm) a_{1i} a_{2j} \ldots a_{np} \tag{3–8}$$

The sum in (3–8) is taken over all permutations of the second subscript. A plus sign precedes a given term if the subscripts $(i, j, \ldots, p)$ are an even permutation† of $(1, 2, \ldots, n)$ and a minus sign is used if the subscripts are an odd permutation. From the definition (3–8) it is easily seen that each term of the summation has n elements if A is an $n \times n$ matrix and that there are $n!$ terms.

Example

$$\begin{vmatrix} a_{11} & a_{12} & a_{13} \\ a_{21} & a_{22} & a_{23} \\ a_{31} & a_{32} & a_{33} \end{vmatrix} = \begin{aligned} & a_{11}a_{22}a_{33} - a_{12}a_{21}a_{33} + a_{12}a_{23}a_{31} \\ & - a_{13}a_{22}a_{31} + a_{13}a_{21}a_{32} - a_{11}a_{23}a_{32} \end{aligned} \tag{3–9}$$

† An even or odd permutation is one in which the number of binary inversions of the natural numbers $1, 2, \ldots, n$ from their natural order is even or odd, i.e., the numbers of pairs of digits that are out of order, is even or odd. For example, (1, 3, 2, 5, 4) has three inversions and is therefore odd.

The first term $a_{11}a_{22}a_{33}$, in which the second subscripts are in their natural order, is considered an even permutation.

The *minor* D_{ij} of the element a_{ij} of a matrix A is the determinant obtained by striking out the ith row and j^{th} column. The *cofactor* A_{ij} of the element a_{ij} of A equals $(-1)^{i+j}D_{ij}$. It can be shown that the determinant of an nth order matrix A can be expressed as:

$$|A| = \sum_{j=1}^{n} a_{ij}A_{ij} \qquad i = 1, 2, \ldots, n \tag{3–10}$$

where A_{ij} is the cofactor of element a_{ij}. It is equally true that:

$$|A| = \sum_{i=1}^{n} a_{ij}A_{ij} \qquad j = 1, 2, \ldots, n \tag{3–11}$$

What (3–10) and (3–11) provide is a method for evaluating a determinant by successive expansions.

Example

We shall evaluate the following determinant by expanding in terms of the first row

$$\begin{vmatrix} 2 & -3 & 7 \\ 1 & 4 & 2 \\ 3 & 5 & 6 \end{vmatrix} = 2\begin{vmatrix} 4 & 2 \\ 5 & 6 \end{vmatrix} -(-3)\begin{vmatrix} 1 & 2 \\ 3 & 6 \end{vmatrix} + 7\begin{vmatrix} 1 & 4 \\ 3 & 5 \end{vmatrix}$$

$$= 2(24 - 10) + 3(6 - 6) + 7(5 - 12) = -21$$

Matrix Inverse: A matrix B is called the inverse of the square matrix A if $AB = I$. The inverse of A is usually denoted A^{-1}. A square matrix A is said to be *non-singular* if $|A| \neq 0$. If $|A| = 0$, the matrix is *singular*. For every non-singular matrix A there exists a unique A^{-1} such that:

$$AA^{-1} = A^{-1}A = I \tag{3–12}$$

Furthermore, it can be shown that if $|A| \neq 0$,

$$A^{-1} = \frac{1}{|A|}R \tag{3–13}$$

where R is the *adjoint matrix* of A and is defined as the matrix of transposed cofactors of the matrix A, i.e.,

$$R = \begin{bmatrix} A_{11} & A_{21} & \cdots & A_{n1} \\ A_{12} & A_{22} & \cdots & A_{n2} \\ \cdot & \cdot & & \cdot \\ \cdot & \cdot & & \cdot \\ \cdot & \cdot & & \cdot \\ A_{1n} & A_{2n} & \cdots & A_{nn} \end{bmatrix} \tag{3–14}$$

Some important properties of the inverse are

$$\begin{aligned} (AB)^{-1} &= B^{-1}A^{-1} \\ (A^{-1})^{-1} &= A \\ (A')^{-1} &= (A^{-1})' \end{aligned} \tag{3–15}$$

Rank of Matrix A: The rank of any matrix A is the order of the largest square array in A whose determinant does not vanish. The rank of A is denoted $r(A)$.

Example

$$A = \begin{bmatrix} 2 & 3 & 1 \\ 4 & 2 & 0 \end{bmatrix} \quad \text{and} \quad r(A) = 2$$

$$\text{Since} \begin{vmatrix} 2 & 3 \\ 4 & 2 \end{vmatrix} = -8 \neq 0$$

$$B = \begin{bmatrix} 2 & 3 & 0 \\ 4 & 6 & 0 \end{bmatrix} \quad \text{and} \quad r(B) = 1$$

since only first order determinants do not vanish.

3–2 VECTORS AND EUCLIDEAN SPACES

Vectors: A matrix with a single row or a single column will be called a vector. We will adopt the convention that all vectors are *column vectors*, i.e., are matrices with a single column. We designate vectors as $\bar{a} = [a_1, a_2, \ldots, a_m]$ or $\bar{x} = (x_1, x_2, \ldots, x_n)$ depending upon the context. However, these will always be column vectors. If we wish to designate a *row* *vector*, we write $\bar{a}'$ or $\bar{x}'$, since the transpose of a matrix with a single column is a single row. Just as (x_1, x_2, x_3) can be considered to be a point in a three-dimensional space, so can a point $\bar{x} = (x_1, x_2, \ldots, x_n)$ be considered to be a point in an n-dimensional space.

Unit Vector: A vector whose i^{th} component is one and all of whose other components are zero is called the unit vector $\bar{e}_i$. If we deal with n component vectors, then there are n such unit vectors

$$\begin{aligned} \bar{e}_1 &= (1, 0, \ldots, 0) \\ \bar{e}_2 &= (0, 1, \ldots, 0) \\ &\cdot \\ &\cdot \\ &\cdot \\ \bar{e}_n &= (0, 0, \ldots, 1) \end{aligned} \tag{3–16}$$

It can be seen that the unit vectors are each single columns of the identity matrix, I.

Null Vector: A vector, all of whose components are zero, is called the null vector. A null vector can be of any dimension. We designate the null vector $\bar{0}$. Therefore,

$$\bar{0} = (0, 0, \ldots, 0) \tag{3–17}$$

Sum Vector: A vector having the value one for each of its components is called the sum vector. It is written as $\bar{1}$. Therefore,

$$\bar{1} = (1, 1, \ldots, 1) \tag{3–18}$$

The name derives from the fact that if we multiply the two matrices $\bar{1}'$ and any column matrix $\bar{c}$ we obtain, e.g.,

$$\bar{1}'\bar{c} = (1, 1, 1)\begin{pmatrix} c_1 \\ c_2 \\ c_3 \end{pmatrix} = c_1 + c_2 + c_3 = \sum_{j=1}^{3} c_j$$

Scalar Product: The scalar product of two vectors $\bar{a}$ and $\bar{b}$ is denoted $\bar{a}'\bar{b}$ and is given by

$$\bar{a}'\bar{b} = \sum_{i=1}^{n} a_i b_i \tag{3–19}$$

Distance: In two- and three-dimensional space, the origin can be considered to be a vector (0, 0) or (0, 0, 0) and any point (x_1, x_2) or (x_1, x_2, x_3)

can be expressed as:

$$(x_1, x_2) = x_1(1, 0) + x_2(0, 1)$$
$$(x_1, x_2, x_3) = x_1(1, 0, 0) + x_2(0, 1, 0) + x_3(0, 0, 1)$$

It will be recognized that vectors such as (1, 0, 0) are simply points located a unit distance on one of the co-ordinate axes. Similarly, a point in an n-dimensional space can be represented

$$\bar{x} = x_1\bar{e}_1 + x_2\bar{e}_2 + \dots + x_n\bar{e}_n \tag{3–20}$$

Hence the unit vectors $\bar{e}_i$ define a co-ordinate system for n-dimensional space.

The *distance* between two points $\bar{x}$ and $\bar{y}$ in n-dimensional space will be denoted $|\bar{x} - \bar{y}|$ and it is defined by

$$|\bar{x} - \bar{y}| = [(\bar{x} - \bar{y})'(\bar{x} - \bar{y})]^{1/2} = \left[\sum_{i=1}^{n} (x_i - y_i)^2\right]^{1/2} \tag{3–21}$$

This definition obviously derives from a generalization of distance in two and three-dimensional spaces.

Euclidean Space: An n-dimensional Euclidean space is the collection of all points or vectors $\bar{x} = (x_1, x_2, \dots, x_n)$. Addition and multiplication by a scalar is defined as for matrices. Associated with any two vectors is a number which we call the distance between the vectors and which is given by (3–21). We denote an n-dimensional Euclidean space by E^n.

Linear Combination: A vector $\bar{x}$ in E^n is a linear combination of the vectors $\bar{x}_1, \bar{x}_2, \dots, \bar{x}_p$ in E^n if

$$\bar{x} = \alpha_1\bar{x}_1 + \alpha_2\bar{x}_2 + \dots + \alpha_p\bar{x}_p \tag{3–22}$$

for some set of scalars α_i, $i = 1, 2, \dots, p$.

Linear Dependence: A set of vectors $\bar{x}_1, \bar{x}_2, \dots, \bar{x}_p$ in E^n is said to be linearly dependent if there exists a set of scalars α_i, $i = 1, 2, \dots, p$, not all of which are zero, such that

$$\alpha_1\bar{x}_1 + \alpha_2\bar{x}_2 + \dots + \alpha_p\bar{x}_p = \bar{0} \tag{3–23}$$

If the only set of α_i for which (3–23) holds is $\alpha_i = 0$, $i = 1, 2, \dots, p$, then the set of vectors is said to be *linearly independent*. A set of vectors $\bar{x}_1, \bar{x}_2, \dots, \bar{x}_p$ in E^n is linearly dependent if and only if some one of the vectors is a linear combination of the other vectors. A vector $\bar{x}$ is said to be linearly dependent on a set of vectors $\bar{x}_1, \bar{x}_2, \dots, \bar{x}_p$ if $\bar{x}$ can be written as a linear combination of $\bar{x}_1, \bar{x}_2, \ldots, \bar{x}_p$; otherwise x is said to be linearly independent of $\bar{x}_1, \bar{x}_2, \ldots, \bar{x}_p$.

If a set of vectors is linearly independent, any subset of these vectors is also linearly independent. On the other hand, if a given set of vectors is linearly dependent, any larger set containing the given set as a subset will also be linearly dependent.

The unit vectors for E^n, i.e., $\bar{e}_1, \bar{e}_2, \ldots, \bar{e}_n$ are an important example of a linearly independent set of vectors since

$$\sum_{i=1}^{n} \alpha_i \bar{e}_i = (\alpha_1, \alpha_2, \ldots, \alpha_n) = \bar{0}$$

implies that $\alpha_i = 0$, $i = 1, 2, \ldots, n$.

Spanning Set of Vectors: A set of vectors $\bar{x}_1, \bar{x}_2, \ldots, \bar{x}_p$ in E^n is said to span the Euclidean space E^n, if every vector in E^n can be written as a linear combination of $\bar{x}_1, \bar{x}_2, \ldots, \bar{x}_p$. It is obvious that the unit vectors $\bar{e}_1, \bar{e}_2, \ldots, \bar{e}_n$ span E^n since any vector $\bar{x}$ can be written

$$\bar{x} = x_1\bar{e}_1 + x_2\bar{e}_2 + \ldots + x_n\bar{e}_n \tag{3-24}$$

It is also clear that the smallest set of vectors that can span E^n must be a linearly independent set.

Basis: A linearly independent set of vectors from E^n which spans the entire space E^n is called a basis for E^n. The unit vectors are a basis for E^n since they are linearly independent. There can be an infinite number of bases for E^n. However, the representation of any vector in terms of a given set of basis vectors is unique. In other words, given a set of basis vectors and some arbitrary vector, this vector can be expressed as a linear combination of the basis vectors in only one way, i.e., the α_i, $i = 1, 2, \ldots, p$ are unique.

It is not difficult to show that any basis for E^n must contain exactly n linearly independent vectors. Hence the dimension of a Euclidean space is characterized both by the number of components of the vectors which make up the space as well as the number of vectors in any basis for the space.

Basic Solution: Suppose we have a set of m simultaneous linear equations in n variables, i.e., $A\bar{x} = \bar{b}$ and $r(A) = m$, $m < n$. Let us set $n - m$ of the variables to zero, so that the resulting set of equations includes m of the columns of the matrix A such that these columns (vectors) are linearly independent. (In effect, we are choosing an $m \times m$ non-singular submatrix from A.) The solution to this new set of m equations in m variables is called a *basic solution.* These m variables which may be different from zero are called *basic variables.* The reason for this name is that the m linearly independent columns of A form a basis for E^m.

If we denote the submatrix of m linearly independent columns of A by B, then since $|B| \neq 0$, we know that the basic solution can be obtained from

$$\bar{x}_B = B^{-1}\bar{b} \tag{3-25}$$

where $\bar{x}_B$ is the vector of basic variables. An upper bound on the number of possible basic solutions is $n!/[m!\,(n - m)!]$. This upper bound canbe attained only if every possible combination of m columns of the matrix A is a linearly independent set.

Degenerate Basic Solution: A basic solution to the m equations in n variables, $A\bar{x} = \bar{b}$ is said to be degenerate if one or more of the basic variables is zero. Otherwise the solution is said to be non-degenerate.

3–3 HYPERPLANES AND CONVEX SETS

In this section we shall extend certain notions which are familiar in two- and three-dimensional spaces to n-dimensional space. We shall use the usual notation of set theory, since a vector can be regarded as a point in E^n. Hence, we shall deal with point sets, i.e., sets whose elements are points in E^n. The expression

$$S = \{\bar{x} \mid 2x_1 + 3x_2 = 10\}$$

will be understood to mean "all points $\bar{x} = (x_1, x_2)$ which lie on the straight line $2x_1 + 3x_2 = 10$." In general, the expression

$$S = \{\bar{x} \mid P(\bar{x})\}$$

will mean the set of points which have the properties described by the expression or expressions $P(\bar{x})$.

Line: A line in E^n is defined most conveniently as a set of points, i.e., a line through two given points $\bar{x}_1$ and $\bar{x}_2$, $\bar{x}_1 \neq \bar{x}_2$ is defined as the set of points

$$L = \{\bar{x} \mid \bar{x} = \alpha\bar{x}_1 + (1 - \alpha)\bar{x}_2, \quad \text{all real } \alpha\} \tag{3-26}$$

Hyperplanes: The analogue of a line in E^2 and a plane in E^3 is a *hyperplane* in E^n. For example, $2x_1 + 3x_2 = 10$ is a line in E^2 and $3x_1 + 2x_2 + 4x_3 = 20$ is a plane in E^3. In E^n, the set of points $\bar{x}$ which satisfy

$$c_1x_1 + c_2x_2 + \ldots + c_nx_n = z \tag{3-27}$$

defines a hyperplane for fixed values of c_i, $i = 1, 2, \ldots, n$ and z. If we define $\bar{c} = (c_1, c_2, \ldots, c_n)$ then we may write the vector equation for a hyperplane as $\bar{c}'\bar{x} = z$. A hyperplane will pass through the origin if and only if $z = 0$. Hence $\bar{c}'\bar{x} = 0$ is a hyperplane through the origin. The vector $\bar{c}$ is normal to the hyperplane $\bar{c}'\bar{x} = z$. Any two hyperplanes such that

$$\bar{c}_1'\bar{x} = z_1, \qquad \bar{c}_2'\bar{x} = z_2 \text{ and } \bar{c}_1 = \alpha\bar{c}_2, \qquad \alpha \neq 0$$

are parallel to each other. From this it is easy to deduce that if $\bar{c}'\bar{x} = z_1$ and $\bar{c}'\bar{x} = z_2$ with $z_2 > z_1$ then these hyperplanes are parallel, and that moving a hyperplane $\bar{c}'\bar{x} = z_1$ parallel to itself is accomplished by increasing (or decreasing) a given value of z.

Half Spaces: A hyperplane $\bar{c}'\bar{x} = z$ divides E^n into three sets of points. These are points on the hyperplane, and points on one or the other side of the hyperplane, i.e.,

$$S_1 = \{\bar{x} \mid \bar{c}'\bar{x} = z\}$$
$$S_2 = \{\bar{x} \mid \bar{c}'\bar{x} < z\}$$
$$S_3 = \{\bar{x} \mid \bar{c}'\bar{x} > z\}$$

The sets S_2 and S_3 are called *open half-spaces.* If we consider all points lying on a hyperplane or on one side of it we obtain sets $S_4 = \{\bar{x} \mid \bar{c}'\bar{x} \leq z\}$, $S_5 = \{\bar{x} \mid \bar{c}'\bar{x} \geq z\}$. These are called *closed half-spaces.*

Convex Combination: Given a set S and that $\bar{x}_1$ and $\bar{x}_2$ are points in S, a convex combination of the two points is defined as $\bar{x} = \alpha\bar{x}_1 + (1 - \alpha)\bar{x}_2$, $0 \leq \alpha \leq 1$. A convex combination of a finite number of points $\bar{x}_1, \bar{x}_2, \ldots, \bar{x}_p$ is defined as

$$\bar{x} = \sum_{i=1}^{p} \beta_i\bar{x}_i, \qquad \beta_i \geq 0 \qquad i = 1, 2, \ldots, p$$

$$\text{where } \sum_{i=1}^{p} \beta_i = 1$$

Convex Set: A set S is convex if for any points $\bar{x}_1$, $\bar{x}_2$ in S, every convex combination of these points is in the set. In other words, a set S is convex if for any points $\bar{x}_1$, $\bar{x}_2$, the line segment joining the points is also in the set.

Example

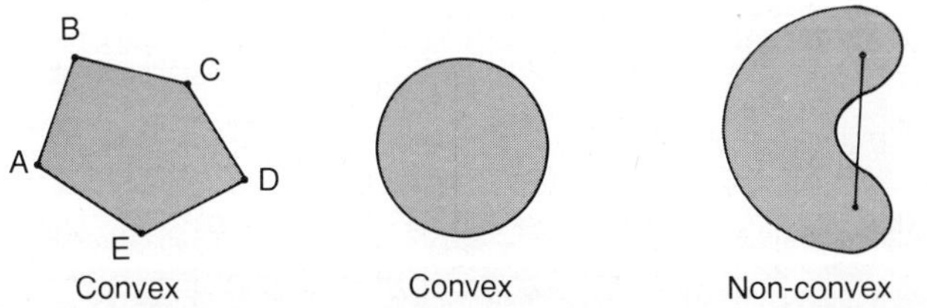

Extreme Point: A point $\bar{x}$ is an extreme point of a convex set S, if and only if it is not possible to represent $\bar{x}$ as a convex combination of some two other points in the set, i.e., there do *not* exist points $\bar{x}_1$, $\bar{x}_2$ such that $\bar{x} = \alpha\bar{x}_1 + (1 - \alpha)\bar{x}_2$, $0 < \alpha < 1$. In the example under "convex set," the points A, B, C, D, E are the only extreme points of the pentagon and its interior. The circle and its interior constitute a convex set. It has an infinite number of extreme points; namely all points on the circle.

Convex Set Properties: The intersection of any number of convex sets is a convex set. It is also easy to show that hyperplanes and open and closed half-spaces are convex sets. Hence, it is clear that the intersection of any number of hyperplanes and/or half-spaces will be a convex set.

The set of all convex combinations of a finite number of points is called the *convex polyhedron* generated by these points. A result of some importance which can be proved is the following: Any point in a closed, strictly bounded convex set S, with a finite number of extreme points, can be written as a convex combination of the extreme points. In other words, S is the set of all convex combinations of its extreme points. A convex polyhedron is a convex set with a finite number of extreme points. However, a convex polyhedron may not be strictly bounded. An example of such a case is shown in Figure 1–3.

PROBLEMS

1. Calculate the products AB and BA when they are defined

$$A = \begin{bmatrix} 1 & 3 & 2 & 5 \\ 2 & 1 & 6 & 0 \\ 0 & 4 & 7 & 2 \end{bmatrix} \qquad B = \begin{bmatrix} 2 & 8 \\ 1 & 0 \\ 0 & 2 \\ 4 & 3 \end{bmatrix}$$

$$A = \begin{bmatrix} 1 & 3 & 0 \\ 4 & 9 & 7 \end{bmatrix} \qquad B = \begin{bmatrix} 1 & 6 \\ 5 & 3 \\ 2 & 1 \end{bmatrix}$$

$$A = \begin{bmatrix} 1 & 0 & 0 \\ 0 & 1 & 0 \\ 0 & 0 & 1 \end{bmatrix} \qquad B = \begin{bmatrix} 7 & 6 & 3 \\ 2 & 1 & 4 \\ 0 & 1 & 3 \end{bmatrix}$$

2. Show that

$$(A + B)'(AB)' = A'B'A' + B'B'A'$$

3. Calculate the inverse of the following matrices

$$A = \begin{bmatrix} 2 & 4 \\ 1 & 3 \end{bmatrix} \qquad B = \begin{bmatrix} 1 & 2 & 3 \\ 3 & 1 & 2 \\ 2 & 1 & 3 \end{bmatrix}$$

4. What is the rank of the following matrices?

$$A = \begin{bmatrix} 1 & 0 \\ -1 & 2 \end{bmatrix}; \qquad B = \begin{bmatrix} 3 & 0 & 1 \\ 0 & -4 & 0 \\ 0 & 0 & 5 \end{bmatrix}$$

$$C = \begin{bmatrix} 2 & 0 \\ 0 & 8 \end{bmatrix} \qquad D = \begin{bmatrix} 3 & 0 & 0 \\ 0 & 0 & 2 \\ 4 & 0 & 5 \end{bmatrix}$$

$$E = \begin{bmatrix} 8 & 4 \\ 1 & \frac{1}{2} \end{bmatrix} \qquad F = \begin{bmatrix} 2 & 1 & 0 \\ -1 & -2 & 1 \end{bmatrix}$$

$$G = \begin{bmatrix} 3 \\ 1 \end{bmatrix} \qquad H = \begin{bmatrix} 1 & 0 & 1 \\ 0 & 1 & 0 \end{bmatrix}$$

5. Are the following sets of vectors linearly dependent or linearly independent?

a) [1, 0, 3], [2, 1, 1], [3, 4, 0]
b) [3, 1, 2], [4, 2, 8], [2, 1, 4]
c) [1, 0, 2], [3, 1, 4]

6. Which of the following sets of vectors form a basis for E^3?

a) [2, 1, 1], [0, 1, 1], [3, 1, 4]
b) [1, −3, 2], [3, 1, 2], [6, 2, 4]
c) [1, 0, 1], [2, 0, 3], [−1, 0, 2]
d) [3, 6, 1], [6, 2, 8], [1, 0, 1]

7. Can the vector $\bar{b} = [3, 4, 2]$ be expressed in terms of the vectors

$$\bar{a}_1 = [3, 0, 6]; \ \bar{a}_2 = [2, 0, 4]; \ \bar{a}_3 = [0, 1, 0]$$

What does your result mean?

8. Find all basic solutions to the following set of equations:

$$\begin{aligned} 4x_1 + 2x_2 + x_3 + 3x_4 + 2x_5 &= 10 \\ 2x_1 + 3x_2 + 3x_3 + 2x_4 + 5x_5 &= 20 \end{aligned}$$

9. Consider a rectangle whose vertices are (0, 0), (1, 0), (1, 1), (0, 1). Express the following points as convex combinations of the four extreme points of the

convex set defined by the rectangle and its interior:

a) (0.4, 0.6)
b) (0.9, 0.3)

10. Which of the following are convex sets?

$$S_1 = \{\bar{x} \mid 2x_1^2 + 2x_1x_2 + x_2^2 \leq 10\}$$
$$S_2 = \{\bar{x} \mid x_1^2 + x_2^2 = 10\}$$
$$S_3 = \{\bar{x} \mid x_1 + x_2 \leq 2, 4x_1 + 3x_2 \geq 12\}$$
$$S_4 = \{\bar{x} \mid x_1 \geq 0, x_2 \geq 0, x_2 \leq 5\}$$

11. Is the non-negative orthant of E^2 a convex set? Justify your answer geometrically. Is all of E^2 a convex set?

REFERENCES

1. Hadley, G.: *Linear Algebra.* Reading, Mass., Addison-Wesley, 1961.
2. Steinberg, D.: *Computational Matrix Algebra*, New York, McGraw-Hill, 1974.

4

The Simplex Method

4-1 INTRODUCTION

In this section we shall state the standard linear programming problem, indicate its relationship to the various ways a "linear programming problem" may be formulated, and provide the basic terminology we shall employ throughout the remainder of the discussion of linear programming.

It will be recalled from the discussion in Chapters 1 and 2 that when an actual linear programming problem is formulated, the constraints may be mixed in the sense that some or all of them may be in the form of inequalities and some or all of them may hold as strict equalities. Nevertheless, we shall always write a linear programming problem in what we shall call the "standard form" as follows:

$$\text{Max } z = \sum_{j=1}^{n} c_j x_j$$

$$\sum_{j=1}^{n} a_{ij} x_j = b_i \qquad i = 1, 2, \ldots, m \tag{4-1}$$

$$x_j \geq 0 \qquad j = 1, 2, \ldots, n$$

or equivalently, in matrix notation as

$$\begin{aligned} \text{Max } z &= \bar{c}'\bar{x} \\ A\bar{x} &= \bar{b} \\ \bar{x} &\geq \bar{0} \end{aligned} \tag{4-2}$$

In order to do this, it is apparent that we have somehow converted those constraints of the original problem that were stated as inequalities into some sort of "equivalent" equality form. Let us indicate how this is done and then provide the theoretical justification for doing so.

Consider a typical inequality:

$$\sum_{j=1}^{p} a_{ij} x_j \leq b_i \tag{4-3}$$

If we add an additional non-negative variable x_{p+1}, to the left hand side of equation (4–3) we then have

$$\sum_{j=1}^{p} a_{ij}x_j + x_{p+1} = b_i \qquad (4\text{–}4)$$
$$x_{p+1} \geq 0$$

The variable x_{p+1} is referred to as a *slack variable*. In the same way a constraint of the form

$$\sum_{j=1}^{p} a_{ij}x_j \geq b_i$$

can be rewritten as

$$\sum_{j=1}^{p} a_{ij}x_j - x_{p+1} = b_i \qquad (4\text{–}5)$$
$$x_{p+1} \geq 0$$

In this case the variable x_{p+1} is called a *surplus variable*.

Let us now show that the addition of slack and surplus variables does not change in any way the set of possible solutions to the original problem statement. We do this in the following theorem.

Theorem 4–1 Given a constraint set of the form

$$\begin{aligned} D\bar{x}\{\leq, =, \geq\}\bar{d} \\ \bar{x} \geq \bar{0} \end{aligned} \qquad (4\text{–}6)$$

(where D is an $m \times p$ matrix, $\bar{d}$ is an m–vector, and $\bar{x}$ is a p–vector) and also the constraint set obtained by adding $n - p$ slack or surplus variables

$$\begin{aligned} A\bar{y} = \bar{d} \\ y \geq \bar{0} \end{aligned} \qquad (4\text{–}7)$$

where A is an $m \times n$ matrix and y is an n–vector. Then every solution to (4–6) corresponds to a solution to (4–7) and, conversely, every solution to (4–7) corresponds to a solution to (4–6).

Proof: We shall partition the matrix $A = (D, S)$ where S is an $m \times (n - p)$ sub-matrix of coefficients corresponding to the new slack and surplus variables, $\bar{x}_s$ and where $\bar{y} = \begin{bmatrix} \bar{x} \\ \bar{x}_s \end{bmatrix}$. Let us now consider any solution $\bar{x}^0$ to (4–6). It is clear that if we substitute $\bar{x}^0$ in (4–7) we can solve for vector $\bar{x}_s^0$ such that a vector $\bar{y}^0$ will be defined which satisfies (4–7). Conversely, given a solution $\bar{y}^1$ to (4–7), this can be partitioned as $\bar{y}^1 = \begin{bmatrix} \bar{x}^1 \\ \bar{x}_s^1 \end{bmatrix}$ and we see that we can satisfy (4–6) even if it is necessary to make some or all of the components of $\bar{x}_s^1 = 0$: Hence, we have established a one-to-one correspondence between the solutions of (4–6) and (4–7).

Since the simplex algorithm makes use of the basic notions of Gaussian elimination or pivoting for solving sets of linear equations, it is convenient to deal with systems of equalities. Theorem 4–1 provides the justification for converting a mixed system into a system of equalities with an enlarged number of non-negative variables. This form is the one we shall deal with henceforth, both theoretically and computationally.

We take as our standard linear programming problem

$$\begin{aligned} \text{Max } z &= \bar{c}'\bar{x} \\ A\bar{x} &= \bar{b} \\ \bar{x} &\geq \bar{0} \end{aligned} \tag{4–8}$$

where $\bar{c}$, $\bar{x}$ are n-vectors, $\bar{b}$ is an m-vector and A is an $m \times n$ matrix. If $\bar{a}_j$ denotes the jth column of A, then the system of equations $A\bar{x} = \bar{b}$ can also be written

$$\bar{a}_1\bar{x}_1 + \bar{a}_2x_2 + \ldots + \bar{a}_nx_n = \bar{b}$$

With this in mind, we now state some definitions which will be used throughout the remainder of this book:

Feasible Solution: A feasible solution to the linear programming problem (4–8) is a vector $\bar{x} = (x_1, x_2, \ldots, x_n)$ which satisfies all the constraints of the problem.

Basic Feasible Solution: A basic feasible solution to the linear programming problem is a feasible solution with no more than m positive x_j and such that the vectors $\bar{a}_j$ associated with these latter x_j are a linearly independent set of vectors.

Nondegenerate Basic Feasible Solution: A nondegenerate basic feasible solution is a basic feasible solution with exactly m positive x_j. Hence a *degenerate* basic feasible solution has fewer than m positive x_j.

4–2 NECESSARY ANALYTICAL RESULTS

The constraint equations of (4–1) or (4–2) correspond in geometric terms, as was indicated graphically in Chapter 1, to hyperplanes in an n-dimensional space. As we noted in Chapter 3, the intersection of these hyperplanes is another hyperplane, of lower dimension. The values of $\bar{x}$ which lie on this intersection hyperplane, that is, which satisfy all the constraints, form a convex set, X_c, of points. This will be proven in Theorem 4–2below. Moreover, this convex set will have a finite number of extreme points or vertices. A point $\bar{x}^*$ corresponding to an extreme point of X_c will have at most m non-zero components, if there are m constraints. This will be proved in Theorem 4–5.The extreme points of X_c are of great importance because, as will be shown in Theorem 4–3 the optimal solution to a linear programming problem, that is, the maximum or minimum value of the objective function, occurs at an extreme point of the convex set, X_c.

In terms of the definitions given in Section 4–1 and the previous discussion, we shall focus our attention on basic solutions to the equations of (4–1), to solutions obtained by allowing $n - m$ of the variables to be zero. In addition we shall concentrate on solutions which are also non-negative, i.e., that satisfy the restriction, $\bar{x} \geq \bar{0}$. Let us now prove the necessary theorems.

Theorem 4–2 The set of points $\bar{x}$ satisfying $A\bar{x} = \bar{b}$, $\bar{x} \geq \bar{0}$ forms a convex set.

Proof. Let $X_c = \{\bar{x} \mid A\bar{x} = \bar{b}, \bar{x} \geq \bar{0}\}$. We wish to show that X_c is a convex set. A set is convex, by definition, if given any two points in the set, $\bar{x}_1$, $\bar{x}_2$, a convex combination is also in the set. Let us show this.

$$\text{Let } \bar{x}_c = \lambda \bar{x}_1 + (1 - \lambda)\bar{x}_2, \qquad 0 \leq \lambda \leq 1$$

In order for $\bar{x}_c$ to be in X_c, it must satisfy $A\bar{x} = \bar{b}$ and $\bar{x} \geq \bar{0}$. Since $\bar{x}_1 \geq \bar{0}$ and $\bar{x}_2 \geq \bar{0}$ by hypothesis, so are $\lambda\bar{x}_1 \geq 0$, $(1 - \lambda)x_2 \geq \bar{0}$, since λ and $1 - \lambda$ are both non-negative. Therefore

$$\bar{x}_c = \lambda \bar{x}_1 + (1 - \lambda)\bar{x}_2 \geq \bar{0} \tag{4–9}$$

Now let us consider the constraints, $A\bar{x} = \bar{b}$. Since $A\bar{x}_1 = \bar{b}$ and $A\bar{x}_2 = \bar{b}$, it follows that

$$\begin{aligned} A\bar{x}_c &= A[\lambda\bar{x}_1 + (1 - \lambda)\bar{x}_2] \\ &= \lambda A\bar{x}_1 + (1 - \lambda)A\bar{x}_2 \\ &= \lambda\bar{b} + (1 - \lambda)\bar{b} = \bar{b} \end{aligned}$$

Therefore $\bar{x}_c \in X_c$ and X_c is a convex set.

It will be recalled that a point, x^* of a convex set X_c is an extreme point if it cannot be expressed as a convex combination of two other points in X_c. Geometrically, this means that no line segment in X_c contains x^* except as an end joint. Analytically, x^* is an extreme point, if there *do not* exist any points x_1, x_2 and a scalar $\lambda > 0$ such that

$$\bar{x}^* = \lambda\bar{x}_1 + (1 - \lambda)\bar{x}_2$$

These extreme points of X_c have an important property which we now prove.

Theorem 4–3 The objective function of a linear programming problem assumes its maximum at an extreme point of the convex set of feasible solutions.

Proof: We assume the number of extreme points is finite. Let us designate these extreme points by $\bar{x}_1^*, \bar{x}_2^*, \ldots, \bar{x}_t^*$. Let us designate the optimal solution by $\bar{x}_m$. If $\bar{x}_m$ is an extreme point then we are done. Suppose, otherwise, that the optimal value of the objective, $z_m = \bar{c}'\bar{x}_m$ occurs for $\bar{x}_m$, not an extreme point. If $\bar{x}_m$ is not an extreme point, it can be expressed as a convex combination of extreme points (see Chapter 3, Section 3-3), i.e.,

$$\bar{x}_m = \sum_{k=1}^{t} \lambda_k \bar{x}_k^* \tag{4–10}$$

where

$$\lambda_k \geq 0, \qquad k = 1, 2, \ldots, t$$

$$\sum_{k=1}^{t} \lambda_k = 1 \tag{4–11}$$

Let us now examine the objective function, $z = \bar{c}'\bar{x}$ for $\bar{x}_m$. We have

$$z_m = \bar{c}'\bar{x}_m = \bar{c}' \sum_{k=1}^{t} \lambda_k \bar{x}_k^* = \sum_{k=1}^{t} \lambda_k \bar{c}'\bar{x}_k^* \tag{4–12}$$

If we now define

$$\bar{c}'\bar{x}_p^* = \max_k \{\bar{c}'\bar{x}_k^*\} \tag{4–13}$$

and substitute $\bar{c}'\bar{x}_p^*$ for each $\bar{c}'\bar{x}_k^*$ in (4–12), we have

$$z_m \leq \sum_{k=1}^{t} \lambda_k \bar{c}'\bar{x}_p^* = \bar{c}'\bar{x}_p^* \sum_{k=1}^{t} \lambda_k = \bar{c}'\bar{x}_p^* = z_p \tag{4–14}$$

However, we assumed that $z_m = \bar{c}'\bar{x}_m$ was the *maximum* value that the objective function could have, i.e., $z_m \geq z = \bar{c}'\bar{x}$, for any $\bar{x}$. From (4–14) we have that $z_p = \bar{c}'\bar{x}_p^* \geq z_m^*$. Therefore, it must be true that

$$z_p = \bar{c}'\bar{x}_p^* = z_m \tag{4–15}$$

and hence, there is an extreme point $\bar{x}_p^*$ where the objective function takes on its maximum value.

In the remaining theorems, we shall make reference to the columns of the matrix A, which was assumed to have n columns. Let us designate these columns $\bar{a}_1, \bar{a}_2, \ldots, \bar{a}_n$. It will be recalled that each of these columns is an m-component vector. We shall require sets of linearly independent vectors or columns of A. Without loss of generality, we will assume that these are the vectors $\bar{a}_1, \bar{a}_2, \ldots, \bar{a}_k$ where $k < n$. It is clear that a renumbering of the vectors can always arrange that the first k vectors are linearly independent.

Theorem 4–4 If there exists a set of $k \leq m$ linearly independent columns of A, namely, $\bar{a}_1, \bar{a}_2, \ldots, \bar{a}_k$, such that

$$x_1\bar{a}_1 + x_2\bar{a}_2 + \ldots + x_k\bar{a}_k = \bar{b} \tag{4–16}$$

with all $x_j \geq 0$, then the point $\bar{x} = (x_1, x_2, \ldots, x_k, 0, \ldots, 0)$ is an extreme point of X_c, the convex set of feasible solutions.

Proof: Suppose $\bar{x}$ is not an extreme point. Then, since it is feasible, it can be written as a convex combination of two other points, $\bar{u} \in X_c$ and $\bar{v} \in X_c$, i.e.,

$$\bar{x} = \lambda\bar{u} + (1 - \lambda)\bar{v} \qquad 0 < \lambda < 1 \tag{4–17}$$

We have assumed that all $x_j \geq 0$. We also note that $0 < \lambda < 1$. Therefore, since the last $n - k$ components of $\bar{x}$ are zero, it follows that the last $n - k$ components of $\bar{u}$ and $\bar{v}$ must also be zero. Hence,

$$\begin{aligned} \bar{u} &= (u_1, u_2, \ldots, u_k, 0, \ldots, 0) \\ \bar{v} &= (v_1, v_2, \ldots, v_k, 0, \ldots, 0) \end{aligned} \tag{4–18}$$

We note that $\bar{x}$, $\bar{u}$, and $\bar{v}$ are all feasible solutions. Therefore, we may write, in view of (4–16) and (4–18),

$$\begin{aligned} x_1\bar{a}_1 + x_2\bar{a}_2 + \ldots + x_k\bar{a}_k &= \bar{b} \\ u_1\bar{a}_1 + u_2\bar{a}_2 + \ldots + u_k\bar{a}_k &= \bar{b} \\ v_1\bar{a}_1 + v_2\bar{a}_2 + \ldots + v_k\bar{a}_k &= \bar{b} \end{aligned} \tag{4–19}$$

If, in the preceding equations, we subtract the second from the first and also subtract the third from the second we obtain

$$\begin{aligned} (x_1 - u_1)\bar{a}_1 + (x_2 - u_2)\bar{a}_2 + \ldots + (x_k - u_k)\bar{a}_k &= \bar{0} \\ (u_1 - v_1)\bar{a}_1 + (u_2 - v_2)\bar{a}_2 + \ldots + (u_k - v_k)\bar{a}_k &= \bar{0} \end{aligned}$$

We assumed, however, that the columns of A, namely, $\bar{a}_1, \bar{a}_2, \ldots, \bar{a}_k$, were a set of linearly independent vectors. This implies that

$$\begin{aligned} x_j - u_j &= 0, \qquad j = 1, 2, \ldots, k \\ u_j - v_j &= 0, \qquad j = 1, 2, \ldots, k \end{aligned} \tag{4–20}$$

since all the coefficients must vanish identically. We therefore have, from equations (4–20)

$$x_j = u_j = v_j, \qquad j = 1, 2, \ldots, k$$

Hence, our assumption that $\bar{x}$ was not an extreme point of X_c, i.e., that $\bar{x}$ could be expressed as a convex combination of $\bar{u} \in X_c$ and $\bar{v} \in X_c$ is not true. Therefore, by definition, if $\bar{x}$ cannot be expressed as a convex combination of two other points in X_c, it must be an extreme point of X_c.

Theorem 4–5 Let $\bar{x}^* = (x_1^*, x_2^*, \ldots, x_n^*)$ be an extreme point of X_c. Then the columns of A which are associated with positive x_j^* are linearly independent, with at most m of the x_j^* being positive.

Proof: We shall assume, without loss of generality, that the first k of the x_j^* are nonzero, i.e., positive. This being the case, we can write

$$x_1^*\bar{a}_1 + x_2^*\bar{a}_2 + \ldots + x_k^*\bar{a}_k = \bar{b} \tag{4–21}$$

We shall prove this theorem by contradiction. We shall therefore assume that the vectors $\bar{a}_j, j = 1, 2, \ldots, k$ are linearly *dependent*. If this is so, then there exists a linear combination of these vectors such that

$$\alpha_1\bar{a}_1 + \alpha_2\bar{a}_2 + \ldots + \alpha_k\bar{a}_k = \bar{0} \tag{4–22}$$

with at least one $\alpha_j \neq 0, j = 1, 2, \ldots, k$. We now multiply equation (4–22) by some scalar, $\beta > 0$ and in one case add the result to equation (4–21) and in the other, subtract the result from equation (4–21). We then obtain the following equations:

$$\begin{aligned} \sum_{j=1}^{k} (x_j^* + \beta\alpha_j)\bar{a}_j &= \bar{b} \\ \sum_{j=1}^{k} (x_j^* - \beta\alpha_j)\bar{a}_j &= \bar{b} \end{aligned} \tag{4–23}$$

From (4–23) we see that we have two vectors as follows:

$$\begin{aligned} \bar{u} &= (x_1^* + \beta\alpha_1, x_2^* + \beta\alpha_2, \ldots, x_k^* + \beta\alpha_k, 0, \ldots, 0) \\ \bar{v} &= (x_1^* - \beta\alpha_1, x_2^* - \beta\alpha_2, \ldots, x_k^* - \beta\alpha_k, 0, \ldots, 0) \end{aligned} \tag{4–24}$$

The vectors $\bar{u}$ and $\bar{v}$ may not be feasible solutions unless β is properly chosen. We can see that if β is chosen so that it is positive and sufficiently small so that all of the first k components of $\bar{u}$ and $\bar{v}$ are positive, then $\bar{u}$ and $\bar{v}$ will be feasible solutions, i.e., $\bar{u}, \bar{v} \in X_c$. This clearly can be done if we let

$$0 < \beta < \min_j \frac{x_j^*}{|\alpha_j|} \qquad \{j \mid \alpha_j \neq 0\}$$

Now, we note also that in addition to $\bar{u}$ and $\bar{v}$ being feasible solutions, we can write

$$\bar{x}^* = \tfrac{1}{2}\bar{u} + \tfrac{1}{2}\bar{v} \tag{4–25}$$

However, equation (4–25) contradicts the assumption that $\bar{x}^*$ is an extreme point. Hence, the assumption of the linear dependence of $\bar{a}_1, \bar{a}_2, \ldots, \bar{a}_k$ has led us to a contradiction. Therefore the vectors $\bar{a}_1, \bar{a}_2, \ldots, \bar{a}_k$ are linearly independent. A well known result from linear algebra states that every set of $p + 1$ vectors in a p-dimensional space is a linearly dependent set. Hence, by the result just obtained and the fact that the $\bar{a}_j$ were assumed to be m-component vectors, we cannot have *more* than m positive x_j. This concludes the proof of the theorem.

Let us now consider the consequences of the above theorems. If we assume that the set of equations $A\bar{x} = \bar{b}$ is to have at least one solution, and if the rank of A is equal to m, then, as was indicated in Chapter 3, there must exist at least one linearly independent set of m vectors from the set $\bar{a}_1, \bar{a}_2, \ldots, \bar{a}_n$. This result, taken in conjunction with Theorem 4–5, allows us to assume that in all cases the set of vectors $\bar{a}_1, \bar{a}_2, \ldots, \bar{a}_n$ contains a set of m linearly independent vectors.† What we have shown then in these theorems may be briefly summarized as follows:

1. The set of vectors $\bar{a}_1, \bar{a}_2, \ldots, \bar{a}_n$ contains at least one set of $m \leq n$ linearly independent vectors (rank (A) = m).
2. Every basic feasible solution to the standard linear programming problem corresponds to an extreme point of the convex set of feasible solutions, X_c (Theorems 4–4 and 4–5).
3. Every extreme point of X_c has m linearly independent vectors from the set $\bar{a}_1, \bar{a}_2, \ldots, \bar{a}_n$ associated with it (Theorem 4–5).
4. There is some extreme point of X_c at which the objective function $z = \bar{c}'\bar{x}$ takes on its maximum value (Theorem 4–3).

The preceding statements clearly indicate that even though the convex set X_c has an infinite number of feasible solutions $\bar{x}$, we need only examine a finite subset of this infinite set. This finite subset consists of these basic feasible solutions, which correspond to extreme points of X_c. Such extreme point solutions are generated by considering sets of m linearly independent vectors from the set of n vectors $\bar{a}_1, \bar{a}_2, \ldots, \bar{a}_n$ which are the columns of the matrix A. It should be noted that even though X_c consists of feasible solutions, as well as its extreme points, there is *not* a one-to-one correspondence between extreme points of X_c and all the sets of m linearly independent vectors out of the total set of n vectors. There are many more of the latter than the former.

† It should be noted that if the rank of A is less than m, i.e., there are fewer than m linearly independent vectors, it will not affect the computational method we are about to develop. It does no harm to have a "degenerate" basis, i.e., one with zero coefficients in a solution. It is assumed here to simplify the theoretical discussion that the rank of A is m. It causes no theoretical problems if the rank of A is less than m.

Sets of m linearly independent vectors give rise to *basic* solutions rather than basic feasible solutions. Nevertheless, every extreme point of X_c corresponds to one of these basic solutions. In short, not all basic solutions are feasible.

Even though the set of basic solutions is finite, it is much too large to examine exhaustively for reasonable values of m and n. The number of basic solutions is simply the number of ways m linearly independent vectors can be chosen from a total set of n. This would also be an upper bound on the number of basic feasible solutions.† This number is

$$\binom{n}{m} = \frac{n!}{m!\,(n - m)!}$$

One can see that for a problem with, for example, $n = 400$, $m = 100$ (a not excessively large linear programming problem in practice), there is no practical possibility of examining this number of basic solutions.

It is fairly clear that what is required is a procedure that will generate not basic solutions but, rather, basic *feasible* solutions and, in addition, not *all* possible basic feasible solutions, since the number is too large. Rather, what we should look for is a method which will generate only a very small subset of the total set of basic feasible solutions, but which contains the *optimal* basic feasible solution. We also require a means by which this optimal basic feasible solution can be readily identified as such. These two important requirements are indeed met by the *simplex method*, which will be developed in detail in the following section. Its basic conception is that of George Dantzig.[2]

4-3 THE SIMPLEX METHOD—THEORETICAL DEVELOPMENT

The theorems in the previous section have given us some insight into the nature of the solution or solutions to a linear programming problem. We will now develop some results which will enable us to solve a linear programming problem by an iterative process which has come to be known as the "simplex method." It has many specific computational embodiments and variations. Its implementation on a digital computer is often very complex and subject to many refinements. However, all versions of the "simplex method" have a certain core of considerations in common. We shall call this the "primal simplex method" or, simply, the "simplex method." We now develop this method. In order to do this we shall require some standard notation, assumptions and definitions.

† More realistic bounds on the number of basic feasible solutions can be found in Grünbaum.[1]

We shall assume that our standard linear programming problem is stated in the form given by equations (4–2):

$$\begin{aligned} \text{Max } z &= \bar{c}'\bar{x} \\ A\bar{x} &= \bar{b} \\ \bar{x} &\geq \bar{0} \end{aligned} \tag{4–2}$$

where A is an $m \times n$ matrix, $\bar{c}$, $\bar{x}$ are n–vectors, and $\bar{b}$ is an m–vector. We shall denote the jth column of A as $\bar{a}_j$, $j = 1, 2, \ldots, n$. The matrix B will consist of m linearly independent columns of A and, therefore, it will serve as a basis for E^m. For this reason, B will be called a *basis matrix*. From the properties of a basis (see Chapter 3, Section 3-2), we know that any vector in E^m can be written as a linear combination of the vectors in B. Therefore, any column of the matrix A can be expressed as a linear combination of the columns in B, i.e.,

$$\bar{a}_j = y_{1j}\bar{a}_1 + y_{2j}\bar{a}_2 + \ldots + y_{mj}\bar{a}_m = \sum_{i=1}^{m} y_{ij}\bar{a}_i \tag{4–26}$$

The y_{ij} in (4–26) are scalar coefficients. We have assumed for notational convenience that the basis B contains the first m columns of A, i.e., $\bar{a}_1$, $\bar{a}_2$, $\ldots$, $\bar{a}_m$. This can be assumed without loss of generality, since the numbering of the vectors is arbitrary.

We can consider an m-component vector $\bar{y}_j$ whose components are the scalar coefficients of (4–26) i.e.,

$$\bar{y}_j = \begin{bmatrix} y_{1j} \\ y_{2j} \\ \cdot \\ \cdot \\ \cdot \\ y_{mj} \end{bmatrix} \tag{4–27}$$

With the notation of (4–27) we can write (4–26) as

$$\bar{a}_j = B\bar{y}_j \tag{4–28}$$

We assumed that B was a basis for E^m and, hence, B must be a nonsingular matrix. Therefore, we may also write

$$\bar{y}_j = B^{-1}\bar{a}_j \tag{4–29}$$

Since B consists of the first m columns of A, we can consider the matrix A as being partitioned into the basis matrix B and a non-basic matrix N of order $m \times (n - m)$. Therefore, we have that $A = (B, N)$ and we may write

$$A\bar{x} = (B, N)\bar{x} = (B, N)\begin{bmatrix} \bar{x}_B \\ \bar{x}_N \end{bmatrix} = \bar{b} \tag{4–30}$$

where $\bar{x}$ has now been partitioned into $\bar{x}_B$ and $\bar{x}_N$ just as A has been partitioned. From (4–30) we have

$$B\bar{x}_B + N\bar{x}_N = \bar{b} \tag{4–31}$$

However, from the definition of a *basic solution* we see that $\bar{x}_N = \bar{0}$. Therefore,

$$B\bar{x}_B = \bar{b} \tag{4–32}$$
$$\bar{x}_B = B^{-1}\bar{b}$$

and $\bar{x}_B = B^{-1}b$ is a basic solution to $A\bar{x} = \bar{b}$. The variables $\bar{x}_B$ are called *basic variables.*

We can equally well partition the vector $\bar{c}'$ as we have partitioned the rows of A and the vector $\bar{x}$. If we do so, we have

$$\bar{c}' = (\bar{c}'_B, \bar{c}'_N) \quad \text{where} \quad \bar{c}'_B = (c_{B1}, c_{B2}, \ldots, c_{Bm})$$

and the objective function, $z = \bar{c}'\bar{x}$ can now be expressed

$$z = \bar{c}'\bar{x} = (\bar{c}'_B, \bar{c}'_N)\begin{bmatrix} \bar{x}_B \\ \bar{x}_N \end{bmatrix} = \bar{c}'_B\bar{x}_B + \bar{c}'_N\bar{x}_N = \bar{c}'_B\bar{x}_B \tag{4–33}$$

One last definition that we shall require later is a scalar quantity associated with each $\bar{y}_j$ vector. It is defined as

$$z_j = \sum_{i=1}^{m} y_{ij}c_{Bi} = \bar{c}'_B\bar{y}_j \tag{4–34}$$

With the preceding notation and definitions we may proceed. The central ideas of the simplex method are quite simple. We shall engage in an iterative process. At each stage of this process we shall have a basic feasible solution to (4–2). We must examine whether this current basic feasible solution is optimal. Hence, we need a test for optimality. If it is optimal, we are done. If the current basic feasible solution is not optimal we need a method by which we can find a better basic feasible solution. The entire iterative process is of necessity finite, since the number of basic feasible solutions is finite. If we never return to a previous solution, we must ultimately reach the optimal solution.

Let us now consider the above somewhat general description in more detail. We must at each stage of the iterative process, show that the current basic feasible solution is optimal or find an improved basic feasible solution. Let us consider the latter problem first.

Suppose we have available to us a basic feasible solution to $A\bar{x} = \bar{b}$. As before, we assume for simplicity that the first m columns of A are in the basis matrix B. This being so, we may write

$$\sum_{i=1}^{m} x_{Bi}\bar{a}_i = \bar{b} \tag{4–35}$$

The x_{Bi} are the components of the solution vector, $\bar{x}_B$. Our objective is to find another basic feasible solution. Clearly what must be done is to replace one or more of the vectors $\bar{a}_i$ which are currently in B and replace them with some of the vectors, $\bar{a}_j$ which are in N. There are a great many different ways that this could be done. In the simplex method of Dantzig, usually, one vector at a time is removed from B and replaced with a vector from N, in order to generate a new basis matrix and an associated basic feasible solution. In some computational algorithms, some exceptions to this can be found. However, it will always be the case that the simplex theory and computational algorithms will involve and can be best explained initially, in terms of replacing a single vector at a time in the basis matrix B.

A simple geometric interpretation can be attached to the process of replacing a single vector in a basis with some other vector. As we have noted previously, each basic feasible solution corresponds to an extreme point of the convex set of feasible solutions, X_c. When we replace one vector at a time, what we are doing is moving from one extreme point of X_c to an *adjacent* extreme point. Hence, what we shall now consider is how to find an improved adjacent extreme point or equivalently, an improved basic feasible solution, by replacing one basis vector with another.

Let us assume that we have a basic feasible solution $\bar{x}_B$ and, hence, that equation (4–35) is satisfied. We know that any vector of A, and in particular those not in the basis, can be expressed in terms of the basis vectors as

$$\bar{a}_j = \sum_{i=1}^{m} y_{ij}\bar{a}_i, \qquad j = m + 1, m + 2, \ldots, n \tag{4–36}$$

We are using the convention that vectors designated $\bar{a}_i$ are in the basis and those designated $\bar{a}_j$ are not in the basis. We wish to have one of these nonbasic vectors $\bar{a}_j$ enter the basis and replace some vector $\bar{a}_r$, where $\bar{a}_r$ is in the basis. Let us not concern ourselves for the present with which particular $\bar{a}_j$ is to enter the basis; for the moment we shall select an $\bar{a}_j$ arbitrarily. Let us then single out some $\bar{a}_r$ in the basis such that $y_{rj} \neq 0$. This will lead us shortly to see how to choose $\bar{a}_r$. Using equation (4–36) we may write

$$\bar{a}_j = y_{rj}\bar{a}_r + \sum_{\substack{i=1 \\ i\neq r}}^{m} y_{ij}\bar{a}_i \tag{4–37}$$

If we solve equation (4–37) for $\bar{a}_r$, we obtain

$$\bar{a}_r = \frac{1}{y_{rj}}\bar{a}_j - \sum_{\substack{i=1 \\ i\neq r}}^{m} \frac{y_{ij}}{y_{rj}}\bar{a}_i \tag{4–38}$$

We now substitute $\bar{a}_r$ from equation (4–38) into the expression for a basic solution given by equation (4–35). This results in

$$\sum_{\substack{i=1 \\ i\neq r}}^{m} x_{Bi}\bar{a}_i + \frac{x_{Br}}{y_{rj}}\bar{a}_j - x_{Br}\sum_{\substack{i=1 \\ i\neq r}}^{m} \frac{y_{ij}}{y_{rj}}\bar{a}_i = \bar{b} \tag{4–39}$$

Upon rearrangement, equation (4–39) yields

$$\sum_{\substack{i=1 \\ i\neq r}}^{m} \left(x_{Bi} - x_{Br}\frac{y_{ij}}{y_{rj}}\right)\bar{a}_i + \frac{x_{Br}}{y_{rj}}\bar{a}_j = \bar{b} \tag{4–40}$$

Equation (4–40) clearly will give us a new *basic* solution. However, although the constraints $A\bar{x} = \bar{b}$ are obviously satisfied, we have not taken any precautions to insure that the new coefficients of $\bar{a}_i$ and $\bar{a}_j$ in equation (4–40) are necessarily non-negative. Let us then examine the new coefficients of (4–40). If we designate the new values of x_{Bi} as $\hat{x}_{Bi}$, we see that they have the form

$$\begin{aligned} \hat{x}_{Bi} &= x_{Bi} - x_{Br}\frac{y_{ij}}{y_{rj}}, \qquad i \neq r \\ \hat{x}_{Br} &= \frac{x_{Br}}{y_{rj}} \end{aligned} \tag{4–41}$$

From an examination of equations (4–41) it can be seen that some of the $\hat{x}_{Bi}$ could certainly be less than zero unless this possibility is explicitly excluded. Hence, we must require that

$$\begin{aligned} \hat{x}_{Bi} &= x_{Bi} - x_{Br}\frac{y_{ij}}{y_{rj}} \geq 0, \qquad i \neq r \\ \hat{x}_{Br} &= \frac{x_{Br}}{y_{rj}} \geq 0 \end{aligned} \tag{4–42}$$

Some simple conclusions can be drawn from the requirements of (4–42). If $x_{Br} > 0$ then we must require that $y_{rj} > 0$. If *all* the $y_{ij} \leq 0$ then the $\hat{x}_{Bi}$ will be non-negative. However, if some of the $y_{ij} > 0$, then whether or not $x_{Bi} - x_{Br} \dfrac{y_{ij}}{y_{rj}}$ will be non-negative will depend upon which $\bar{a}_r$ was chosen to be removed from the basis. This, then, is the decisive criterion in choosing the $\bar{a}_r$ to be removed, viz., the necessity to retain, at each step, a basic *feasible* solution. Let us see how to choose the correct $\bar{a}_r$. If some $y_{ij} > 0$, we can divide the first set of inequalities of (4–42) by y_{ij} to obtain

$$\frac{x_{Bi}}{y_{ij}} - \frac{x_{Br}}{y_{rj}} \geq 0 \tag{4–43}$$

If we wish to insure that (4–43) will always hold, then we can simply choose the column r which will be removed from the basis B as follows:

$$\frac{x_{Br}}{y_{rj}} = \min_i \left\{ \frac{x_{Bi}}{y_{ij}} \,\middle|\, y_{ij} > 0 \right\} \tag{4–44}$$

If r is chosen according to (4–44), then it is clear that none of the new coefficients, $\hat{x}_{Bi}$ can be negative. Hence (4–44) gives us the rule to use to prevent our new solution from becoming infeasible because of the negativity of one or more of the x_{Bi}.

There are two points which relate to the presence of degeneracy in a basic feasible solution which should be mentioned. The first is concerned with how degeneracy can arise in a basic feasible solution. It can be seen from the relationship of (4–44) that if the minimum is not unique more than one basis variable will be forced to zero. Hence, one or more variables in the new basic solution will be zero and, therefore, we will have a *degenerate* basic feasible solution. The second point deals with what may happen if we start with a degenerate solution. For this case, if x_{Br} was zero, the new basic solution would also be degenerate since $\hat{x}_{Br} = \dfrac{x_{Br}}{y_{rj}} = 0$. However, it is also possible to start with a degenerate solution and *not* have the new solution be degenerate. This can occur because x_{Br} is selected from those x_{Bi} that have $y_{ij} > 0$. If the $x_{Bi} = 0$ do not have associated $y_{ij} > 0$, then x_{Br} would be determined from the non-degenerate x_{Bi} and hence, in this case, the new solution might not be degenerate. Neither of these points relating to degeneracy affects the use of relation (4–44) in practice. However, as we shall see later, the presence of degeneracy is the source of some theoretical difficulties which require detailed consideration and resolution.

We can now summarize what we have shown with respect to determining the vector $\bar{a}_r$ which is to leave the basis, B. What we need to do is simply compute $\frac{x_{Br}}{y_{rj}}$ from equation (4–44); this tells us which column r of the basis matrix B is to be removed. We then have a new basis matrix $\hat{B}$ consisting of the columns $\bar{a}_i$, $i \neq r$ and a new column $\bar{a}_j$ which was selected to enter the basis. The new basic feasible solution is given by $\hat{\bar{x}}_B = \hat{B}^{-1}\bar{b}$, and the new values of the basic variables are given by equations (4–41).

In the preceding, we assumed that some vector $\bar{a}_j$ was to enter the basis, then we considered what needed to be done to insure a new basic feasible solution. We saw that no matter what vector was selected to enter the basis, it was always possible, providing at least one $L_{ij} > 0$, to remove a vector in such a way, using equation (4–44) as to have a new basic feasible solution. We turn now to the question of selecting the vector to enter the basis. The entire rationale for the stagewise process of moving from one basic feasible solution to another is to do so in such a way that we improve the value of the objective function at each stage, if that is possible, or receive some indication that the current solution and the current value of the objective function is optimal. Let us now examine how this may be done.

In terms of the original basic feasible solution, the objective function is given by

$$z = \bar{c}'\bar{x} = \bar{c}_B'\bar{x}_B = \sum_{i=1}^{m} c_{Bi}x_{Bi} \tag{4–45}$$

Using the previously introduced notation, the new objective function is given by

$$\hat{z} = \bar{c}'\hat{\bar{x}} = \hat{\bar{c}}_B'\hat{\bar{x}}_B = \sum_{i=1}^{m} \hat{c}_{Bi}\hat{x}_{Bi} \tag{4–46}$$

From the preceding two equations, it follows that since $\bar{a}_r$ was removed from the basis and $\bar{a}_j$ entered the basis, we have that

$$\begin{aligned} \hat{c}_{Bi} &= c_{Bi}, \qquad i \neq r \\ \hat{c}_{Br} &= c_j \end{aligned} \tag{4–47}$$

If we now substitute values of $\hat{x}_{Bi}$ from equation (4–41) and values of $\hat{c}_{Bi}$ from equation (4–47) into equation (4–46) we obtain

$$\hat{z} = \sum_{\substack{i=1 \\ i \neq r}}^{m} c_{Bi}\left(x_{Bi} - x_{Br}\frac{y_{ij}}{y_{rj}}\right) + \frac{x_{Br}}{y_{rj}}c_j \tag{4–48}$$

In order to simplify equation (4–48) we may note that the missing term in the summation is

$$c_{Br}\left(x_{Br} - x_{Br}\frac{y_{rj}}{y_{rj}}\right) = 0$$

and therefore we can add this to the summation to yield

$$\hat{z} = \sum_{i=1}^{m} c_{Bi}\left(x_{Bi} - x_{Br}\frac{y_{ij}}{y_{rj}}\right) + \frac{x_{Br}}{y_{rj}} c_j \tag{4–49}$$

Let us now rearrange equation (4–49) as follows:

$$\hat{z} = \sum_{i=1}^{m} c_{Bi}x_{Bi} - \frac{x_{Br}}{y_{rj}} \sum_{i=1}^{m} c_{Bi}y_{ij} + \frac{x_{Br}}{y_{rj}} c_j$$

$$= z - \frac{x_{Br}}{y_{rj}} z_j + \frac{x_{Br}}{y_{rj}} c_j$$

Upon slight rearrangement we have

$$\hat{z} = z + \frac{x_{Br}}{y_{rj}} (c_j - z_j) \tag{4–50}$$

Equation (4–50) will give us the information that we have been seeking. We know that in the absence of degeneracy, $\frac{x_{Br}}{y_{rj}} > 0$. Therefore, if we select a vector $\bar{a}_j$ such that $c_j - z_j > 0$ (or $z_j - c_j < 0$ as it is more often written), we will have insured that $\hat{z} > z$ and, hence, we have a larger value of the objective function. Even if degeneracy was present, it is still true that $\hat{z} \geq z$. If we choose the vector $\bar{a}_j$ to enter the basis such that $\frac{x_{Br}}{y_{rj}}(c_j - z_j)$ is the largest possible for those $\bar{a}_j, j = m + 1, m + 2, \ldots, n$, then we will have obtained the largest possible increase in the objective function z. In practice, in order to minimize the amount of computation involved, a vector $\bar{a}_j$ is selected such that $z_j - c_j$ is the smallest of the $z_j - c_j < 0$. This seems to be equally satisfactory and greatly reduces the amount of calculation required at each iteration, since it can be calculated without any reference to the vector leaving the basis. This point is not known with certainty, however. The reader may wish to consult reference 3 for one study of this matter. To summarize, we shall determine the vector to enter the basis as follows:

$$z_k - c_k = \min_j \{z_j - c_j \mid z_j - c_j < 0\} \tag{4–51}$$

The subscript k in (4–51) indicates the vector to enter the basis.

Having described the two parts of each stage of the iteration process of the simplex method, we need now to consider when and how to terminate the calculation. We have seen from equation (4–50), that in the absence of degeneracy, as long as there is at least one vector $\bar{a}_j$ not in the basis with a value of $z_j - c_j < 0$, then if we insert that vector into the basis we will obtain an increase in the objective function. In fact this is clearly a monotone process in which each value of the objective function is greater than the previous value. Hence, we can continue this process until there are no vectors $\bar{a}_j$ with $z_j - c_j < 0$. This means that for all vectors $\bar{a}_j$, $z_j - c_j \geq 0$. Suppose that this occurs at a point $\bar{x}_B$ with a corresponding value of the objective function equal to z_0. We shall now show that when $z_j - c_j \geq 0, j = m + 1, m + 2, \ldots, n$, z_0 is the maximum value of the objective function.

Let $x'_j \geq 0$, $j = 1, 2, \ldots, n$ be any feasible solution to $A\bar{x} = \bar{b}$. We therefore have

$$\sum_{j=1}^{n} x'_j \bar{a}_j = \bar{b}, \qquad x'_j \geq 0 \qquad j = 1, 2, \ldots, n \tag{4-52}$$

The corresponding value of the objective function is

$$z = c_1 x'_1 + c_2 x'_2 + \ldots + c_n x'_n \tag{4-53}$$

We know further that we can express any vector $\bar{a}_j$ in A as a linear combination of the basis vectors:

$$\bar{a}_j = \sum_{i=1}^{m} y_{ij} \bar{a}_i \qquad j = 1, 2, \ldots, n \tag{4-54}$$

We now substitute from equations (4–54) into the first equation of (4–52):

$$x'_1 \sum_{i=1}^{m} y_{i1} \bar{a}_i + x'_2 \sum_{i=1}^{m} y_{i2} \bar{a}_i + \ldots + x'_n \sum_{i=1}^{m} y_{in} \bar{a}_i = \bar{b} \tag{4-55}$$

Equation (4–55) can be rearranged to yield

$$\left[\sum_{j=1}^{n} x'_j y_{1j}\right] \bar{a}_1 + \left[\sum_{j=1}^{n} x'_j y_{2j}\right] \bar{a}_2 + \ldots + \left[\sum_{j=1}^{n} x'_j y_{mj}\right] \bar{a}_m = \bar{b} \tag{4-56}$$

Equation (4–56) expresses the vector $\bar{b}$ in terms of the basis vectors, $\bar{a}_i$. The coefficients in this expression are $\sum_{j=1}^{n} x'_j y_{ij}$, $i = 1, 2, \ldots, m$. However, it will be recalled from Chapter 3 that the representation of any vector in terms of a single set of basis vectors is unique. If we therefore compare equations (4–52) and (4–56) we see that

$$x_{Bi} = \sum_{j=1}^{n} x'_j y_{ij} \qquad i = 1, 2, \ldots, m \tag{4-57}$$

Let us now consider the objective function, z'. We have assumed that at the point $\bar{x}_B$, $z_j - c_j \geq 0$ for all $\bar{a}_j$ not in the basis. For those columns of A that are in the basis we have

$$\bar{y}_j = \bar{y}_i = B^{-1}\bar{a}_i = \bar{e}_i \qquad j = 1, 2, \ldots, m \tag{4–58}$$

where $\bar{e}_i$ is the unit vector with a "1" in the ith row and zeros elsewhere. Equations (4–58) enable us to calculate z_j as follows:

$$z_j = \bar{c}_B'\bar{y}_j = \bar{c}_B'\bar{e}_i = c_j \qquad j = 1, 2, \ldots, m \tag{4–59}$$

From equation (4–59) we see that $z_j - c_j = 0$ for $j \leq m$. Combining this with the previous result that $z_j - c_j \geq 0, j = m + 1, m + 2, \ldots, n$ we have that $z_j - c_j \geq 0$ for all j. Since $z_j \geq c_j, j = 1, 2, \ldots, n$ we can now substitute z_j for c_j in (4–53) to obtain

$$z_1x_1' + z_2x_2' + \ldots + z_nx_n' \geq z' \tag{4–60}$$

If we again make use of the definition, $z_j = \sum_{i=1}^{m} y_{ij}c_{Bi}$ and substitute into (4–60) we have that

$$x_1' \sum_{i=1}^{m} y_{i1}c_{Bi} + x_2' \sum_{i=1}^{m} y_{i2}c_{Bi} + \ldots + x_n' \sum_{i=1}^{m} y_{in}c_{Bi} \geq z'$$

which can be rearranged to give

$$\left[\sum_{j=1}^{n} x_j'y_{1j}\right]c_{B1} + \left[\sum_{j=1}^{n} x_j'y_{2j}\right]c_{B2} + \ldots + \left[\sum_{j=1}^{n} x_j'c_{mj}\right]c_{Bm} \geq z' \tag{4–61}$$

If we now substitute from equation (4–57) into (4–61) we obtain

$$c_{B1}x_{B1} + c_{B2}x_{B2} + \ldots + c_{Bm}x_{Bm} = \bar{c}_B'\bar{x}_B = z_0 \geq z' \tag{4–62}$$

Equation (4–62) tells us that when we have reached a solution $\bar{x}_B$ for which all $z_j - c_j \geq 0$ its value of z_0 is at least as large as any other feasible solution z'. Hence, z_0 is the maximum value of the objective function.

In the absence of degeneracy, we appear to have a finite computational algorithm. Degeneracy will be discussed in more detail in the next chapter. However, there is one assumption we have made which requires further attention. We have assumed that if we were not yet optimal; that is, if one or more $z_j - c_j < 0$, we could always replace one vector in the basis with another. However, if we consider the criterion for removal of a vector from the basis:

$$\frac{x_{Br}}{y_{rj}} = \min_i \left\{ \frac{x_{Bi}}{y_{ij}} \middle| \, y_{ij} > 0 \right\}$$

the successful removal of a vector from the basis requires that at least one $y_{ij} > 0$ for $i = 1, 2, \ldots, m$. Is it possible that when we try to insert some vector into the basis with $z_j - c_j < 0$, all the associated $y_{ij} \leq 0$? We shall see, if this turns out to be the case, that this is an indication of an unbounded solution; i.e., the objective function can be made arbitrarily large for a maximization problem.

Let us assume that we have a basic feasible solution $\bar{x}_B$ and, therefore,

$$\sum_{i=1}^{m} x_{Bi}\bar{a}_i = \bar{b} \tag{4–63}$$

If we both add and subtract $\phi\bar{a}_k$ to (4–63) we do not change it and we have

$$\sum_{i=1}^{m} x_{Bi}a_i - \phi\bar{a}_k + \phi\bar{a}_k = \bar{b} \tag{4–64}$$

In equation (4–64) ϕ is any scalar and $\bar{a}_k$ is the vector about to enter the basis. $\bar{a}_k$ can be expressed in terms of the basis vectors as follows:

$$\bar{a}_k = \sum_{i=1}^{m} y_{ik}\bar{a}_i \tag{4–65}$$

If we substitute (4–65) into (4–64) we have

$$\sum_{i=1}^{m} x_{Bi}\bar{a}_i - \phi \sum_{i=1}^{m} y_{ik}\bar{a}_i + \phi\bar{a}_k = \bar{b}$$

which can be rearranged to

$$\sum_{i=1}^{m} (x_{Bi} - \phi y_{ik})\bar{a}_i + \phi\bar{a}_k = \bar{b} \tag{4–66}$$

Equation (4–66) is instructive. If, as we assumed, $y_{ik} \leq 0$, $i = 1, 2, \ldots, m$ and $\phi > 0$, then we have a feasible but non-basic, solution since $m + 1$ variables are positive. The objective function for the solution under these conditions is

$$\hat{z} = \sum_{i=1}^{m} c_{Bi}(x_{Bi} - \phi y_{ik}) + c_k\phi$$

or

$$\hat{z} = z + \phi(c_k - z_k) \tag{4–67}$$

It can be seen from equation (4–67) that if $z_k - c_k < 0$ and $\phi > 0$, then $\hat{z}$ can be made arbitrarily large as ϕ increases. In this case, we see that the given linear programming problem has an unbounded solution. As was indicated in Chapter 1, no properly formulated linear programming problem, purporting to deal with the real world, can have an unbounded maximum. Normally, such an indication, i.e., that all $y_{ij} \leq 0$ for some j, for which $z_j - c_j < 0$ is a sign that an error has been made in problem formulation or in data preparation for the calculation.

4–4 SUMMARY OF THE PRIMAL SIMPLEX ALGORITHM

What we have been discussing in the previous section is usually referred to, somewhat loosely, as the "primal simplex algorithm." We will reserve the computational details for the next chapter. However, let us state the algorithm itself in a systematic fashion.

Primal Simplex Algorithm

1. Determine an initial basic feasible solution, $\bar{x}_0$ by some method. (This is discussed in Chapter 5.)

2. For those vectors not in the basis, $j \in J_N$ calculate $z_j - c_j$. If all $z_j - c_j \geq 0$, we have an optimal solution.

3. If one or more $z_j - c_j < 0$, $j \in J_N$, we select a vector to enter the basis $\bar{a}_k$, using the criterion

$$z_k - c_k = \min_{j \in J_N} \{z_j - c_j \mid z_j - c_j < 0\}$$

4. If all $y_{ik} \leq 0$, an unbounded maximum exists. If at least one $y_{ik} > 0$, we select the vector $\bar{a}_r$ to be removed from the basis $\bar{a}_r$, by the criterion

$$\frac{x_{Br}}{y_{rk}} = \min_i \left\{ \frac{x_{Bi}}{y_{ik}} \middle| \; y_{ik} > 0 \right\}$$

5. With the new basis B_1 formed from B_0 by removing $\bar{a}_r$ and replacing it with a_k, we compute the new solution $\bar{x}_1$, new values of y_{ij}, $z_j - c_j$, and z.

6. Return to Step 2.

The foregoing steps comprise the necessary individual tasks that must be performed. However, the details of how, for example, an initial basic feasible solution is chosen or how data is organized and manipulated for computational purposes, is extremely important. These matters are discussed in Chapter 5. Because of these differences in detail one could probably justifiably speak of many different primal simplex algorithms. The algorithmic steps given in the foregoing paragraph are the quintessential characteristics that are shared by such detailed computational algorithms.

4-5 AN EXAMPLE OF THE PRIMAL SIMPLEX PROCESS

In the next chapter a comprehensive discussion of the computational details of the primal simplex algorithm will be presented. What we shall give here is a simple example of what is done in terms of the six steps described in the previous section. The various computational formats used do not show this easily. The following example will hopefully make this process transparent.

Consider the following linear programming problem:

$$\begin{aligned} \text{Max } z &= 2x_1 - 3x_2 \\ 2x_1 + 5x_2 &\geq 10 \\ 3x_1 + 8x_2 &\leq 24 \\ x_1, x_2 &\geq 0 \end{aligned} \tag{4-68}$$

First we convert the problem stated in equations (4–68) into equality form by adding slack and surplus variables:

$$\begin{aligned} \text{Max } z &= 2x_1 - 3x_2 \\ 2x_1 + 5x_2 - x_3 &= 10 \\ 3x_1 + 8x_2 + x_4 &= 24 \\ x_1, x_2, x_3, x_4 &\geq 0 \end{aligned} \tag{4-69}$$

The matrix A and its component column vectors $\bar{a}_j$ are as follows:

$$A = \begin{bmatrix} 2 & 5 & -1 & 0 \\ 3 & 8 & 0 & 1 \end{bmatrix}$$

$$\bar{a}_1 = \begin{bmatrix} 2 \\ 3 \end{bmatrix} \qquad \bar{a}_2 = \begin{bmatrix} 5 \\ 8 \end{bmatrix} \qquad \bar{a}_3 = \begin{bmatrix} -1 \\ 0 \end{bmatrix} \qquad \bar{a}_4 = \begin{bmatrix} 0 \\ 1 \end{bmatrix}$$

The vector $\bar{b}$ is:

$$\bar{b} = \begin{bmatrix} 10 \\ 24 \end{bmatrix}$$

We can choose as our initial basis a matrix B using columns $\bar{a}_1$ and $\bar{a}_4$ since these are linearly independent. This implies that

$$x_1\bar{a}_1 + x_4\bar{a}_4 = \bar{b}$$

$$x_1\begin{bmatrix}2\\3\end{bmatrix} + x_4\begin{bmatrix}0\\1\end{bmatrix} = \begin{bmatrix}10\\24\end{bmatrix}$$

From the foregoing equation we conclude that

$$x_1 = 5, \qquad x_4 = 9, \qquad x_2 = x_3 = 0$$

Hence, our initial basic feasible solution is $\bar{x}_B = \begin{bmatrix}5\\9\end{bmatrix}$ and $B = \begin{bmatrix}2 & 0\\3 & 1\end{bmatrix}$.

Now we shall express the vectors not in the basis in terms of the basis vectors.

$$\bar{a}_2 = y_{12}\bar{a}_1 + y_{42}\bar{a}_4$$
$$\bar{a}_3 = y_{13}\bar{a}_1 + y_{43}\bar{a}_4$$

Writing out the foregoing equations, we have

$$\begin{bmatrix}5\\8\end{bmatrix} = y_{12}\begin{bmatrix}2\\3\end{bmatrix} + y_{42}\begin{bmatrix}0\\1\end{bmatrix}$$

$$\begin{bmatrix}-1\\0\end{bmatrix} = y_{13}\begin{bmatrix}2\\3\end{bmatrix} + y_{43}\begin{bmatrix}0\\1\end{bmatrix}$$

The first of these vector equations yields

$$\bar{y}_2' = [y_{12}, y_{42}] = [\tfrac{5}{2}, \tfrac{1}{2}]$$

The second yields

$$\bar{y}_3' = [y_{13}, y_{43}] = [-\tfrac{1}{2}, \tfrac{3}{2}]$$

We can now compute the values of z_j for $\bar{a}_2$ and $\bar{a}_3$.

$$z_j = \bar{c}_B'\bar{y}_j \qquad \bar{c}_B' = (2, 0)$$

$$z_2 = (2, 0)\begin{bmatrix}\frac{5}{2}\\ \frac{1}{2}\end{bmatrix} = 5$$

$$z_3 = (2, 0)\begin{bmatrix}-\frac{1}{2}\\ \frac{3}{2}\end{bmatrix} = -1$$

$$z_2 - c_2 = 5 - (-3) = 8$$

$$z_3 - c_3 = -1 - 0 = -1$$

Since the only $z_j - c_j, j = 2, 3$ that is negative is $z_3 - c_3$, we shall choose $\bar{a}_3$ to enter the basis. We must now determine the vector to leave. The criterion is

$$\operatorname*{Min}_i \left\{ \frac{x_{Bi}}{y_{i3}} \middle| \; y_{i3} > 0 \right\}$$

In this small example the choice would be between $\dfrac{x_1}{y_{13}}$ and $\dfrac{x_4}{y_{43}}$. However, since, $y_{13} < 0$ we have no choice and $\bar{a}_4$ must leave the basis.

The original value of the objective function was

$$z = \bar{c}_B' \bar{x}_B = (2, 0)\begin{bmatrix} 5 \\ 9 \end{bmatrix} = 10$$

According to equation (4–50):

$$\hat{z} = z + \frac{x_4}{y_{43}}(c_3 - z_3) = 10 + \frac{9}{\frac{3}{2}}(1) = 16$$

and we see that $\hat{z} > z$. We can check this by direct computation of the new value of $\bar{x}_B$

$$\hat{x}_1 \bar{a}_1 + \hat{x}_3 \bar{a}_3 = \bar{b}$$

$$\hat{x}_1 \begin{bmatrix} 2 \\ 3 \end{bmatrix} + \hat{x}_3 \begin{bmatrix} -1 \\ 0 \end{bmatrix} = \begin{bmatrix} 10 \\ 24 \end{bmatrix}$$

Therefore,

$$\hat{x}_1 = 8,\ \hat{x}_3 = 6 \text{ and } \hat{\bar{x}}_B = \begin{bmatrix} 8 \\ 6 \end{bmatrix}$$

$$\hat{z} = \hat{\bar{c}}_B' \hat{\bar{x}}_B = (2, 0)\begin{bmatrix} 8 \\ 6 \end{bmatrix} = 16$$

We will now repeat the entire process. We express the vectors not in the basis $\bar{a}_2, \bar{a}_4$ in terms of $\bar{a}_1, \bar{a}_3$ which are the vectors in the basis.

$$\bar{a}_2 = y_{12}\bar{a}_1 + y_{32}\bar{a}_3$$

$$\bar{a}_4 = y_{14}\bar{a}_1 + y_{34}\bar{a}_3$$

This gives us

$$\begin{bmatrix} 5 \\ 8 \end{bmatrix} = y_{12}\begin{bmatrix} 2 \\ 3 \end{bmatrix} + y_{32}\begin{bmatrix} -1 \\ 0 \end{bmatrix}$$

from which we obtain

$$\bar{y}_2' = [y_{12}, y_{32}] = [\tfrac{8}{3}, \tfrac{1}{3}]$$

and

$$\begin{bmatrix} 0 \\ 1 \end{bmatrix} = y_{14}\begin{bmatrix} 2 \\ 3 \end{bmatrix} + y_{34}\begin{bmatrix} -1 \\ 0 \end{bmatrix}$$

from which we have

$$\bar{y}_4' = [y_{14}, y_{34}] = [\tfrac{1}{3}, \tfrac{2}{3}]$$

$$z_2 = (2, 0)\begin{bmatrix} \frac{8}{3} \\ \frac{1}{3} \end{bmatrix} = \tfrac{16}{3}$$

$$z_4 = (2, 0)\begin{bmatrix} \frac{1}{3} \\ \frac{2}{3} \end{bmatrix} = \tfrac{2}{3}$$

Therefore,

$$z_2 - c_2 = \tfrac{16}{3} - (-3) = \tfrac{25}{3}$$

$$z_4 - c_4 = \tfrac{2}{3} - 0 = \tfrac{2}{3}$$

Since both $z_2 - c_2 > 0$ and $z_4 - c_4 > 0$, we have obtained the optimal solution, which is

$$x_1 = 8, \qquad x_2 = 0, \qquad x_3 = 6, \qquad x_4 = 0 \quad \text{with} \quad z = 16$$

The above example is given only to illustrate the ideas previously discussed. It is not suggested as a guide to practical computation. That subject, in addition to other complications, is reserved for the next chapter.

PROBLEMS

1. Solve the following problem by examining all basic solutions:

$$\begin{aligned} \text{Max } z &= 3x_1 + 2x_2 + x_3 \\ 2x_1 + x_2 + 3x_3 &= 20 \\ x_1 + 2x_2 + x_3 &= 10 \\ x_1, x_2, x_3 &\geq 0 \end{aligned}$$

2. Prove that if there is a feasible solution to $A\bar{x} = \bar{b}$, $\bar{x} \geq \bar{0}$ then there is a basic feasible solution, assuming that $r(A) = m$.

3. If an optimal basic solution is nondegenerate and $z_j - c_j > 0$ for every vector not in the final basis, prove that the optimal solution is unique.

4. Would it be possible to devise an algorithm which inserted two vectors into the basis at each iteration instead of one? Discuss how this might be done. What is the geometric interpretation of such a method?

5. Suppose the vector to be removed from the basis as determined by (4-44) is unique. If this is the case, can the new basic feasible solution ever be degenerate?

6. Consider the linear programming problem given by:

$$\text{Max } z = 3x_1 + x_2 + 2x_3 + 4x_4$$

$$\begin{aligned} 2x_1 + 3x_2 + x_3 + 5x_4 &= 30 \\ x_1 + 2x_2 + 3x_3 + 4x_4 &= 40 \\ 3x_1 + x_2 + 4x_3 + x_4 &= 25 \\ x_1, x_2, x_3, x_4 &\geq 0 \end{aligned}$$

Choose some initial basis and then find the optimal solution as in Section 4.5.

REFERENCES

1. B. Grünbaum: *Convex Polytopes*. New York, Interscience Publishers, 1967.
2. G. Dantzig: Maximization of a Linear Function of Variables Subject to Linear Inequalities. In: *Activity Analysis of Production and Allocation*. T. C. Koopmans, Ed., New York, John Wiley and Sons, 1951.
3. H. W. Kuhn and R. E. Quandt: An Experimental Study of the Simplex Method. In: *Proceedings of Symposia of Applied Mathematics*. Vol. XV, Providence, Rhode Island, American Mathematical Society, 1963.

5

The Simplex Method: Computational Details

5–1 INTRODUCTION

In the previous chapter, a general statement of the primal simplex algorithm was given. In addition, an example of the simplex process was given, in the sense that at each stage all necessary quantities were calculated from their definitions. The point of that example was to illustrate the most important aspects of the process, namely, the selection of the vector to enter the basis, the determination of the vector to leave the basis and, finally, the recognition that an optimal solution has been obtained. In actual practice, we always seek to minimize the amount of computation that is required. The first step toward this goal is the development of simple formulae by which all quantities of interest at iteration $k + 1$ can be calculated from those at iteration k. In addition, we shall discuss an organizational format, the so-called "simplex tableau," for arranging all the information we require in a compact format.

In connection with the general computational details we shall also discuss some of the fine points related to such matters as obtaining an initial solution, redundancy and degeneracy.

5–2 TRANSFORMATION FORMULAE

It will be recalled that for each vector that is not in the basis, it is necessary to have a value of $z_j - c_j$, and that this value depends upon the various y_{ij}. Hence it is of interest to us to find a simple method to transform the components of each $\bar{y}_j$ at each stage.

Let us first establish our notational conventions. The columns of A, i.e., the m columns $\bar{a}_j$ of A, that are in the basis will be represented as an $m \times m$

matrix B. These various $\bar{a}_j$ can be in B in any order. We will represent B as

$$B = [\bar{b}_1, \bar{b}_2, \ldots, \bar{b}_m]$$

Which particular $\bar{a}_j$ is represented by $\bar{b}_i$ has to be specified independently. The subscript i on the $\bar{b}_i$ should not be confused with the subscript j on the $\bar{a}_j$. Hence, in terms of our development in Chapter 4, we see that any column $\bar{a}_j$ of A can be expressed as a linear combination of the columns of B, i.e.,

$$\bar{a}_j = y_{1j}\bar{b}_1 + y_{2j}\bar{b}_2 + \ldots + y_{mj}\bar{b}_m = \sum_{i=1}^{m} y_{ij}\bar{b}_i = B\bar{y}_j \tag{5–1}$$

Let us now return to the problem of the transformation of the y_{ij}.

Let us assume that $\bar{a}_k$ is to enter the basis and $\bar{b}_r$ is to be removed from the basis. We know that

$$\bar{a}_j = \sum_{i=1}^{m} y_{ij}\bar{b}_i \qquad j = 1, 2, \ldots, n \tag{5–2}$$

We can rewrite (5–2) for the case of $\bar{a}_k$ as follows:

$$\bar{a}_k = y_{rk}\bar{b}_r + \sum_{\substack{i=1\\ i\neq r}}^{m} y_{ik}\bar{b}_i \tag{5–3}$$

Since $\bar{b}_r$ is to be removed from the basis, we know that $y_{rk} \neq 0$. Hence, we can solve (5–3) for $\bar{b}_r$ to obtain

$$\bar{b}_r = \frac{1}{y_{rk}}\bar{a}_k - \sum_{\substack{i=1\\ i\neq r}}^{m} \frac{y_{ik}}{y_{rk}}\bar{b}_i \tag{5–4}$$

Let us now write (5–2) in a format similar to (5–3) as

$$\bar{a}_j = y_{rj}\bar{b}_r + \sum_{\substack{i=1\\ i\neq r}}^{m} y_{ij}\bar{b}_i \qquad j = 1, 2, \ldots, n \tag{5–5}$$

and substitute (5–4) into (5–5) to obtain

$$\bar{a}_j = \sum_{\substack{i=1\\ i\neq r}}^{m} \left(y_{ij} - \frac{y_{rj}y_{ik}}{y_{rk}}\right)\bar{b}_i + \frac{y_{rj}}{y_{rk}}\bar{a}_k \qquad j = 1, 2, \ldots, n \tag{5–6}$$

Equations (5–2) expressed any vector $\bar{a}_j$ in terms of the vectors $\bar{b}_i$ in the current basis B. Then we inserted $\bar{a}_k$ and removed $\bar{b}_r$. Therefore, equations

(5–6) give the representation of any vector $\bar{a}_j$ in terms of the new basis vectors, $\bar{b}_i$, $i = 1, 2, \ldots, m$; $i \neq r$ and $\bar{a}_k$. We now rewrite (5–6) as

$$\bar{a}_j = \sum_{i=1}^{m} \hat{y}_{ij}\hat{\bar{b}}_i \qquad j = 1, 2, \ldots, n \tag{5–7}$$

where we define: $\hat{\bar{b}}_i = \bar{b}_i$, $i \neq r$, $\hat{\bar{b}}_r = \bar{a}_k$. If we compare equations (5–6) and (5–7) we see that

$$\begin{aligned} \hat{y}_{ij} &= y_{ij} - \frac{y_{rj}y_{ik}}{y_{rk}} \qquad i \neq r \\ \hat{y}_{rj} &= \frac{y_{rj}}{y_{rk}} \end{aligned} \tag{5–8}$$

Equations (5–8) enable us to compute, at each stage, a new set of y_{ij} from the previous set, once we know which vectors are to enter and leave the basis.

It is also necessary at each iteration, to transform the values of $z_j - c_j$. It is more efficient to do this than to recalculate the $\hat{z}_j - c_j$ from their definition. First we note that, by definition (see Equation (4-34))

$$\hat{z}_j - c_j = \sum_{i=1}^{m} \hat{\bar{c}}_{Bi}\hat{y}_{ij} - c_j \qquad j = 1, 2, \ldots, n \tag{5–9}$$

where $\hat{c}_{Bi} = c_{Bi}$, $i \neq r$ and $\hat{c}_{Br} = c_k$. If we substitute from (5–8) into (5–9) we have

$$\hat{z}_j - c_j = \sum_{\substack{i=1 \\ i \neq r}}^{m} c_{Bi}\left(y_{ij} - \frac{y_{rj}y_{ik}}{y_{rk}}\right) + \frac{y_{rj}}{y_{rk}} c_k - c_j \qquad j = 1, 2, \ldots, n \tag{5–10}$$

It is apparent that the missing term in the summation in (5–10) is

$$c_{Br}\left(y_{rj} - \frac{y_{rj}y_{rk}}{y_{rk}}\right) = 0 \tag{5–11}$$

Hence, we can combine (5–10) and (5–11) to obtain

$$\begin{aligned} \hat{z}_j - c_j &= \sum_{i=1}^{m} c_{Bi}\left(y_{ij} - \frac{y_{rj}y_{ik}}{y_{rk}}\right) + \frac{y_{rj}}{y_{rk}} c_k - c_j \qquad j = 1, 2, , \ldots, n \\ &= \sum_{i=1}^{m} c_{Bi}y_{ij} - \frac{y_{rj}}{y_{rk}} \sum_{i=1}^{m} c_{Bi}y_{ik} + \frac{y_{rj}}{y_{rk}} c_k - c_j \\ &= \sum_{i=1}^{m} c_{Bi}y_{ij} - c_j - \frac{y_{rj}}{y_{rk}}\left[\sum_{i=1}^{m} c_{Bi}y_{ik} - c_k\right] \\ &= z_j - c_j - \frac{y_{rj}}{y_{rk}}(z_k - c_k) \end{aligned}$$

Therefore,

$$\hat{z}_j - c_j = z_j - c_j - \frac{y_{rj}}{y_{rk}}(z_k - c_k) \qquad j = 1, 2, \ldots, n \tag{5-12}$$

We already know how to transform the values of x_{Bi}, $i = 1, 2, \ldots, m$. These transformation equations were developed in Chapter 4 in connection with the development of the criterion for determining the vector to leave the basis. They are given by equations (4–41) and are restated here in terms of our current notation:

$$\begin{aligned} \hat{x}_{Bi} &= x_{Bi} - x_{Br}\frac{y_{ik}}{y_{rk}} \qquad i \neq r \\ \hat{x}_{Br} &= \frac{x_{Br}}{y_{rk}} \end{aligned} \tag{5-13}$$

Similarly, the transformation of the value of the objective function z was also derived in Chapter 4 and is given by equation (4–50), which in terms of our current notation is

$$\hat{z} = z + \frac{x_{Br}}{y_{rk}}(c_k - z_k) \tag{5-14}$$

Equations (5–8), (5–12), (5–13), and (5–14) will enable us to transform all the quantities of interest at each stage of the simplex calculation. It is convenient to use *one* set of transformation equations for all the aforementioned quantities. This can be done by what amounts to adding an extra row and column to the y_{ij} array. If we define

$$\begin{aligned} y_{i0} &\equiv x_{Bi}, \qquad i = 1, 2, \ldots, m \\ y_{m+1,0} &\equiv z \\ y_{m+1,j} &\equiv z_j - c_j, \qquad j = 1, 2, \ldots, n \end{aligned} \tag{5-15}$$

Then the following equations transform the y_{ij}, $z_j - c_j$, x_{Bi}, z for all i, j:

$$\begin{aligned} \hat{y}_{ij} &= y_{ij} - \frac{y_{ik}y_{rj}}{y_{rk}} \qquad j = 0, 1, \ldots, n \quad i = 1, 2, \ldots, m+1; \quad i \neq r \\ \hat{y}_{rj} &= \frac{y_{rj}}{y_{rk}} \qquad j = 0, 1, \ldots, n \end{aligned} \tag{5-16}$$

Equations (5–16) are often written in the form of vector equations as follows. We define

$$\bar{w}_j = [\bar{y}_j, y_{m+1,j}] \qquad j = 0, 1, \ldots, n \tag{5-17}$$

$$\bar{\gamma} = \left[-\frac{y_{1k}}{y_{rk}}, -\frac{y_{2k}}{y_{rk}}, \ldots, -\frac{y_{r-1,k}}{y_{rk}}, \frac{1}{y_{rk}} - 1, -\frac{y_{r+1,k}}{y_{rk}}, \ldots, -\frac{y_{mk}}{y_{rk}}\right]$$

Then the transformation equation is

$$\overline{\overline{w}}_j = \overline{w}_j + y_{rj}\bar{\gamma} \qquad j = 0, 1, \ldots, n \tag{5-18}$$

Equation (5–18) transforms all variables of interest.

5-3 SIMPLEX TABLEAU

A simplex tableau is a format for arranging the array of numbers y_{ij}, all i, j so that the necessary steps of the simplex algorithm can be conveniently carried out when one is performing calculations manually. What is shown in the tableau is the current basic feasible solution $\bar{x}_B$, which vectors $\bar{a}_j$ are in the basis, the values $\bar{y}_j$ which allow one to express each $\bar{a}_j$ in terms of the vectors in the basis, the current value of the objective function z and the values of $z_j - c_j$ for each $\bar{a}_j$. Hence, we have at each stage all the necessary information to determine whether or not the current solution is optimal. If it is not, we can determine the vector to enter and the vector to leave the basis. Furthermore, as equations (5–16) and (5–18) indicate, all the necessary quantities needed to construct the succeeding simplex tableau are also available for the required calculations.

We will leave the problem of how to find an initial basic feasible solution to the next section. We assume then that either initially or at some subsequent stage, we know which vectors are in the basis. There are, then, many possible different ways to put the simplex tableau together. Variants will be found in different books and papers. However, they all contain the same information. The tableau given in Figure 5–1 is the one we shall use.

It will be noted in Figure 5–1 that under the column headed "Basis" we list which vectors $\bar{a}_j$ are currently in the basis. For example, $\bar{b}_1$ might be $\bar{a}_5$, $\bar{b}_2$ might be $\bar{a}_3$, etc. The values of x_{Bi} in the next column give the values of x_5, x_3, etc., that are in the basic feasible solution $\bar{x}_B \geq \bar{0}$ and the values of c_{Bi} in the first column are the corresponding prices of the basic variables. The remaining columns give the components y_{ij} of the vectors $\bar{y}_j$, where $\bar{y}_j = B^{-1}\bar{a}_j$. At the top of each $\bar{y}_j$ the c_j are noted for convenience. Under each $\bar{y}_j$ we record the corresponding $z_j - c_j$ and under the $\bar{x}_B$ column, the current value of the objective function is recorded.

It is obvious that all the information we require to carry out the simplex process described in the previous chapter is contained in the simplex tableau. We can immediately tell if the solution contained in the tableau is optimal. If examination of the $z_j - c_j$ reveals that $z_j - c_j \geq 0$ $j = 1, 2, \ldots, n$ we know the solution is optimal. If one or more of the $z_j - c_j$ are negative then we determine the vector to enter the basis by

$$z_k - c_k = \min_j \{z_j - c_j \mid z_j - c_j < 0\} \tag{5-19}$$

			x_1	c_2	$\cdots$	c_ℓ	$\cdots$	c_n
$\bar{c}_B$	*Basis*	$\bar{x}_B$	$\bar{y}_1$	$\bar{y}_2$	$\cdots$	$\bar{y}_\ell$	$\cdots$	$\bar{y}_n$
c_{B1}	$\bar{b}_1$	x_{B1}	y_{11}	y_{12}	$\cdots$	$y_{1\ell}$	$\cdots$	y_{1n}
c_{B2}	$\bar{b}_2$	x_{B2}	y_{21}	y_{22}	$\cdots$	$y_{2\ell}$	$\cdots$	y_{2n}
$\vdots$	$\vdots$	$\vdots$	$\vdots$	$\vdots$	$\vdots$	$\vdots$	$\vdots$	$\vdots$
c_{Bm}	$\bar{b}_m$	x_{Bm}	y_{m1}	y_{m2}	$\cdots$	$y_{m\ell}$	$\cdots$	y_{mn}
		z	$z_1 - c_1$	$z_2 - c_2$	$\cdots$	$z_\ell - c_\ell$	$\cdots$	$z_n - c_n$

Figure 5–1 Simplex tableau.

The vector to leave the basis is found by

$$\frac{x_{Br}}{y_{rk}} = \min_i \left\{ \frac{x_{Bi}}{y_{ik}} \,\middle|\, y_{ik} > 0 \right\} \tag{5–20}$$

These values are easily found from the third column of the tableau $\bar{x}_B$ and the particular $\bar{y}_k$ which corresponds to the minimum $z_j - c_j$.

Let us assume then that $\bar{b}_r$ leaves the basis and $\bar{a}_k$ replaces it and see what the transformed simplex tableau looks like after transforming all the quantities in the tableau using equations (5–16). We present some sample conversions and then the entire transformed tableau in Figure 5–2. For example,

$$\hat{y}_{11} = y_{11} - \frac{y_{1k}y_{r1}}{y_{rk}} \qquad r \neq 1, \qquad k \neq 1$$

$$\hat{y}_{23} = y_{23} - \frac{y_{2k}y_{r3}}{y_{rk}} \qquad r \neq 2, \qquad k \neq 3 \tag{5–21}$$

$$\hat{y}_{r3} = \frac{y_{r3}}{y_{rk}}$$

$$\hat{x}_{B2} = x_{B2} - y_{2k}\frac{x_{Br}}{y_{rk}}$$

$$\hat{z}_3 - c_3 = z_3 - c_3 - \frac{y_{r3}}{y_{rk}}(z_k - c_k)$$

It is easily seen that $\hat{\bar{y}}_k$ will *always* be a vector of the form $\bar{e}_r$ because

$$\hat{y}_{ik} = y_{ik} - y_{ik}\frac{y_{rk}}{y_{rk}} = 0 \qquad i \neq r$$

$$\hat{y}_{rk} = \frac{y_{rk}}{y_{rk}} = 1 \tag{5–22}$$

			c_1	c_2		c_k		c_n
$\bar{c}_B$	Basis	$\bar{x}_B$	$\bar{y}_1$	$\bar{y}_2$	$\cdots$	$\bar{y}_k$	$\cdots$	$\bar{y}_n$
c_{B1}	$\bar{b}_1$	$x_{B1} - y_{1k}\dfrac{x_{Br}}{y_{rk}}$	$y_{11} - y_{1k}\dfrac{y_{r1}}{y_{rk}}$	$y_{12} - y_{1k}\dfrac{y_{r2}}{y_{rk}}$	$\cdots$	0	$\cdots$	$y_{1n} - y_{1k}\dfrac{y_{rn}}{y_{rk}}$
c_{B2}	$\bar{b}_2$	$x_{B2} - y_{2k}\dfrac{x_{Br}}{y_{rk}}$	$y_{21} - y_{2k}\dfrac{y_{r1}}{y_{rk}}$	$y_{22} - y_{2k}\dfrac{y_{r2}}{y_{rk}}$	$\cdots$	0	$\cdots$	$y_{2n} - y_{2k}\dfrac{y_{rn}}{y_{rk}}$
$\vdots$	$\vdots$	$\vdots$	$\vdots$	$\vdots$	$\cdot$	$\cdot$	$\cdot$	$\vdots$
c_{Br}	$\hat{b}_r = \bar{a}_k$	$\dfrac{x_{Br}}{y_{rk}}$	$\dfrac{y_{r1}}{y_{rk}}$	$\dfrac{y_{r2}}{y_{rk}}$		1	$\cdot$	$\dfrac{y_{rn}}{y_{rk}}$
$\vdots$	$\vdots$	$\vdots$	$\vdots$	$\vdots$	$\cdot$	$\vdots$	$\cdot$	$\vdots$
c_{Bm}	$\bar{b}_m$	$x_{Bm} - y_{mk}\dfrac{x_{Br}}{y_{rk}}$	$y_{m1} - y_{mk}\dfrac{y_{r1}}{y_{r1}}$	$y_{m2} - y_{mk}\dfrac{y_{r2}}{y_{r2}}$	$\cdots$	0	$\cdots$	$y_{mn} - y_{mk}\dfrac{y_{rn}}{y_{rk}}$
		$z - \dfrac{x_{Br}}{y_{rk}}(z_k - c_k)$	$(z_1 - c_1) - \dfrac{y_{r1}}{y_{rk}}(z_k - c_k)$	$(z_2 - c_2) - \dfrac{y_{r2}}{y_{rk}}(z_k - c_k)$	$\cdots$	0	$\cdots$	$(z_n - c_n) - \dfrac{y_{rn}}{y_{rk}}(z_k - c_k)$

Figure 5–2 Transformed simplex tableau.

Hence there is never any need to calculate $\bar{y}_k$. The column of the simplex tableau that corresponds to $\bar{y}_k$ is simply changed to a unit vector, $\bar{e}_r$. Similarly, $\hat{z}_k - c_k$ is calculated to be zero from (5–16) or (5–12). With the new tableau we again examine the $z_j - c_j$ to see if the new solution is optimal. If not we continue the simplex process.

5–4 DETERMINATION OF AN INITIAL BASIC FEASIBLE SOLUTION

In order to have an initial simplex tableau we must know which vectors $\bar{a}_j$ are in B and, hence, since $\bar{x}_B = B^{-1}\bar{b}$, what our initial basic feasible solution is. Since a basis consists of m linearly independent vectors, it is not always immediately obvious how to select m such linearly independent vectors. We shall discuss here several methods for finding such an initial basic feasible solution or of obtaining an indication that the problem under consideration has no feasible solution.

The simplest case, for which an initial solution can be determined with virtually no effort, is the case in which the original problem constraint set consisted entirely of inequalities of the form

$$\sum_{j=1}^{p} a_{ij}x_j \leq b_i \qquad i = 1, 2, \ldots, m \tag{5–23}$$

and also that $b_i \geq 0$, $i = 1, 2, \ldots, m$. For this case we would have to add a slack variable to each of the m inequalities of (5–23) to obtain

$$\begin{aligned} \sum_{j=1}^{p} a_{ij}x_j + x_{p+i} &= b_i \qquad i = 1, 2, \ldots, m \\ x_{p+i} &\geq 0 \qquad i = 1, 2, \ldots, m \end{aligned} \tag{5–24}$$

Hence the matrix A for this constraint set (see Theorem 4–1) consists of

$$A = (D, S) = (D, I) \tag{5–25}$$

where D is the submatrix of coefficients of the x_j, and I is an identity matrix which is the submatrix of coefficients of the x_{p+i}. If we now partition

$$\bar{x} = [\bar{x}_D, \bar{x}_s] \tag{5–26}$$

where $\bar{x}_s$ is the vector of slack variables, then it is easy to see how to obtain an initial basic feasible solution. Since

$$A\bar{x} = (D, I)\begin{bmatrix} \bar{x}_D \\ \bar{x}_s \end{bmatrix} = \bar{b}$$

we have that

$$D\bar{x}_D + I\bar{x}_s = \bar{b} \tag{5–27}$$

If we let $\bar{x}_D = \bar{0}$, then we have that $\bar{x}_s = \bar{b}$. Since by hypothesis $\bar{b} \geq \bar{0}$, we see that

$$\bar{x}_s = \bar{b} \geq \bar{0} \tag{5–28}$$

is a basic feasible solution. This solution has a particularly convenient basis to work with, since $B = I$ and therefore $B^{-1} = I$. Hence,

$$\bar{y}_j = B^{-1}\bar{a}_j = I\bar{a}_j = \bar{a}_j \qquad j = 1, 2, \ldots, n \tag{5–29}$$

and since $\bar{x}_s$ consists entirely of slack variables, we see that $\bar{c}_B = \bar{0}$, since slack variables have zero values of c_j. Therefore,

$$z_j - c_j = \bar{c}'_B\bar{y}_j - c_j = -c_j \qquad j = 1, 2, \ldots, n \tag{5–30}$$

and

$$z = \bar{c}'_B\bar{x}_B = 0 \tag{5–31}$$

Hence we have all the quantities necessary for the initial tableau without the need for any calculation.

Unfortunately, very few problems encountered in practice satisfy the assumptions of the foregoing paragraph as described by (5–23). However, it may still be the case that when one examines the matrix A, it is found to contain an $m \times m$ identity submatrix, i.e., it contains m vectors (slack, surplus, or otherwise) which are unit vectors $\bar{e}_i$ and which together comprise an identity matrix I.† This case is almost as simple as the first one. Not all the c_{Bi} will be zero and hence the only change is that we must calculate the $z_j - c_j$ from

$$z_j - c_j = \bar{c}'_B\bar{y}_j - c_j = \bar{c}'_B\bar{a}_j - c_j \qquad j = 1, 2, \ldots, n \tag{5–32}$$

Since this is a simple calculation, the present case is also a very convenient one for which to determine an initial basic feasible solution.

We now turn our attention to the general case, one in which the A matrix does *not* contain an $m \times m$ identity submatrix. At this point, there are two divergent paths we might choose. The first would be to try to find a way to determine some m linearly independent vectors present in A to form our initial basis. This path was not the one chosen by the early practitioners of linear programming although there are some indications that this may well be a perfectly reasonable approach (see Hasegawa[1]). Instead, another approach

† We shall continue to assume throughout this chapter that $\bar{b} \geq \bar{0}$. This is not a limitation in practice since if one or more of the inequalities or equations had $b_i < 0$, we could multiply by -1 to convert to the assumed form. This would, of course, change the direction of the inequality involved.

has been uniformly taken. This consists of adopting an identity basis, whether or not one is present in the matrix A. In other words, because of the simplicity that results from having $B = I$, we decide we will start with such a basis. Since we have assumed that the matrix A did not contain such a submatrix, it is obvious that we must augment A and its associated variables in some way. Instead of dealing with the original set of constraints $A\bar{x} = \bar{b}$, we make use of an augmented set of constraint equations:

$$A\bar{x} + I\bar{x}_a = \bar{b} \tag{5–33}$$

What we have done is create a new matrix $F = (A, I)$ and a new set of variables $[\bar{x}, \bar{x}_a]$ by adding an $m \times m$ identity submatrix to A, and m additional variables $\bar{x}_a = [x_{a1}, x_{a2}, \ldots, x_{am}]$, whose associated vectors are $\bar{e}_i$. The additional variables x_{ai} are called *artificial variables*. The $\bar{e}_i$ which are associated with these variables will be designated $\bar{a}_{n+i}$, $i = 1, 2, \ldots, m$, and are called *artificial vectors*.

We have now obtained an initial basis $B = I$ and such that $\bar{x} = \bar{0}$ and $\bar{x}_a = \bar{b} \geq \bar{0}$. The only difficulty is that what we have is a basic feasible solution to (5–33). However, it is most assuredly *not* a basic feasible solution to the original constraints $A\bar{x} = \bar{b}$. This is obvious because no solution to $A\bar{x} = \bar{b}$ can contain any $x_{ai} > 0$, yet the initial basic feasible solution has *all* $x_{ai} > 0$, unless one or more $b_i = 0$. Hence, what we must do is somehow convert the "solution" we have to one which contains no positive values of x_{ai}, if this is possible. To do this we use the simplex method to insert vectors into the basis and so, one by one, remove the artificial vectors $\bar{a}_{n+i}$ from the initial basis until we finally have a basic feasible solution to the original problem $A\bar{x} = \bar{b}$. Then we can continue the simplex process until we arrive at the optimal basic feasible solution to the original linear programming problem.

In order to use the simplex process to insert and remove vectors from the basis, we assign extremely unfavorable prices to the variables x_{ai}, so that it is always possible to improve the value of the objective function z by continuing to remove any remaining artificial vectors $\bar{a}_{n+i}$ from the basis. Formally, this is done as follows. If c_{ai}, $i = 1, 2, \ldots, m$ are prices associated with the x_{ai} we then define

$$c_{ai} = -M, \qquad M > 0, \qquad i = 1, 2, \ldots, m \tag{5–34}$$

assuming that we are dealing with a maximization problem. (If we are minimizing z, then $c_{ai} = M$ for all artificial variables.) Since the value of $z_j - c_j$ determines which vector is to enter the basis, let us calculate $z_j - c_j$ for those variables not in the initial basis. We then have

$$z_j - c_j = \bar{c}'_B\bar{y}_j - c_j = \sum_{i=1}^{n} c_{ai}y_{ij} - c_j = -M\sum_{i=1}^{m} y_{ij} - c_j \tag{5–35}$$

We know that not all $y_{ij} \leq 0$; otherwise the objective function would be unbounded. Furthermore, if all $z_j - c_j \geq 0$, i.e., $-M \sum_{i=1}^{m} y_{ij} - c_j \geq 0$, we would have the optimal solution to

$$\text{Max } z = \sum_{j=1}^{n} c_j x_j - M \sum_{i=1}^{m} x_{ai} \tag{5–36}$$

But we know if M is sufficiently large that this cannot occur unless all $x_{ai} = 0$. Hence, at least one $z_j - c_j < 0$ for M sufficiently large, no matter what the relative magnitudes of $\sum_{i=1}^{m} y_{ij}$ and c_j. Hence, for some $j = \ell$,

$$z_\ell - c_\ell = -M \sum_{i=1}^{m} y_{i\ell} - c_\ell < 0 \tag{5–37}$$

and some $\bar{a}_\ell$ will enter the basis at each stage until all the artificial vectors are removed. It is never necessary to specify M when performing manual calculations. (This method is no longer used in computer codes.) We merely assume that M is sufficiently large.

In summary, then, we are trying to solve the problem:

$$\text{Max } z = \bar{c}'\bar{x}$$

subject to

$$\begin{aligned} A\bar{x} &= \bar{b} \\ \bar{x} &\geq \bar{0} \end{aligned} \tag{5–38}$$

In order to have a convenient basis $B = I$ to start with, we begin our computation by solving a modified problem:

$$\text{Max } z = \bar{c}'\bar{x} - M\bar{1}'\bar{x}_a$$

subject to

$$\begin{aligned} A\bar{x} + I\bar{x}_a &= \bar{b} \\ \bar{x} &\geq \bar{0} \\ \bar{x}_a &\geq \bar{0} \end{aligned} \tag{5–39}$$

For (5–39), our initial basic feasible solution is $\bar{x}_a = \bar{b}$, $\bar{x} = \bar{0}$. By applying the simplex method to (5–39) we ultimately arrive at a basic feasible solution which no longer has any artificial variables which are positive. We then continue with the simplex method until an optimal basic feasible solution is obtained. Once an artificial vector has been removed from the basis, it may be discarded, since it may never be a candidate for re-entry into the basis when M is sufficiently large.

It should be noted that it is not necessary in all cases to append a full artificial basis to begin the calculation. If there are present in the matrix A one or more vectors $\bar{e}_i$, then we need add only those artificial vectors $\bar{a}_{n+i} \equiv \bar{e}_i$ which are required to make a full identity matrix I. Hence, we do not make the problem size greater than necessary nor do we perform more iterations than necessary to drive the artificial vectors out of the basis.

The method we have been discussing is usually attributed to Charnes[2] although it appears it was suggested earlier by Dantzig.[3] In any case, the method is of historical interest only. The "two-phase" method to be described in the following paragraphs is less cumbersome and, at least for part of the calculation, would lead to the same sequence of vectors entering and leaving the basis.

We shall now describe a process for finding an initial basic feasible solution which is usually referred to as the *two-phase method.* The name refers to the fact that the total number of iterations required to solve a linear programming problem may be separated into two distinct phases. In Phase I, we carry out a process of eliminating any artificial variables in the basis. Phase I normally ends with either no artificial variables in the basis, or if any remain they will have zero values. In this case we initiate Phase II, which is the usual simplex process. If one or more artificial variables remain at the end of Phase I with positive values, then there is no feasible solution to the original linear programming problem. It can be seen that the basic approach of the two-phase method is similar to that of the previous method, in which we used prices of $-M$. Let us now examine this two-phase method in greater detail.

We proceed in *Phase I* by adding a set of m additional artificial variables, $\bar{x}_a = [x_{a_1}, x_{a_2}, \ldots, x_{a_m}]$ as before. The associated vectors are $\bar{a}_{n+1} \equiv \bar{e}_i$, $i = 1, 2, \ldots, m$. We then assign Phase I prices to all the variables as follows:

$$\begin{aligned} c_j &= 0 \qquad & j &= 1, 2, \ldots, n \\ c_{ai} &= -1 \qquad & i &= 1, 2, \ldots, m \end{aligned} \tag{5-40}$$

Hence the Phase I objective function is reduced to

$$z_a = \bar{c}'\bar{x} - \bar{1}'\bar{x}_a = -\sum_{i=1}^{m} x_{ai} \tag{5-41}$$

It is quite clear, since we are maximizing z_a, that the maximum value z_a can attain is zero. Furthermore, it is clear that there are only two ways this can occur. The first is if all the artificial vectors originally in the basis have been replaced by legitimate vectors and hence $z_a = 0$. The second way is if those artificial vectors remaining in the basis have corresponding values of $x_{ai} = 0$. On the other hand, whenever $z_a < 0$ at least one artificial variable, corresponding to an artificial vector in the basis, has a positive value, i.e., $x_{ai} > 0$.

Whenever this occurs, we do not have an initial basic feasible solution to the original linear programming problem.

Starting with a basis consisting of artificial vectors we will apply the simplex method, and at each iteration a legitimate non-basic vector will become basic. Legitimate vectors will replace artificial vectors for the same reason they did in the $-M$ method we discussed previously. Hence the value of z_a will increase at each iteration. We note that once an artifical vector has been removed from the basis during Phase I it can be discarded, since it will not be permitted to re-enter. This would defeat the purpose of Phase I.

Phase I iterations are terminated when all $z_j - c_j \geq 0$ for non-basis vectors. When this occurs there are three possibilities:

1. $z_a < 0$ and one or more artificial vectors remain in the basis with $x_{ai} > 0$.
2. $z_a = 0$ and all artificial vectors have been removed from the basis.
3. $z_a = 0$ and one or more artificial vectors remain in the basis with $x_{ai} = 0$.

If the simplex tableau at the end of Phase I (all $z_j - c_j \geq 0$) gives rise to Case 1 above then, clearly, there is no feasible solution to the original problem. If the final simplex tableau shows Case 2 or 3, we have found an initial basic feasible solution to the original problem and we can proceed with Phase II. However, some additional precautions will have to be taken if the final tableau ends in Case 3. These will be discussed presently.

Phase II proceeds with the usual simplex calculation. First we restore the original prices to the legitimate variables and assign zero prices to any artificial variables remaining in the basis at a zero level. Hence, with these changed prices, we recompute the $z_j - c_j$, $j = 1, 2, \ldots, n$. Otherwise, the final tableau of Phase I becomes the initial tableau for Phase II. We then proceed with the simplex calculations.

If Phase I ended in Case 2, there are no artificial vectors in the basis and we have a normal simplex calculation. However, if Phase I ended in Case 3 there is always the possibility that one or more of the artificial variables may become positive in some subsequent iteration. If this occurred we would no longer have a basic feasible solution. Hence, we must take steps to avoid this. There are several approaches that can be taken. First, however, let us see under what circumstances an artificial variable left in the basis at a zero level could become positive at some later stage.

Let $i \in I_a$ be the set of indices of these artificial vectors in the basis at a zero level. Now suppose some vector $\bar{a}_k$ is to enter the basis. There are several possibilities.

Suppose that $y_{ik} > 0$ for all $i \in I_a$. For this case one of the artificial vectors could be removed, although it would not necessarily be removed. If it is removed, however, the vector $\bar{a}_k$ will obviously enter at a zero level, since

$$\hat{x}_{Br} = \frac{x_{Br}}{y_{rk}} = \frac{0}{y_{rk}} = 0 \tag{5-42}$$

Furthermore, the values of all other variables will remain unchanged, since

$$\hat{x}_{Bi} = x_{Bi} - (0)\frac{y_{ik}}{y_{rk}} = x_{Bi} \tag{5–43}$$

Hence, any other artificial variables in the basis at a zero level will remain at a zero level. Therefore, this case causes no difficulty.

Now suppose $y_{ik} = 0$ for all $i \in I_a$. For this case, no artificial vector can be removed. Furthermore, the values of the artificial variables in the basis will remain zero, since

$$\hat{x}_{Bi} = x_{Bi} - x_{Br}\frac{y_{ik}}{y_{rk}} = 0 - \frac{x_{Br}(0)}{y_{rk}} = 0, \qquad i \in I_a \tag{5–44}$$

Again, this case causes no problems.

Finally, let us suppose that $y_{ik} < 0$ for at least one $i \in I_a$. Clearly, those artificial vectors with $y_{ik} < 0$ will remain in the basis since they do not satisfy the requirement for removal, viz., $y_{rk} > 0$. However, we can see that the values of the variables corresponding to those $y_{ik} < 0$, $i \in I_a$ will become *positive* (if $x_{Br} > 0$) since

$$\hat{x}_{Bi} = x_{Bi} - x_{Br}\frac{y_{ik}}{y_{rk}} = -\frac{x_{Br}y_{ik}}{y_{rk}} > 0, \qquad i \in I_a \tag{5–45}$$

Hence, it is this case that we must prevent from occurring, because we cannot allow an artificial variable to become positive in Phase II. Three ways have been suggested for dealing with this problem. We will consider two now and leave the third for Chapter 6.

One approach to avoiding the difficulty we have described is to remove an artificial vector from the basis if $y_{ik} < 0$ for one or more $i \in I_a$, instead of using the usual criterion, i.e., that given by equation (5–20). If there is more than one artificial vector with $y_{ik} < 0$, $i \in I_a$ then we can arbitrarily choose one of them. If $\bar{b}_r$ is the artificial vector we choose to remove, then,

$$\hat{x}_{Br} = \frac{x_{Br}}{y_{rk}} = 0 \tag{5–46}$$

and the values of all the other variables are:

$$\hat{x}_{Bi} = x_{Bi} - x_{Br}\frac{y_{ik}}{y_{rk}} = x_{Bi} - (0)\frac{y_{ik}}{y_{rk}} = x_{Bi} \tag{5–47}$$

Therefore, all the other variables remain the same. Hence, we have avoided having the artificial variables become positive. Whenever faced with the situation of one or more $y_{ik} < 0$, $i \in I_a$ we can resort to this deviation. Otherwise, we follow the usual simplex criterion for determining the vector to be removed. Note that the value of the objective function stays the same whenever it is necessary to use this rule.

A second approach to preventing artificial variables from becoming positive in Phase II is as follows. Again we assume that Phase I has concluded with $z_a = 0$, but one or more of the artificial variables x_{ai} are in the basis with $x_{ai} = 0$. If we examine the basic equation for how the objective function changes from iteration to iteration, i.e.,

$$\hat{z}_a = z_a - \frac{x_{Br}}{y_{rk}}(z_k - c_k) \tag{5–48}$$

we can note the following. At the end of Phase I, $z_a = 0$. Hence,

$$\hat{z}_a = -\frac{x_{Br}}{y_{rk}}(z_k - c_k) \tag{5–49}$$

for any vector which enters the basis at the conclusion of Phase I. Note that the $z_k - c_k$ refer to the Phase I prices and hence at the conclusion of Phase I all $z_j - c_j \geq 0$. It therefore follows from (5–49) that if any vector for which $z_j - c_j > 0$ enters the basis in Phase II, $\hat{z}_a < 0$ and hence one or more artificial variables will become positive. On the other hand, if any vector enters the basis for which $z_j - c_j = 0$, this causes no change, that is, $\hat{z}_a = 0$. Therefore, none of the artificial variables could be positive.

The above analysis suggests the following procedure for avoidance of artificial variables becoming positive in Phase II. We divide the vectors into two classes. In class 1 we place those having $z_j - c_j > 0$ at the conclusion of Phase I. In class 2 we place those vectors having $z_j - c_j = 0$ at the conclusion of Phase I. Then we do not permit any vector in class 1 to enter the basis during Phase II. With this restriction, it is not possible for any remaining artificial variables to become positive during Phase II.

5–5 EXAMPLES OF SIMPLEX CALCULATIONS

In this section we will present several examples illustrating each of the methods of initiating and carrying forward the calculations we have previously discussed.

EXAMPLE 1

Identity submatrix present in matrix A.

$$\begin{aligned} \text{Max } z &= 3x_1 + 4x_2 + 2x_3 \\ x_1 - 2x_2 + 4x_3 &\leq 36 \\ 2x_1 + 3x_2 - 5x_3 &\leq 40 \\ 3x_1 + 2x_2 + x_3 &\leq 28 \\ x_1, x_2, x_3 &\geq 0 \end{aligned} \tag{5–50}$$

We convert the problem (5–50) into the standard form as follows:

$$\begin{aligned} \text{Max } z &= 3x_1 + 4x_2 + 2x_3 \\ x_1 - 2x_2 + 4x_3 + x_4 &= 36 \\ 2x_1 + 3x_2 - 5x_3 + x_5 &= 40 \\ 3x_1 + 2x_2 + x_3 + x_6 &= 28 \\ x_j \geq 0, \qquad j &= 1, 2, \ldots, 6 \end{aligned} \tag{5–51}$$

We take as our initial basis $B = (\bar{b}_1, \bar{b}_2, \bar{b}_3) = (\bar{a}_4, \bar{a}_5, \bar{a}_6) = I$, since these constitute a built-in identity submatrix. Hence, our initial tableau is as given in Figure 5–3.

From Tableau I we see that not all $z_j - c_j \geq 0$. Hence, we do not have the optimal solution. We choose the vector to enter the basis from

$$z_k - c_k = \min_j \{z_j - c_j < 0\} = \min(-3, -4, -2) = -4 = z_2 - c_2$$

Hence $\bar{a}_2$ enters the basis. We determine the vector to leave the basis from

$$\frac{x_{Br}}{y_{rk}} = \min_i \left\{ \frac{x_{Bi}}{y_{ik}} \middle| y_{ik} > 0 \right\} = \min \left(\frac{40}{3}, \frac{28}{2} \right) = \frac{40}{3} = \frac{x_{B2}}{y_{22}}$$

Hence $\bar{a}_5$ leaves the basis. Using the transformation equations (5–16) we can determine all the quantities we need for the next tableau. We have circled element $y_{rk} = y_{22}$ in Tableau I. This is called the pivot element because of the analogy with a similar process in the Gaussian elimination method of solving linear equations. The circle serves to remind us that this pivot must be divided by itself to become "1" in the next tableau. A few sample calcula-

			3	4	2	0	0	0
$\bar{c}_B$	Basis	$\bar{x}_B$	$\bar{y}_1$	$\bar{y}_2$	$\bar{y}_3$	$\bar{y}_4$	$\bar{y}_5$	$\bar{y}_6$
0	$\bar{a}_4$	36	1	−2	4	1	0	0
0	$\bar{a}_5$	40	2	③	−5	0	1	0
0	$\bar{a}_6$	28	3	2	1	0	0	1
		0	−3	−4	−2	0	0	0

Figure 5–3 Tableau I.

tions are as follows:

$$r = 2,\ k = 2 \qquad y_{rk} = y_{22} = 3$$

$$\hat{y}_{21} = \frac{y_{21}}{y_{22}} = \frac{2}{3}$$

$$\hat{y}_{20} = \hat{x}_{B2} = \frac{x_2}{y_{22}} = \frac{40}{3}$$

$$\hat{y}_{13} = y_{13} - \frac{y_{12}y_{23}}{y_{22}} = 4 - \frac{(-2)(-5)}{3} = -2$$

$$\hat{y}_{35} = y_{35} - \frac{y_{32}y_{25}}{y_{22}} = 0 - \frac{(2)(1)}{3} = -\frac{2}{3}$$

$$\hat{y}_{44} = \hat{z}_4 - c_4 = y_{44} - \frac{y_{42}y_{24}}{y_{22}} = 0 - \frac{(-4)(0)}{3} = 0$$

The fully transformed tableau is given in Figure 5–4.

We see in Tableau II that $\hat{z} > z$ but we are still not optimal since not all $z_j - c_j \geq 0$. Hence, we continue the simplex process. The vector to enter the basis is determined from

$$z_k - c_k = \min\left(-\tfrac{1}{3}, -\tfrac{26}{3}\right) = -\tfrac{26}{3} = z_3 - c_3$$

Therefore $\bar{a}_3$ enters the basis. The vector to leave the basis is determined from

$$\frac{x_{Br}}{y_{r3}} = \min\left(\tfrac{188}{3}/\tfrac{2}{3}, \tfrac{4}{3}/\tfrac{13}{3}\right) = \min\left(94, \tfrac{4}{13}\right) = \tfrac{4}{13} = \frac{x_{B3}}{y_{33}}$$

Again we make use of the transformation formulae and we obtain Tableau III in Figure 5–5.

We see in Tableau III that all $z_j - c_j \geq 0$. Hence we have obtained the optimal solution: $x_1 = 0$, $x_2 = 13.85$, $x_3 = 0.3077$, $x_4 = 62.46$, $x_5 = 0$, $x_6 = 0$ and $z = 56$. We note, as will often be the case, that one of the slack

			3	4	2	0	0	0
$\bar{c}_B$	Basis	$\bar{x}_B$	$\bar{y}_1$	$\bar{y}_2$	$\bar{y}_3$	$\bar{y}_4$	$\bar{y}_5$	$\bar{y}_6$
0	$\bar{a}_4$	$\frac{188}{3}$	$\frac{7}{3}$	0	$\frac{2}{3}$	1	$\frac{2}{3}$	0
4	$\bar{a}_2$	$\frac{40}{3}$	$\frac{2}{3}$	1	$-\frac{5}{3}$	0	$\frac{1}{3}$	0
0	$\bar{a}_6$	$\frac{4}{3}$	$\frac{5}{3}$	0	$\frac{13}{3}$ (circled)	0	$-\frac{2}{3}$	1
		$\frac{160}{3}$	$-\frac{1}{3}$	0	$-\frac{26}{3}$	0	$\frac{4}{3}$	0

Figure 5–4 Tableau II.

			3	4	2	0	0	0
$\bar{c}_B$	Basis	$\bar{x}_B$	$\bar{y}_1$	$\bar{y}_2$	$\bar{y}_3$	$\bar{y}_4$	$\bar{y}_5$	$\bar{y}_6$
0	$\bar{a}_4$	$\frac{812}{13}$	$\frac{27}{13}$	0	0	1	$\frac{10}{13}$	$-\frac{2}{13}$
4	$\bar{a}_2$	$\frac{180}{13}$	$\frac{17}{13}$	1	0	0	$\frac{1}{13}$	$\frac{5}{13}$
2	$\bar{a}_3$	$\frac{4}{13}$	$\frac{5}{13}$	0	1	0	$-\frac{2}{13}$	$\frac{3}{13}$
		56	3	0	0	0	0	2

Figure 5–5 Tableau III.

variables, $x_4 > 0$. This means that one of our "resources" is not being fully allocated in order to achieve an optimal solution.

EXAMPLE 2

The use of artificial variables: $-M$ method.

$$\begin{aligned} \text{Max } z &= x_1 + 4x_2 + 3x_3 \\ 2x_1 - x_2 + 5x_3 &= 40 \\ x_1 + 2x_2 - 3x_3 &\geq 22 \\ 3x_1 + x_2 + 2x_3 &= 30 \\ x_1, x_2, x_3 &\geq 0 \end{aligned} \tag{5–52}$$

First we put problem (5–52) into standard form:

$$\begin{aligned} \text{Max } z &= x_1 + 4x_2 + 3x_3 \\ 2x_1 - x_2 + 5x_3 &= 40 \\ x_1 + 2x_2 - 3x_3 - x_4 &= 22 \\ 3x_1 + x_2 + 2x_3 &= 30 \\ x_1, x_2, x_3, x_4 &\geq 0 \end{aligned} \tag{5–53}$$

We then add artificial vectors $\bar{a}_5, \bar{a}_6, \bar{a}_7$, to form an initial artificial basis, and take as our objective function:

$$z = x_1 + 4x_2 + 3x_3 - Mx_{a1} - Mx_{a2} - Mx_{a3} \tag{5–54}$$

Our initial basis is $B = (\bar{b}_1, \bar{b}_2, \bar{b}_3) = (\bar{a}_5, \bar{a}_6, \bar{a}_7)$ and the initial tableau is given in Figure 5–6.

			1	4	3	0	$-M$	$-M$	$-M$
$\bar{c}_B$	Basis	$\bar{x}_B$	$\bar{y}_1$	$\bar{y}_2$	$\bar{y}_3$	$\bar{y}_4$	$\bar{y}_5$	$\bar{y}_6$	$\bar{y}_7$
$-M$	$\bar{a}_5$	40	2	-1	5	0	1	0	0
$-M$	$\bar{a}_6$	22	1	2	-3	-1	0	1	0
$-M$	$\bar{a}_7$	30	③	1	2	0	0	0	1
		$-92M$	$-6M-1$	$-2M-4$	$-4M-3$	M	0	0	0

Figure 5–6 Tableau I.

If we examine Tableau I we see that:

$$z_k - c_k = \text{Min}\,(-6M-1, -2M-4, -4M-3) = -6M-1$$
$$= z_1 - c_1$$

Therefore, $\bar{a}_1$ enters the basis. The vector to leave the basis is:

$$\frac{x_{Br}}{y_{r1}} = \min\left(\tfrac{40}{2}, \tfrac{22}{1}, \tfrac{30}{3}\right) = 10 = \frac{x_{B3}}{y_{31}}$$

Hence $y_{rk} = y_{31}$. We use the usual transformation formulae to obtain Tableau II in Figure 5–7. Note that $\bar{a}_7$ is removed; it is not necessary to carry it along in Tableau II since it can never be a candidate for re-entry into the basis. From Tableau II we see that we are not yet optimal and that:

$$z_k - c_k = \min\left(-\tfrac{11}{3}, -\tfrac{7}{3}\right) = -\tfrac{11}{3} = z_2 - c_2$$

Therefore, $\bar{a}_2$ enters the basis. The vector to leave the basis is:

$$\frac{x_{Br}}{y_{r2}} = \min\left(\frac{12}{\frac{5}{3}}, \frac{10}{\frac{1}{3}}\right) = \frac{36}{5} = \frac{x_{B2}}{y_{22}}$$

			1	4	3	0	$-M$	$-M$
$\bar{c}_B$	Basis	$\bar{x}_B$	$\bar{y}_1$	$\bar{y}_2$	$\bar{y}_3$	$\bar{y}_4$	$\bar{y}_5$	$\bar{y}_6$
$-M$	$\bar{a}_5$	20	0	$-\frac{5}{3}$	$\frac{11}{3}$	0	1	0
$-M$	$\bar{a}_6$	12	0	$\textcircled{\frac{5}{3}}$	$-\frac{11}{3}$	-1	0	1
1	$\bar{a}_1$	10	1	$\frac{1}{3}$	$\frac{2}{3}$	0	0	0
		$-32M$ $+10$	0	$-\frac{11}{3}$	$-\frac{7}{3}$	M	0	0

Figure 5–7 Tableau II.

			1	4	3	0	$-M$
$\bar{c}_B$	Basis	$\bar{x}_B$	$\bar{y}_1$	$\bar{y}_2$	$\bar{y}_3$	$\bar{y}_4$	$\bar{y}_5$
$-M$	$\bar{a}_5$	32	0	0	0	-1	1
4	$\bar{a}_2$	$\frac{36}{5}$	0	1	$-\frac{11}{5}$	$-\frac{3}{5}$	0
1	$\bar{a}_1$	$\frac{38}{5}$	1	0	$\textcircled{\frac{7}{5}}$	$\frac{1}{5}$	0
		$-32M + \frac{182}{5}$	0	0	$-\frac{52}{5}$	$M - \frac{11}{5}$	0

Figure 5–8 Tableau III.

Hence, $y_{rk} = y_{22}$. We transform Tableau II to obtain Tableau III. We drop $\bar{a}_6$ since it is no longer required. In Tableau III we note that we still have one artificial vector in the basis and we are still not optimal. We have only one choice for entry into the basis and that is $\bar{a}_3$. Similarly, only $\bar{a}_1$ can be removed, because only $y_{33} > 0$ in $\bar{y}_3$. We note that at this iteration an artificial vector is not removed. Transforming Tableau III we obtain Tableau IV in Figure 5–9. Tableau IV has some interesting information for us. All $z_j - c_j \geq 0$, yet we still have an artificial vector in the basis. Hence, the original problem has *no* feasible solution. While this should not occur in practical problems, errors of formulation or errors in data can result in this conclusion. The fact that the optimality criteria is satisfied while an artificial vector is in the basis at a positive level tells us that there are no feasible solutions to our problem.

			1	4	3	0	$-M$
$\bar{c}_B$	Basis	$\bar{x}_B$	$\bar{y}_1$	$\bar{y}_2$	$\bar{y}_3$	$\bar{y}_4$	$\bar{y}_5$
$-M$	$\bar{a}_5$	32	0	0	0	-1	1
4	$\bar{a}_2$	$\frac{134}{7}$	$\frac{11}{7}$	1	0	$-\frac{2}{7}$	0
3	$\bar{a}_3$	$\frac{38}{7}$	$\frac{5}{7}$	0	1	$\frac{1}{7}$	0
		$-32M + \frac{650}{7}$	$\frac{52}{7}$	0	0	$M - \frac{5}{7}$	0

Figure 5–9 Tableau IV.

EXAMPLE 3: TWO-PHASE CALCULATION

$$\begin{aligned} \text{Max } z &= -2x_1 - 3x_2 - 5x_3 \\ 2x_1 + 3x_2 + 2x_3 &\geq 10 \\ x_1 + 4x_2 + 5x_3 &\geq 15 \\ 3x_1 + x_2 + x_3 &\geq 20 \\ x_1, x_2, x_3 &\geq 0 \end{aligned} \tag{5–55}$$

			0	0	0	0	0	0	−1	−1	−1
$\bar{c}_B$	Basis	$\bar{x}_B$	$\bar{y}_1$	$\bar{y}_2$	$\bar{y}_3$	$\bar{y}_4$	$\bar{y}_5$	$\bar{y}_6$	$\bar{y}_7$	$\bar{y}_8$	$\bar{y}_9$
−1	$\bar{a}_7$	10	2	(3)	2	−1	0	0	1	0	0
−1	$\bar{a}_8$	15	1	4	5	0	−1	0	0	1	0
−1	$\bar{a}_9$	20	3	1	1	0	0	−1	0	0	1
		−45	−6	−8	−8	1	1	1	0	0	0

Figure 5–10 Phase I—Tableau I.

We convert (5–55) to standard form which is as follows:

$$
\begin{aligned}
&\text{Max } z = -2x_1 - 3x_2 - 5x_3\\
&2x_1 + 3x_2 + 2x_3 - x_4 = 10\\
&x_1 + 4x_2 + 5x_3 - x_5 = 15\\
&\quad 3x_1 + x_2 + x_3 - x_6 = 20\\
&x_j \geq 0, \qquad j = 1, 2, \ldots, 6
\end{aligned}
\tag{5–56}
$$

For Phase I of the calculation we define a new objective function:

$$z_a = -x_{a_1} - x_{a_2} - x_{a_3} \tag{5–57}$$

The artificial vectors corresponding to x_{a_1}, x_{a_2}, x_{a_3} are $\bar{a}_7 = \bar{e}_1$, $\bar{a}_8 = \bar{e}_2$, $\bar{a}_9 = \bar{e}_3$. The first tableau for Phase I is shown in Figure 5–10. From Tableau I in Figure 5–10 we determine the vector to enter the basis from

$$z_k - c_k = \min(-6, -8, -8) = -8 = z_2 - c_2$$

Therefore, $\bar{a}_2$ enters the basis. The vector to leave is determined from

$$\frac{x_{Br}}{y_{r2}} = \min\left(\tfrac{10}{3}, \tfrac{15}{4}, \tfrac{20}{1}\right) = \tfrac{10}{3} = \frac{x_{B1}}{y_{12}}$$

Therefore, $\bar{a}_7$ leaves the basis. We now transform Tableau I to obtain Tableau II in Figure 5–11. It is not necessary to transform $\bar{y}_7$. We note from Tableau

			0	0	0	0	0	0	−1	−1
$\bar{c}_B$	Basis	$\bar{x}_B$	$\bar{y}_1$	$\bar{y}_2$	$\bar{y}_3$	$\bar{y}_4$	$\bar{y}_5$	$\bar{y}_6$	$\bar{y}_8$	$\bar{y}_9$
0	$\bar{a}_2$	$\frac{10}{3}$	$\frac{2}{3}$	1	$\frac{2}{3}$	$-\frac{1}{3}$	0	0	0	0
−1	$\bar{a}_8$	$\frac{5}{3}$	$-\frac{5}{3}$	0	($\frac{7}{3}$)	$\frac{4}{3}$	−1	0	1	0
−1	$\bar{a}_9$	$\frac{50}{3}$	$\frac{7}{3}$	0	$\frac{1}{3}$	$\frac{1}{3}$	0	−1	0	1
		$-\frac{55}{3}$	$-\frac{2}{3}$	0	$-\frac{8}{3}$	$-\frac{5}{3}$	1	1	0	0

Figure 5–11 Phase I—Tableau II.

II that $z_a \neq 0$ and that artificial vectors $\bar{a}_8$ and $\bar{a}_9$ are still in the basis. Therefore, we continue. The vector to enter the basis is

$$z_k - c_k = \min\left(-\tfrac{2}{3}, -\tfrac{8}{3}, -\tfrac{5}{3}\right) = -\tfrac{8}{3} = z_3 - c_3$$

Therefore $\bar{a}_3$ enters the basis. The vector to leave the basis is

$$\frac{x_{Br}}{y_{r3}} = \min\left(\tfrac{10}{3}\big/\tfrac{2}{3}, \tfrac{5}{3}\big/\tfrac{7}{3}, \tfrac{50}{3}\big/\tfrac{1}{3}\right) = \min\left(5, \tfrac{5}{7}, 50\right) = \tfrac{5}{7} = \frac{x_{B2}}{y_{23}}$$

Therefore, $\bar{a}_8$ leaves the basis. We transform Tableau II and obtain Tableau III in Figure 5–12. Tableau III still has $z_a \neq 0$ and one artificial vector in the

			0	0	0	0	0	0	−1
$\bar{c}_B$	Basis	$\bar{x}_B$	$\bar{y}_1$	$\bar{y}_2$	$\bar{y}_3$	$\bar{y}_4$	$\bar{y}_5$	$\bar{y}_6$	$\bar{y}_9$
0	$\bar{a}_2$	$\frac{20}{7}$	($\frac{8}{7}$)	1	0	$-\frac{5}{7}$	$\frac{2}{7}$	0	0
0	$\bar{a}_3$	$\frac{5}{7}$	$-\frac{5}{7}$	0	1	$\frac{4}{7}$	$-\frac{3}{7}$	0	0
−1	$\bar{a}_9$	$\frac{115}{7}$	$\frac{18}{7}$	0	0	$\frac{1}{7}$	$\frac{1}{7}$	−1	1
		$-\frac{115}{7}$	$-\frac{18}{7}$	0	0	$-\frac{1}{7}$	$-\frac{1}{7}$	1	0

Figure 5–12 Phase I—Tableau III.

basis. We determine the vector to enter the basis from

$$z_k - c_k - \min\left(-\tfrac{18}{7}, -\tfrac{1}{7}, -\tfrac{1}{7}\right) = -\tfrac{18}{7} = z_1 - c_1$$

Therefore, $\bar{a}_1$ enters the basis. The vector to leave is

$$\frac{x_{Br}}{y_{r1}} = \min\left(\tfrac{20}{7}\big/\tfrac{8}{7}, \tfrac{115}{7}\big/\tfrac{18}{7}\right) = \tfrac{5}{2} = \frac{x_{B1}}{y_{11}}$$

Therefore $\bar{a}_2$ leaves the basis. We transform Tableau III and obtain Tableau IV in Figure 5–13. We now allow $\bar{a}_4$ to enter the basis since it is the only

			0	0	0	0	0	0	−1
$\bar{c}_B$	Basis	$\bar{x}_B$	$\bar{y}_1$	$\bar{y}_2$	$\bar{y}_3$	$\bar{y}_4$	$\bar{y}_5$	$\bar{y}_6$	$\bar{y}_9$
0	$\bar{a}_1$	$\frac{5}{2}$	1	$\frac{7}{8}$	0	$-\frac{5}{8}$	$\frac{1}{4}$	0	0
0	$\bar{a}_3$	$\frac{5}{2}$	0	$\frac{5}{8}$	1	$\frac{1}{8}$	$-\frac{1}{4}$	0	0
−1	$\bar{a}_9$	10	0	$-\frac{9}{4}$	0	($\frac{7}{4}$)	$-\frac{1}{2}$	−1	1
		−10	0	$\frac{9}{4}$	0	$-\frac{7}{4}$	$\frac{1}{2}$	1	0

Figure 5–13 Phase I—Tableau IV.

			0	0	0	0	0	0
$\bar{c}_B$	Basis	$\bar{x}_B$	$\bar{y}_1$	$\bar{y}_2$	$\bar{y}_3$	$\bar{y}_4$	$\bar{y}_5$	$\bar{y}_6$
0	$\bar{a}_1$	$\frac{85}{14}$	1	$\frac{1}{14}$	0	0	$\frac{1}{14}$	$-\frac{5}{14}$
0	$\bar{a}_3$	$\frac{25}{14}$	0	$\frac{11}{14}$	1	0	$-\frac{3}{14}$	$\frac{1}{14}$
0	$\bar{a}_4$	$\frac{40}{7}$	0	$-\frac{9}{7}$	0	1	$-\frac{2}{7}$	$-\frac{4}{7}$
		0	0	0	0	0	0	0

Figure 5–14 Phase I—Tableau V.

vector with $z_j - c_j < 0$. We determine the vector to leave from

$$\frac{x_{Br}}{y_{r4}} = \min\left(\tfrac{5}{2}/\tfrac{1}{8},\ 10/\tfrac{7}{4}\right) = \tfrac{40}{7}$$

Therefore $\bar{a}_9$ leaves the basis. We transform Tableau IV and obtain Tableau V in Figure 5–14. In Tableau V we see that $z_a = 0$. This concludes Phase I. We then restore the original prices and recalculate the $z_j - c_j$ and z to begin Phase II. The remainder of the tableau remains the same. The first tableau for Phase II is shown in Figure 5–15. The only vector eligible to enter the basis is $\bar{a}_2$. The vector to leave the basis is

$$\frac{x_{Br}}{y_{r2}} = \min\left(\tfrac{85}{14}/\tfrac{1}{14},\ \tfrac{25}{14}/\tfrac{11}{14}\right) = \min\left(85,\ \tfrac{25}{11}\right) = \tfrac{25}{11} = \frac{x_{B2}}{y_{22}}$$

Hence $\bar{a}_3$ leaves the basis. We transform Tableau VI to obtain Tableau VII in Figure 5–16. We see from Tableau VII that all $z_j - c_j \geq 0$. Hence, we have the optimal solution which is $x_1 = 5.909$, $x_2 = 2.273$, $x_3 = 0$, $x_4 = 8.636$, $x_5 = x_6 = 0$ and $z = -18.64$.

			−2	−3	−5	0	0	0
$\bar{c}_B$	Basis	$\bar{x}_B$	$\bar{y}_1$	$\bar{y}_2$	$\bar{y}_3$	$\bar{y}_4$	$\bar{y}_5$	$\bar{y}_6$
−2	$\bar{a}_1$	$\frac{85}{14}$	1	$\frac{1}{14}$	0	0	$\frac{1}{14}$	$-\frac{5}{14}$
−5	$\bar{a}_3$	$\frac{25}{14}$	0	$\boxed{\frac{11}{14}}$	1	0	$-\frac{3}{14}$	$\frac{1}{14}$
0	$\bar{a}_4$	$\frac{40}{7}$	0	$-\frac{9}{7}$	0	1	$-\frac{2}{7}$	$-\frac{4}{7}$
		$-\frac{295}{14}$	0	$-\frac{15}{14}$	0	0	$\frac{13}{14}$	$\frac{5}{14}$

Figure 5–15 Phase II—Tableau VI.

			−2	−3	−5	0	0	0
$\bar{c}_B$	Basis	$\bar{x}_B$	$\bar{y}_1$	$\bar{y}_2$	y_3	$\bar{y}_4$	y_5	$\bar{y}_6$
−2	$\bar{a}_1$	$\frac{65}{11}$	1	0	$-\frac{1}{11}$	0	$\frac{1}{11}$	$-\frac{4}{11}$
−3	$\bar{a}_2$	$\frac{25}{11}$	0	1	$\frac{14}{11}$	0	$-\frac{3}{11}$	$\frac{1}{11}$
0	$\bar{a}_4$	$\frac{95}{11}$	0	0	$\frac{18}{11}$	1	$-\frac{7}{11}$	$-\frac{5}{11}$
		$-\frac{205}{11}$	0	0	$\frac{15}{11}$	0	$\frac{7}{11}$	$\frac{5}{11}$

Figure 5–16 Phase II—Tableau VII.

EXAMPLE 4: TWO-PHASE CALCULATION—UNBOUNDED SOLUTION

$$\begin{aligned} \text{Max } z &= x_1 - x_2 + 2x_3 \\ -x_1 + x_2 + 2x_3 &\leq 5 \\ -2x_1 + 5x_2 - x_3 &\geq 10 \\ 2x_1 - x_2 + x_3 &\geq 4 \\ x_1, x_2, x_3 &\geq 0 \end{aligned} \tag{5–58}$$

We convert (5–58) to standard form, which is:

$$\begin{aligned} \text{Max } z &= x_1 - x_2 + 2x_3 \\ -x_1 + x_2 + 2x_3 + x_4 &= 5 \\ -2x_1 + 5x_2 - x_3 - x_5 &= 10 \\ 2x_1 - x_2 + x_3 - x_6 &= 4 \\ x_j \geq 0, \quad j &= 1, 2, \ldots, 6 \end{aligned} \tag{5–59}$$

For Phase I of the calculation we define a new objective function:

$$z_a = -x_{a2} - x_{a3} \tag{5–60}$$

The artificial vectors corresponding to x_{a2} and x_{a3} are $\bar{a}_7 = \bar{e}_2$, $\bar{a}_8 = \bar{e}_3$. We have not added an artificial variable x_{a1} to the first equation in (5–59) because a unit vector $\bar{e}_1$ is already present, i.e., $\bar{a}_4 = \bar{e}_1$. The first tableau in Phase I is given in Figure 5–17. Examining Tableau I, we find

$$z_k - c_k = z_2 - c_2 = -4$$

Therefore $\bar{a}_2$ enters the basis. We calculate

$$\frac{x_{Br}}{y_{r2}} = \min\left(\tfrac{5}{1}, \tfrac{10}{5}\right) = 2 = \frac{x_{B2}}{y_{22}}$$

			0	0	0	0	0	0	−1	−1
$\bar{c}_B$	Basis	$\bar{x}_B$	$\bar{y}_1$	$\bar{y}_2$	$\bar{y}_3$	$\bar{y}_4$	$\bar{y}_5$	$\bar{y}_6$	$\bar{y}_7$	$\bar{y}_8$
0	$\bar{a}_4$	5	−1	1	2	1	0	0	0	0
−1	$\bar{a}_7$	10	−2	(5)	−1	0	−1	0	1	0
−1	$\bar{a}_8$	4	2	−1	1	0	0	−1	0	1
		−14	0	−4	0	0	1	1	0	0

Figure 5–17 Phase I—Tableau I.

Therefore, $\bar{a}_7$ leaves the basis. We transform Tableau I to obtain Tableau II in Figure 5–18. It is not necessary to transform $\bar{y}_7$. From Tableau II we see that $z_a \neq 0$. Hence we continue. The vector to enter the basis is

$$z_k - c_k = \min\left(-\tfrac{8}{5}, -\tfrac{4}{5}\right) = -\tfrac{8}{5} = z_1 - c_1$$

The vector to leave the basis is $\bar{a}_8$, since y_{31} is the only y_{i1} that is positive. We transform Tableau II to obtain Tableau III in Figure 5–19. We see from

			0	0	0	0	0	0	−1
$\bar{c}_B$	Basis	$\bar{x}_B$	$\bar{y}_1$	$\bar{y}_2$	$\bar{y}_3$	$\bar{y}_4$	$\bar{y}_5$	$\bar{y}_6$	$\bar{y}_8$
0	$\bar{a}_4$	3	$-\frac{3}{5}$	0	$\frac{11}{5}$	1	$\frac{1}{5}$	0	0
0	$\bar{a}_2$	2	$-\frac{2}{5}$	1	$-\frac{1}{5}$	0	$-\frac{1}{5}$	0	0
−1	$\bar{a}_8$	6	($\frac{8}{5}$)	0	$\frac{4}{5}$	0	$-\frac{1}{5}$	−1	1
		−6	$-\frac{8}{5}$	0	$-\frac{4}{5}$	0	$\frac{1}{5}$	1	0

Figure 5–18 Phase I—Tableau II.

Tableau III that $z_a = 0$. Hence, we have removed all the artificial variables. We have now found a feasible basis for the problem and so we can restore the original prices and proceed with Phase II. In order to do this we use the

			0	0	0	0	0	0
$\bar{c}_B$	Basis	$\bar{x}_B$	$\bar{y}_1$	$\bar{y}_2$	$\bar{y}_3$	$\bar{y}_4$	$\bar{y}_5$	$\bar{y}_6$
0	$\bar{a}_4$	$\frac{21}{4}$	0	0	$\frac{19}{8}$	1	$\frac{1}{8}$	$-\frac{3}{8}$
0	$\bar{a}_2$	$\frac{7}{2}$	0	1	0	0	$-\frac{1}{4}$	$-\frac{1}{4}$
0	$\bar{a}_1$	$\frac{15}{4}$	1	0	$\frac{1}{2}$	0	$-\frac{1}{8}$	$-\frac{5}{8}$
		0	0	0	0	0	0	0

Figure 5–19 Phase I—Tableau III.

			1	−1	2	0	0	0
$\bar{c}_B$	Basis	$\bar{x}_B$	$\bar{y}_1$	$\bar{y}_2$	$\bar{y}_3$	$\bar{y}_4$	$\bar{y}_5$	$\bar{y}_6$
0	$\bar{a}_4$	$\frac{21}{4}$	0	0	$\frac{19}{8}$	1	$\frac{1}{8}$	$-\frac{3}{8}$
−1	$\bar{a}_2$	$\frac{7}{2}$	0	1	0	0	$-\frac{1}{4}$	$-\frac{1}{4}$
1	$\bar{a}_1$	$\frac{15}{4}$	1	0	$\frac{1}{2}$	0	$-\frac{1}{8}$	$-\frac{5}{8}$
		$\frac{1}{4}$	0	0	$-\frac{3}{2}$	0	$\frac{1}{8}$	$-\frac{3}{8}$

Figure 5–20 Phase II—Tableau IV.

last tableau of Phase I, which is Tableau III, with the original prices and the recalculated values of $z_j - c_j$. This is shown in Tableau IV in Figure 5–20. It can be seen from Tableau IV that for $\bar{a}_6$, $y_{i6} \leq 0 \; i = 1, 2, 3$. Hence, there is an unbounded solution and the calculation terminates.

5–6 REDUNDANCY IN PROBLEM CONSTRAINTS—TWO PHASE METHOD

It is possible when formulating a linear programming problem that redundant constraints may be included in the problem statement inadvertently. What this means is that one or more of the constraints place no new restrictions on the variables than the other constraints and hence could be omitted. It is possible to discover when this situation arises fairly easily. We will discuss this within the context of the two-phase method. It should be noted that one need not be concerned about the presence of redundant constraints, since a problem with them can be solved with little more additional effort. There are times, however, when this knowledge may be of use to the problem formulator and analyst.

The case of redundancy may be examined most conveniently at the end of Phase I in the two-phase method. It is clear that if Phase I ended in Case 1 in Section 5–4, then there is no feasible solution to the original problem and the constraints are inconsistent. Further, if Phase I ended in Case 2, we have found a basic feasible solution. Hence the original constraints are obviously consistent and no redundancy is present.

Now let us consider Case 3, i.e., when one or more of the artificial vectors are in the basis with their corresponding variables equal to zero and the optimality criterion is satisfied. Since all the artificial variables are zero, it is clear that we have a consistent, feasible solution to the original problem. However, we shall see that in this case it is possible, but not necessary, that there may be one or more redundant constraints. It will be found useful to consider this case as follows. Let us examine the y_{ij} corresponding to those columns of the basis which contain artificial vectors and such that $x_{Bi} = 0$.

Let us designate these $i \in I_a$. It is clear that there are two possibilities. Either

$$y_{ij} = 0, \qquad \text{all } j,\ i \in I_a$$

or (5–61)

$$y_{ij} \neq 0, \qquad \text{at least one } j,\ i \in I_a$$

Let us consider the second possibility first, and suppose that for $j = l$, $y_{il} \neq 0$, $i \in I_a$. We know from simplex theory that an artificial vector corresponding to some $i \in I_a$ can be removed and replaced by some $\bar{a}_j$. We will have a new basis with the new variable entering at a zero level; the new solution will be feasible and optimal. This is readily seen from (5–13) and (5–14) since

$$\begin{aligned} \hat{x}_{Bi} &= x_{Bi} - (0)\frac{y_{il}}{y_{rl}} = x_{Bi} \\ \hat{x}_{Br} &= \frac{0}{y_{rl}} = 0 \\ \hat{z} &= z + \frac{0}{y_{rl}}(c_l - z_l) = z \end{aligned} \tag{5–62}$$

If we can continue this process until all the artificial vectors present in the basis at a zero level are removed, we will obviously obtain a degenerate basic feasible solution which now contains only columns of A. Hence we see that, if this can be done, no redundancy is present in the original constraints.

If we cannot remove all artificial vectors from the basis, we have obviously reached the state where $y_{ij} = 0$, all j, $i \in I_a$, which was the other possibility we considered in (5–61). Otherwise, we would have been able to continue removing artificial vectors. Let us now examine this case. What the fact that $y_{ij} = 0$, all j, $i \in I_a$ means is that we *cannot* remove an artificial vector from the basis and replace it with some $\bar{a}_j$, while still maintaining a basis. It is evident from this fact that we have redundant constraints. We see this as follows. Every column $\bar{a}_j$ of A can be written as a linear combination of the basis vectors, which are a subset of the columns of A. However, the artificial vectors are clearly not required to represent a given vector in terms of the basis vectors since

$$\bar{a}_j = \sum_{i=1}^{m} y_{ij}\bar{b}_i \tag{5–63}$$

and

$$y_{ij} = 0, \qquad \text{all } j,\ i \in I_a$$

Therefore, every column of A can be written as a linear combination of $m - q$ linearly independent vectors where q is the number of artificial vectors remaining in the basis at a zero level. This is equivalent to stating that $r(A) = m - q$ and, hence, q of the original constraints were redundant.

If it is desired to discover which constraint or constraints are redundant this is relatively easy to do. For simplicity, suppose $q = 1$ and therefore, the uth column of the basis contains the artificial vector $\bar{e}_u$ at a zero level, i.e.,

$x_{Bu} = 0$. Therefore, we know that $y_{uj} = 0$, all j. Since $r(A) = m - 1$ we see that $r(A, \bar{e}_u) = m$. Let us designate the rows of A as $\bar{a}_i'$, i.e., the columns of the transpose of A. Hence, we see that the rows of $(A, \bar{e}_u)$ can be written as

$$\begin{array}{ll} (\bar{a}_i', 0) & i \neq u \\ (\bar{a}_i', 1) & i = u \end{array} \tag{5-64}$$

Since $r(A) = m - 1$ we know that there exists a set of scalars γ_i, not all of which are zero, such that

$$\sum_{i=1}^{m} \gamma_i \bar{a}_i' = \bar{0} \tag{5-65}$$

We shall now show that $\gamma_u \neq 0$ and therefore that $\bar{a}_u'$ can be written as a linear combination of the remaining rows of A. Suppose $\gamma_u = 0$. Then, since $r(A, \bar{e}_u) = m$ the only way the expression of linear independence

$$\sum_{\substack{i=1 \\ i \neq u}}^{m} \lambda_i(\bar{a}_i', 0) + \lambda_u(\bar{a}_u', 1) = \bar{0} \tag{5-66}$$

could be satisfied is that $\lambda_i = 0$, $i = 1, 2, \ldots, m$. However, if we set $\lambda_i = \gamma_i$, we obtain a contradiction since at least one $\gamma_i \neq 0$. Hence $\gamma_u \neq 0$. What this indicates is that the constraint corresponding to $i = u$ is redundant and can be dropped from the original set of constraints. However, it is not necessary to do so. A solution with or without the redundant constraint can and will be obtained. The above argument did not depend in any way on the fact that there was only one redundant constraint. Hence, if there are a number of artificial vectors $\bar{e}_i$ in the basis at a zero level and their corresponding $y_{ij} = 0$ for all j, then for each such $i \in I_a$, the ith constraint can be removed from the original set of constraints because it is redundant.

5-7 DEGENERACY—THE PROBLEM AND ITS RESOLUTION

In developing the theory of the simplex method in Chapter 4, we showed that, in the absence of degeneracy, it was possible by changing one vector in the basis at each iteration, to effect a monotonic increase in the objective function at each iteration. In other words, as long as no degeneracy occurred it was the case that at each iteration, $\hat{z} > z$. Since the number of basic feasible solutions for a linear programming problem is finite, we were certain that we would, in a finite number of steps, reach the optimal solution.

It is the case, however, that degeneracy is a common occurrence in linear programming problems. The presence of degeneracy casts serious doubt on the previous argument for the following reason. Consider the expression for the value of the objective function in terms of the value at the previous stage, i.e.,

$$\hat{z} = z + \frac{x_{Br}}{y_{rk}}(c_k - z_k) \tag{5–67}$$

If $x_{Br} = 0$ and $y_{rk} > 0$, then $\hat{z} = z$. We see then that the key argument that $\hat{z} > z$ at each iteration is violated and hence we can no longer be sure that some basis will not be repeated. In fact, as will be shown by example, it is possible to repeat a sequence of basic feasible solutions endlessly and so cycle indefinitely without ever reaching the optimal solution. While cycling is not thought to be a problem in practice, it is necessary to consider whether or not we can in fact "repair" the simplex theory to avoid the possibility of endless cycling. Hence we must see if we can find a method for generating a sequence of basic feasible solutions which, in the presence of degeneracy, will never repeat a basis and hence will lead to the optimal basic feasible solution. There will also be additional comments in a subsequent section on the "practical" aspects of the problem of degeneracy and cycling.

Several methods have been proposed for resolving the problem of degeneracy. The earliest was that of Charnes.[2,4] A second method proposed by Dantzig Orden, and Wolfe[5] is mathematically equivalent and leads to the same rules of procedure for avoiding repetition of bases. A third method has also been proposed by Wolfe.[6] We shall describe only one of these, the original perturbation method put forth by Charnes.

First, let us recall the origin of degeneracy. Degeneracy in linear programming arises whenever the vector to leave the basis in a simplex iteration is not uniquely determined. It will be recalled that the vector to leave the basis is determined from

$$\frac{x_{Br}}{y_{rk}} = \min_i \left\{ \frac{x_{Bi}}{y_{ik}} \,\middle|\, y_{ik} > 0 \right\} \tag{5–68}$$

If the minimum value of $\dfrac{x_{Bi}}{y_{ik}}$ occurs for more than one i, then a tie results in determining which vector is to leave the current basis. Whichever one is chosen to leave the basis will have a corresponding value of x_{Br} (which will be some x_j) equal to zero in the next solution. However, the other values of x_{Bi} for which $\dfrac{x_{Bi}}{y_{ik}}$ was minimum will also be zero. We see this from

$$\hat{x}_{Bi} = x_{Bi} - x_{Br}\frac{y_{ik}}{y_{rk}} \qquad i \neq r \tag{5–69}$$

Suppose $\dfrac{x_{Bs}}{y_{sk}} = \dfrac{x_{Br}}{y_{rk}}$. Then we have that

$$\hat{x}_{Bs} = x_{Bs} - \frac{x_{Bs}}{y_{sk}}y_{sk} = 0 \tag{5–70}$$

Hence, even though x_{Bs} is still in the basis, it is zero. It is evident from this last observation that if we can find a way to determine the vector to leave the basis in some *unique* fashion, i.e., to devise a rule by which it is not possible to have a tie develop, we should be able to avoid the problem of cycling. This is precisely what the Charnes perturbation method does.

Our original linear programming problem is

$$\begin{aligned} \text{Max } z &= \sum_{j=1}^{n} c_j x_j \\ \sum_{j=1}^{n} x_j \bar{a}_j &= \bar{b} \\ x_j \geq 0 \qquad & j = 1, 2, \ldots, n \end{aligned} \tag{5–71}$$

Now let us consider the following problem, which is referred to as the perturbed problem of the one given in (5–71). It is

$$\begin{aligned} \text{Max } z &= \sum_{j=1}^{n} c_j x_j \\ \sum_{j=1}^{n} x_j \bar{a}_j &= \bar{b} + \sum_{j=1}^{n} \varepsilon^j \bar{a}_j \\ x_j \geq 0, \qquad & j = 1, 2, \ldots, n \end{aligned} \tag{5–72}$$

In (5–72), ε is assumed to be a positive number that can be made as small as necessary for an argument to be presented subsequently. Just as $\bar{x}_B = B^{-1}\bar{b}$, we have for the problem of (5–72) that

$$\begin{aligned} \bar{x}_B(\varepsilon) = B^{-1}\left[\bar{b} + \sum_{j=1}^{n} \varepsilon^j \bar{a}_j\right] &= B^{-1}\bar{b} + \sum_{j=1}^{n} \varepsilon^j B^{-1}\bar{a}_j \\ \therefore\ \bar{x}_B(\varepsilon) &= \bar{x}_B + \sum_{j=1}^{n} \varepsilon^j \bar{y}_j \end{aligned} \tag{5–73}$$

In component form (5–73) can be written

$$x_{Bi}(\varepsilon) = x_{Bi} + \sum_{j=1}^{n} \varepsilon^j y_{ij} \qquad i = 1, 2, \ldots, m \tag{5–74}$$

Similarly, we may compute the perturbed objective function as

$$\begin{aligned} z(\varepsilon) &= \sum_{i=1}^{m} c_{Bi} x_{Bi}(\varepsilon) = \sum_{i=1}^{m} c_{Bi}\left(x_{Bi} + \sum_{j=1}^{n} \varepsilon^j y_{ij}\right) \\ &= \sum_{i=1}^{m} c_{Bi} x_{Bi} + \sum_{j=1}^{n} \sum_{i=1}^{m} \varepsilon^j c_{Bi} y_{ij} = z + \sum_{j=1}^{n} \varepsilon^j z_j \end{aligned} \tag{5–75}$$

For the present, let us assume that we have available to us a basic feasible solution for the perturbed problem. We assume further that this is a non-degenerate solution. We will later show how such an initial basic feasible solution can always be obtained. For this basic feasible solution to the perturbed problem, one of the following conditions must hold:

1. $z_j - c_j \geq 0 \qquad j = 1, 2, \ldots, n$
2. For some k, $z_k - c_k < 0$ and $y_{ik} \leq 0, \qquad i = 1, 2, \ldots, m$
3. For some k, $z_k - c_k < 0$ and at least one $y_{ik} > 0$

If the first condition holds, the current solution is optimal. If the second condition holds, the original problem has an unbounded solution. If the third condition holds, the current solution is not optimal. Hence, we require a change of basis. However, we must do this in such a way that the vector to leave the basis is uniquely determined. First, we shall describe the procedure for accomplishing this and then show that this procedure will always lead to a unique minimum.

Suppose for the perturbed problem that

$$\frac{x_{Br}}{y_{rk}} = \min_i \left\{ \frac{x_{Bi}}{y_{ik}} \middle| y_{ik} > 0 \right\} \tag{5–76}$$

and that $\frac{x_{Br}}{y_{rk}}$ is unique. It can be shown from (5–73) that in this case the new basic feasible solution will be non-degenerate. Now let us consider the case when the minimum in (5–76) is not unique. Suppose that

$$\min_i \left\{ \frac{x_{Bi}}{y_{ik}} \middle| y_{ik} > 0 \right\} = \frac{x_{Bl}}{y_{lk}}, \qquad l = 1, 2, \ldots, p < n \tag{5–77}$$

Then we would attempt to break the tie by computing

$$\min_l \left\{ \frac{x_{Bl}}{y_{lk}} + \varepsilon \frac{y_{l1}}{y_{lk}} \right\}, \qquad l = 1, 2, \ldots, p \tag{5–78}$$

If the minimum is unique in (5–78) we then would make the basis change and continue. However, it is quite possible that two or more of the ratios $\frac{y_{l1}}{y_{lk}}$ would be equal. Hence, there would still be no unique minimum in (5–78). Suppose this is the case (with no loss in generality) for $l = 1, 2, \ldots s < p$. Then we would attempt to break the tie by examining

$$\min_l \left\{ \frac{x_{Bl}}{y_{kl}} + \varepsilon \frac{y_{l1}}{y_{lk}} + \varepsilon^2 \frac{y_{l2}}{y_{lk}} \right\}, \qquad l = 1, 2, \ldots, s \tag{5–79}$$

Again, if the ratios $\dfrac{y_{l2}}{y_{lk}}$ have a unique minimum for $l = 1, 2, \ldots, s$ we are done. If not, we can continue the above process for these values of l for which the ratios are equal. The procedure we have just described for breaking ties will lead to a unique minimum. We will prove this assertion in the next paragraph. For the present assume that $\dfrac{x_{B_r}(\varepsilon)}{y_{rk}}$ was uniquely determined for ε^q where $q < n$. Then we can see that it is not necessary to continue the calculation further, i.e., calculate

$$\frac{x_{B_r}(\varepsilon)}{y_{rk}} = \min_l \left\{\frac{x_{Bl}}{y_{lk}} + \varepsilon\frac{y_{l1}}{y_{lk}} + \varepsilon^2\frac{y_{l2}}{y_{lk}} + \ldots + \varepsilon^n\frac{y_{ln}}{y_{lk}}\right\}$$

we do not need to include terms with $j > q$. All we need do to assure that this is the case is assume that ε is sufficiently small to insure that if the tie was broken for ε^q with $l = 1, 2, \ldots, t$ and, supposing that $l = L$ was the value that led to the unique minimum $\dfrac{y_{Lq}}{y_{Lk}}$, then the following holds:

$$\left|\varepsilon^q\left[\frac{y_{Lq}}{y_{Lk}} - \sum_{\substack{l=1\\ l\neq L}}^{t}\frac{y_{lq}}{y_{lk}}\right]\right| < \left|\sum_{j=q+1}^{u}\varepsilon^j\left[\frac{y_{Lj}}{y_{Lk}} - \sum_{\substack{l=1\\ l\neq L}}^{t}\frac{y_{lj}}{y_{lk}}\right]\right| \qquad (5\text{–}80)$$

$$u = q+1, q+2, \ldots, n$$

We can see that as long as ε is below some maximum value, a unique minimum will have been assured.

Let us now prove that the procedure we have just described will always lead to a unique minimum for $\dfrac{x_{B_r}(\varepsilon)}{y_{rk}}$. We can easily prove this by contradiction. Suppose there was no unique minimum, say, for $l \in L_m$ where $L_m = \{l_1, l_2, \ldots, l_t\}$. Then, we see that

$$\varepsilon^j\frac{y_{l_1j}}{y_{l_1k}} = \varepsilon^j\frac{y_{l_2j}}{y_{l_2k}} = \ldots = \varepsilon^j\frac{y_{l_tj}}{y_{l_tk}} \qquad j = 1, 2, \ldots, n \qquad (5\text{–}81)$$

From (5–81) we see that, for example,

$$y_{l_1j} = y_{l_1k}\frac{y_{l_2j}}{y_{l_2k}} \qquad j = 1, 2, \ldots, n \qquad (5\text{–}82)$$

What (5–82) says is that the l_1^{th} row and the l_2^{th} row of the y_{ij} matrix are *linearly dependent*. This is clearly not possible since the matrix $[y_{ij}]$ contains an identity submatrix and hence linear dependence is impossible. Therefore, we are led to a contradiction and, hence, we will always obtain a unique minimum.

Since we are using the simplex method on the perturbed problem, we are assured that we can proceed from basic feasible solution to basic feasible solution and, since there can be no degeneracy in the perturbed problem, that $\hat{z}(\varepsilon) > z(\varepsilon)$ at each iteration. Hence, no basis can be repeated and we must ultimately arrive at an optimal solution or an indication of an unbounded solution. Since we can always find an optimum solution for the perturbed problem, if one exists, we can obtain the optimal solution to the original problem by letting $\varepsilon = 0$. Since the $z_j - c_j$ are the same for both the original and the perturbed problem, we will then have the optimal solution to the original problem since all $z_j - c_j \geq 0$.

We assumed in the foregoing that we always had available to us an initial non-degenerate basic feasible solution to the perturbed problem. Let us see how this can be accomplished. Suppose we arbitrarily number the vectors $\bar{a}_j$ so that the identity submatrix is contained in the vectors $\bar{a}_1, \bar{a}_2, \ldots, \bar{a}_m$. (We are, of course, assuming that we start with an identity matrix as the initial basis to the unperturbed problem.) If this is the case, we can express the basic variables of the perturbed problem as

$$x_{Bi}(\varepsilon) = x_{Bi} + \varepsilon^i + \sum_{j=m+1}^{n} \varepsilon^j y_{ij} \qquad i = 1, 2, \ldots, m \tag{5–83}$$

If ε is sufficiently small so that

$$|\varepsilon^i| > \left| \sum_{j=m+1}^{n} \epsilon^j y_{ij} \right| \qquad i = 1, 2, \ldots, m \tag{5–84}$$

then since $x_{Bi} \geq 0$, $i = 1, 2, \ldots, m$ it is apparent $x_{Bi}(\varepsilon) > 0$, $i = 1, 2, \ldots, m$. Hence, we have an initial solution for the perturbed problem which is non-degenerate.

One last point is how we actually determine the value of ε to use, if we wish to carry out the procedure we have described. It turns out that we never have to explicitly use ε, and hence do not need to assign a numerical value to it. We see this as follows. Since at some stage we determine a unique minimum of $\dfrac{x_{Bi}(\varepsilon)}{y_{rk}}$, this implies that for some value, $l = L$:

$$\varepsilon^q \frac{y_{Lq}}{y_{Lk}} < \varepsilon^q \frac{y_{Ll}}{y_{Lk}} \qquad l = 1, 2, \ldots, t;\ l \neq L \tag{5–85}$$

Therefore,

$$\frac{y_{Lq}}{y_{Lk}} < \frac{y_{Ll}}{y_{Lk}} \qquad l = 1, 2, \ldots, t;\ l \neq L \tag{5–86}$$

Hence we merely need to compare the ratios in (5–86) without reference to the value of ε. All the data we require is present in the simplex tableau. If a tie is present in some of the minimum values of $\frac{x_{Bi}}{y_{ik}}$ for $y_{ik} > 0$, we next consider ratios of $\frac{y_{i1}}{y_{ik}}$ for the values of i that led to the tie for the minimum. If no unique minimum occurs, we next consider ratios of $\frac{y_{i2}}{y_{ik}}$, etc., until we break the tie. As the preceding arguments have shown, this will solve the perturbed problem without repeating a basis and hence will also solve the original problem.

The following example of how cycling can occur is derived from Beale.[7]†

$$\begin{aligned} \text{Min } z &= -\tfrac{3}{4}x_4 + 20x_5 - \tfrac{1}{2}x_6 + 6x_7 \\ x_1 \quad &+ \tfrac{1}{4}x_4 - 8x_5 - x_6 + 9x_7 = 0 \\ x_2 \quad &+ \tfrac{1}{2}x_4 - 12x_5 - \tfrac{1}{2}x_6 + 3x_7 = 0 \\ x_3 \quad &+ x_6 = 1 \\ x_j \geq 0, &\quad j = 1, 2, \ldots, 7 \end{aligned} \tag{5–87}$$

We present the initial tableau in Figure 5–21.

From Tableau I we see that the most positive $z_j - c_j$ is $z_4 - c_4 = \frac{3}{4}$. Hence $\bar{a}_4$ enters the basis. If we compute

$$\frac{x_{Br}}{y_{rk}} = \min_i \left\{\frac{0}{\frac{1}{4}}, \frac{0}{\frac{1}{2}}\right\}$$

we see that there is a tie for the vector to leave the basis. If we arbitrarily use the rule of the lowest index (something a computer program might do) then $\bar{a}_1$ leaves the basis and we get the tableau shown in Figure 5–22. The subsequent tableaus are presented in Figures 5–23 to 5–27.

			0	0	0	$-\frac{3}{4}$	20	$-\frac{1}{2}$	6
$\bar{c}_B$	Basis	$\bar{x}_B$	$\bar{y}_1$	$\bar{y}_2$	$\bar{y}_3$	$\bar{y}_4$	$\bar{y}_5$	$\bar{y}_6$	$\bar{y}_7$
0	$\bar{a}_1$	0	1	0	0	$\textcircled{\frac{1}{4}}$	-8	-1	9
0	$\bar{a}_2$	0	0	1	0	$\frac{1}{2}$	-12	$-\frac{1}{2}$	3
0	$\bar{a}_3$	1	0	0	1	0	0	1	0
		0	0	0	0	$\frac{3}{4}$	-20	$\frac{1}{2}$	-6

Figure 5–21 Tableau I.

† It will be recalled that a minimization problem can be solved directly without conversion of the objective function. In the optimal tableau, all $z_j - c_j \leq 0$.

			0	0	0	$-\frac{3}{4}$	20	$-\frac{1}{2}$	6
$\bar{c}_B$	Basis	$\bar{x}_B$	$\bar{y}_1$	$\bar{y}_2$	$\bar{y}_3$	$\bar{y}_4$	$\bar{y}_5$	$\bar{y}_6$	$\bar{y}_7$
$-\frac{3}{4}$	$\bar{a}_4$	0	4	0	0	1	-32	-4	36
0	$\bar{a}_2$	0	-2	1	0	0	(4)	$\frac{3}{2}$	-15
0	$\bar{a}_3$	1	0	0	1	0	0	1	0
		0	-3	0	0	0	4	$\frac{7}{2}$	-33

Figure 5–22 Tableau II.

			0	0	0	$-\frac{3}{4}$	20	$-\frac{1}{2}$	6
$\bar{c}_B$	Basis	$\bar{x}_B$	$\bar{y}_1$	$\bar{y}_2$	$\bar{y}_3$	$\bar{y}_4$	$\bar{y}_5$	$\bar{y}_6$	$\bar{y}_7$
$-\frac{3}{4}$	$\bar{a}_4$	0	-12	8	0	1	0	(8)	-84
20	$\bar{a}_5$	0	$-\frac{1}{2}$	$\frac{1}{4}$	0	0	1	$\frac{3}{8}$	$-\frac{15}{4}$
0	$\bar{a}_3$	1	0	0	1	0	0	1	0
		0	-1	-1	0	0	0	2	-18

Figure 5–23 Tableau III.

It can be seen in Tableaus I to VII that whenever a choice has to be made as to which vector would leave the basis, the one with the lowest index i was arbitrarily chosen. We see that Tableau VII is identical with Tableau I and we have indeed cycled without ever becoming optimal and that this was caused by arbitrarily breaking the tie with respect to the vector leaving the basis. It will also be noted that all solutions are degenerate. Let us now

			0	0	0	$-\frac{3}{4}$	20	$-\frac{1}{2}$	6
$\bar{c}_B$	Basis	$\bar{x}_B$	$\bar{y}_1$	$\bar{y}_2$	$\bar{y}_3$	$\bar{y}_4$	$\bar{y}_5$	$\bar{y}_6$	$\bar{y}_7$
$-\frac{1}{2}$	$\bar{a}_6$	0	$-\frac{3}{2}$	1	0	$\frac{1}{8}$	0	1	$-\frac{21}{2}$
20	$\bar{a}_5$	0	$\frac{1}{16}$	$-\frac{1}{8}$	0	$-\frac{3}{64}$	1	0	$(\frac{3}{16})$
0	$\bar{a}_3$	1	$\frac{3}{2}$	-1	1	$-\frac{1}{8}$	0	0	$\frac{21}{2}$
		0	2	-3	0	$-\frac{1}{4}$	0	0	3

Figure 5–24 Tableau IV.

			0	0	0	$-\frac{3}{4}$	20	$-\frac{1}{2}$	6
$\bar{c}_B$	Basis	$\bar{x}_B$	$\bar{y}_1$	$\bar{y}_2$	$\bar{y}_3$	$\bar{y}_4$	$\bar{y}_5$	$\bar{y}_6$	$\bar{y}_7$
$-\frac{1}{2}$	$\bar{a}_6$	0	(2)	-6	0	$-\frac{5}{2}$	56	1	0
6	$\bar{a}_7$	0	$\frac{1}{3}$	$-\frac{2}{3}$	0	$-\frac{1}{4}$	$\frac{16}{3}$	0	1
0	$\bar{a}_3$	1	-2	6	1	$\frac{5}{2}$	-56	0	0
		0	1	-1	0	$\frac{1}{2}$	-16	0	0

Figure 5–25 Tableau V.

			0	0	0	$-\frac{3}{4}$	20	$-\frac{1}{2}$	6
$\bar{c}_B$	Basis	$\bar{x}_B$	$\bar{y}_1$	$\bar{y}_2$	$\bar{y}_3$	$\bar{y}_4$	$\bar{y}_5$	$\bar{y}_6$	$\bar{y}_7$
0	$\bar{a}_1$	0	1	-3	0	$-\frac{5}{4}$	28	$\frac{1}{2}$	0
6	$\bar{a}_7$	0	0	$\textcircled{\frac{1}{3}}$	0	$\frac{1}{6}$	-4	$-\frac{1}{6}$	1
0	$\bar{a}_3$	1	0	0	1	0	0	1	0
		0	0	2	0	$\frac{7}{4}$	-44	$-\frac{1}{2}$	0

Figure 5–26 Tableau VI.

$\bar{c}_B$	Basis	$\bar{x}_B$	$\bar{y}_1$	$\bar{y}_2$	$\bar{y}_3$	$\bar{y}_4$	$\bar{y}_5$	$\bar{y}_6$	$\bar{y}_7$
0	$\bar{a}_1$	0	1	0	0	$\frac{1}{4}$	-8	-1	9
0	$\bar{a}_2$	0	0	1	0	$\textcircled{\frac{1}{2}}$	-12	$-\frac{1}{2}$	3
0	$\bar{a}_3$	1	0	0	1	0	0	1	0
		0	0	0	0	$\frac{3}{4}$	-20	$\frac{1}{2}$	-6

Figure 5–27 Tableau VII.

consider the use of the procedure we previously developed for breaking ties and show how it avoids cycling. From Tableau VII, which is identical to Tableau I, we calculate

$$\text{Min}\left\{\frac{y_{11}}{y_{14}}, \frac{y_{21}}{y_{24}}\right\} = \text{Min}\left\{\frac{1}{\frac{1}{4}}, \frac{0}{\frac{1}{2}}\right\}$$

Therefore we replace $\bar{a}_2$ with $\bar{a}_4$ instead of replacing $\bar{a}_1$ with $\bar{a}_4$ as we did in Tableau I. We then obtain Tableau VIII in Figure 5–28.

From Tableau VIII it is apparent that $\bar{a}_6$ must enter the basis and $\bar{a}_3$ leave. The transformed tableau is given in Figure 5–29.

$\bar{c}_B$	Basis	$\bar{x}_B$	$\bar{y}_1$	$\bar{y}_2$	$\bar{y}_3$	$\bar{y}_4$	$\bar{y}_5$	$\bar{y}_6$	$\bar{y}_7$
0	$\bar{a}_1$	0	1	$-\frac{1}{2}$	0	0	-2	$-\frac{3}{4}$	$\frac{15}{2}$
$-\frac{3}{4}$	$\bar{a}_4$	0	0	2	0	1	-24	-1	6
0	$\bar{a}_3$	1	0	0	1	0	0	$\textcircled{1}$	0
		0	0	$-\frac{3}{2}$	0	0	-2	$\frac{5}{4}$	$-\frac{21}{2}$

Figure 5–28 Tableau VIII.

$\bar{c}_B$	Basis	$\bar{x}_B$	$\bar{y}_1$	$\bar{y}_2$	$\bar{y}_3$	$\bar{y}_4$	$\bar{y}_5$	$\bar{y}_6$	$\bar{y}_7$
0	$\bar{a}_1$	$\frac{3}{4}$	1	$-\frac{1}{2}$	$\frac{3}{4}$	0	-2	0	$\frac{15}{2}$
$-\frac{3}{4}$	$\bar{a}_4$	1	0	2	1	1	-24	0	6
$-\frac{1}{2}$	$\bar{a}_6$	1	0	0	1	0	0	1	0
		$-\frac{5}{4}$	0	$-\frac{3}{2}$	$-\frac{5}{4}$	0	-2	0	$-\frac{21}{2}$

Figure 5–29 Tableau IX.

We see that Tableau IX contains the optimal solution which was obtained in two iterations. Cycling no longer exists.

It is often said that cycling is not a problem of practical importance because it does not *seem* to occur in the solution of linear programming problem of real-world models. While it is true that cycling does not seem to prevent us from ultimately reaching an optimal solution, it is quite possible that cycling does occur in large problems of practical significance and increases the total number of iterations required to reach a solution. What may happen is that accumulations of round-off error during successive cycles may lead to choices of variables that would not have been chosen to leave the basis in previous iterations, i.e., very small accumulations of round-off error will break a tie for the vector to leave the basis and hence lead to a different succession of bases and thence to the optimal solution. While this does not seem to be crippling in terms of time to effect a solution, it may lead to increased time over what would be required to use the tie breaking procedure we have described. However, it is fair to say that to the extent of our knowledge, degeneracy and cycling are not major problems.†

PROBLEMS

1. Determine under what circumstances a vector which is inserted into the basis at one iteration can be removed at the next iteration.

2. Prove that if a vector is removed from the basis at one iteration of the simplex method, it cannot immediately re-enter at the next iteration.

3. Explain why it would be desirable, from the point of view of finding an initial feasible solution, to convert inequalities of the form

$$\sum_{j=1}^{n} a_{ij}x_j \geq 0$$

to the form

$$\sum_{j=1}^{n} (-a_{ij})x_j \leq 0$$

4. Consider how the simplex method could be used to find the inverse, if it exists, of any given square matrix.

5. Solve the following problem without introducing artificial variables.

$$\begin{aligned} \text{Max } z &= 3x_1 + 4x_2 + 5x_3 + x_4 \\ 2x_1 + x_2 + 3x_3 + x_4 &\leq 18 \\ x_1 + 2x_2 + 4x_3 + 2x_4 &\leq 26 \\ 3x_1 + 2x_2 + x_3 + x_4 &\leq 30 \\ x_1, x_2, x_3, x_4 &\geq 0 \end{aligned}$$

† A colleague of one of the authors, Prof. T. C. Kotiah of Southern Illinois University, has recently discovered (to his dismay!) a problem which did in fact cycle, in Phase I. A minor change in one coefficient resulted in a problem which was then solved without encountering cycling.

6. Solve the following problem using the $-M$ method.

$$\begin{aligned} \text{Min } z &= 3x_1 + x_2 + 4x_3 + 7x_4 \\ 3x_1 - x_2 + 2x_3 - 3x_4 &\geq 3 \\ 4x_1 + 2x_2 - 4x_3 - x_4 &\geq 6 \\ -2x_1 + 5x_2 + 4x_3 + 6x_4 &\geq 7 \\ x_1, x_2, x_3, x_4 &\geq 0 \end{aligned}$$

7. Solve the following problem using the two-phase method.

$$\begin{aligned} \text{Max } z &= x_1 + 2x_2 + 4x_3 + 6x_4 \\ 2x_1 + 3x_2 + 4x_3 + x_4 &= 26 \\ x_1 + 5x_2 + x_3 + 4x_4 &\leq 20 \\ 3x_1 + x_2 + 3x_3 + 3x_4 &\leq 30 \\ x_1, x_2, x_3, x_4 &\geq 0 \end{aligned}$$

8. Solve the following problem by any method.

$$\begin{aligned} \text{Min } z &= 5x_1 + 3x_2 + 4x_3 + 2x_4 \\ 3x_1 + 2x_2 + x_3 + x_4 &= 10 \\ x_1 + x_2 + 3x_3 + 4x_4 &= 18 \\ 2x_1 + 3x_2 + 2x_3 + 3x_4 &= 16 \\ x_1, x_2, x_3, x_4 &\geq 0 \end{aligned}$$

9. Show how to solve the following problem and solve it, using a 3×3 basis.

$$\begin{aligned} \text{Max } z &= 3x_1 + 4x_2 + 5x_3 + 8x_4 \\ 2x_1 + x_2 + 5x_3 + 8x_4 &\leq 48 \\ x_1 + 3x_2 + 7x_3 + 5x_4 &\leq 50 \\ 3x_1 + 6x_2 + x_3 + 6x_4 &\leq 76 \\ x_1 \geq 2, \quad x_2 \geq 3, \quad x_3 &\geq 1, \quad x_4 \geq 4 \end{aligned}$$

REFERENCES

1. Hasegawa, H.: A Study of the Dual Simplex Algorithm. M.S. Thesis, Department of Applied Mathematics and Computer Science, Washington University, St. Louis, 1965.
2. Charnes, A., W. W. Cooper, and A. Henderson: *An Introduction to Linear Programming*. New York, John Wiley and Sons, 1953.
3. Dantzig, G.: Maximization of a Linear Function of Variables Subject to Linear Inequalities. In *Activity Analysis of Production and Allocation*. Edited by T. C. Koopmans, New York, John Wiley and Sons, 1951.
4. Charnes, A.: Optimality and Degeneracy in Linear Programming. Econometrica **20**:160, 1952.
5. Dantzig, G., A. Orden, and P. Wolfe: The Generalized Simplex Method for Minimizing a Linear Form under Linear Inequality Constraints. Pacific Jour. Math. **5**:2, 1955.
6. Wolfe, P.: A Technique for Resolving Degeneracy in Linear Programming. Jour. Soc. Ind. and Appl. Math. **11**:205, 1963.
7. Beale, E. M. L.: Cycling in The Dual Simplex Algorithm. Nav. Res. Logist. Quart. **2**:4, 1955.

6

The Revised Simplex Method

6–1 INTRODUCTION

The revised simplex method uses exactly the same theoretical framework to solve a linear programming problem as does the simplex method presented in previous chapters. The "revised" aspect of the revised simplex method concerns which quantities are stored and transformed in each iteration. Otherwise, we are in fact using the "simplex method," i.e., we proceed from one basic feasible solution to another by replacing one vector in the current basis with another. The criteria for selecting the vector to enter the basis and to leave the basis in the revised simplex method are precisely those used in the simplex method. Hence the conceptual basis of the revised simplex algorithm is the same as that of the simplex algorithm.

In the simplex method at each iteration we transform the quantities $\bar{y}_j$, $\bar{x}_B$, $z_j - c_j$ and z. The bulk of the computational labor is involved with transforming the $\bar{y}_j$, $j = 1, 2, \ldots, n$ when in fact the only one of these that is of interest at each iteration is $\bar{y}_k$. Bearing this in mind we may note that if, at each stage, we knew the current value of B^{-1} we could calculate all quantities of interest, since

$$\begin{aligned} \bar{y}_k &= B^{-1}\bar{a}_k \\ \bar{x}_B &= B^{-1}\bar{b} \\ z_j - c_j &= \bar{c}'_B B^{-1}\bar{a}_j - c_j \\ z &= \bar{c}'_B \bar{x}_B \end{aligned} \tag{6–1}$$

Hence, if some way could be found to transform B^{-1} at each iteration, we need not transform all the $\bar{y}_j$. There are other motivations for desiring to do this, which will be discussed in a later section.

There are two versions of the revised simplex method. The first version assumes that an identity matrix is present after slack and surplus variables have been added. For this case, it is not necessary to add artificial variables.

If it is necessary to add artificial variables in order to obtain an initial basis, we will use the second version of the revised simplex method. However, before we discuss this, it will be useful to review some material relating to the computation of matrix inverses.

6–2 THE PRODUCT FORM OF THE INVERSE

In the revised simplex method it is of interest to compute the inverse of a matrix for which one column differs from that of a matrix whose inverse is already known. It would be desirable to have a simple and efficient method for computing the new inverse from the first inverse. Suppose then that we have a non-singular matrix $B = (\bar{b}_1, \bar{b}_2, \ldots, \bar{b}_r, \ldots, \bar{b}_m)$ and that B^{-1} is known. Now we will replace column $\bar{b}_r$ of B with $\bar{a}_k$ yielding the matrix $\hat{B} = (\bar{b}_1, \bar{b}_2, \ldots, \bar{b}_{r-1}, \bar{a}_k, \bar{b}_{r+1}, \ldots, \bar{b}_m)$. We now wish to compute the inverse $\hat{B}^{-1}$ from B^{-1}. Since B constitutes a basis for E^m we can express $\bar{a}_k$ as a linear combination of the columns of B. Hence,

$$\bar{a}_k = \sum_{i=1}^{m} y_{ik}\bar{b}_i \tag{6–2}$$

Only if $y_{rk} \neq 0$ will $\hat{B}$ be non-singular and hence have an inverse. Therefore if we rearrange (6–2) we have

$$\bar{b}_r = \frac{1}{y_{rk}}\bar{a}_k - \sum_{\substack{i=1 \\ i\neq r}}^{m} \frac{y_{ik}}{y_{rk}}\bar{b}_i \tag{6–3}$$

If we rewrite (6–3) in the order of the vectors in $\hat{B}$ we have

$$\begin{aligned} \bar{b}_r = & -\frac{y_{1k}}{y_{rk}}\bar{b}_1 - \frac{y_{2k}}{y_{rk}}\bar{b}_2 - \cdots - \frac{y_{r-1,k}}{y_{rk}}\bar{b}_{r-1} \\ & + \frac{1}{y_{rk}}\bar{a}_k - \frac{y_{r+1,k}}{y_{rk}}\bar{b}_{r+1} - \cdots - \frac{y_{mk}}{y_{rk}}\bar{b}_m \end{aligned} \tag{6–4}$$

We can write (6–4) more conveniently in matrix notation as

$$\bar{b}_r = \hat{B}\bar{\mu}_k \tag{6–5}$$

where

$$\bar{\mu}_k = \left[-\frac{y_{1k}}{y_{rk}}, -\frac{y_{2k}}{y_{rk}}, \ldots, -\frac{y_{r-1,k}}{y_{rk}}, \frac{1}{y_{rk}}, -\frac{y_{r+1,k}}{y_{rk}}, \ldots, -\frac{y_{mk}}{y_{rk}}\right] \tag{6–6}$$

From equations (6–5) and (6–6) we see that the relationship between B and

$\hat{B}$ can be represented as

$$B = \hat{B}J \tag{6–7}$$

where

$$J = (\bar{e}_1, \bar{e}_2, \ldots, \bar{e}_{r-1}, \bar{\mu}_k, \bar{e}_{r+1}, \ldots, \bar{e}_m) \tag{6–8}$$

If we take the inverse of equation (6–7) we have that

$$B^{-1} = J^{-1}\hat{B}^{-1} \tag{6–9}$$

and multiplying (6–9) by J yields

$$\hat{B}^{-1} = JB^{-1} \tag{6–10}$$

Equation (6–10) gives us $\hat{B}^{-1}$ in terms of B^{-1}. The matrix J is simply an identity matrix with the rth column replaced by $\bar{\mu}_k$. Since $\bar{y}_k = B^{-1}\bar{a}_k$ we can easily compute the components of $\bar{\mu}_k$.

The relationship represented in equation (6–10), and the associated results discussed in the preceding paragraph, can be used to compute the inverse of any non-singular matrix. Suppose we have some non-singular matrix:

$$D = (\bar{d}_1, \bar{d}_2, \ldots, \bar{d}_p) \tag{6–11}$$

We begin with an identity matrix since $I_p^{-1} = I_p$. Now we can proceed to insert the columns of D into I one at a time. There are many ways in which this can be done. For example, we might insert $\bar{d}_1$ into the first column of I to yield D_1. Then we could insert the second column of D into D_1, etc. If we continue in this way we have

$$\begin{aligned} D_1^{-1} &= J_1 I = J_1 \\ D_2^{-1} &= J_2 D_1^{-1} = J_2 J_1 \\ &\cdots \\ D^{-1} &= D_p^{-1} = J_p J_{p-1} \ldots J_2 J_1 \end{aligned} \tag{6–12}$$

Therefore we have the general result that

$$D^{-1} = J_p J_{p-1} \ldots J_1 \tag{6–13}$$

Where

$$J_i = (\bar{e}_1, \bar{e}_2, \ldots, \bar{e}_{i-1}, \bar{\mu}_i, \bar{e}_{i+1}, \ldots, \bar{e}_p) \tag{6–14}$$

$$\bar{\mu}_i = \left[-\frac{y_{1i}}{y_{ii}}, -\frac{y_{2i}}{y_{ii}}, \ldots, -\frac{y_{i-1,i}}{y_{ii}}, \frac{1}{y_{ii}}, -\frac{y_{i+1,i}}{y_{ii}}, \ldots, -\frac{y_{pi}}{y_{ii}}\right] \tag{6–15}$$

and also

$$\begin{aligned} \bar{y}_i &= D_{i-1}^{-1}\bar{d}_i \\ D_i^{-1} &= J_i J_{i-1} \ldots J_1 = J_i D_{i-1}^{-1} \qquad i = 1, 2, \ldots, p \\ D_0^{-1} &= I^{-1} = I \end{aligned} \tag{6–16}$$

The relationship given by equation (6–13) is called the product form of the inverse because D^{-1} is represented as a product of matrices J_i. It is possible to insert the columns of D into I in many different orders. In fact this may even be necessary in some cases. We have assumed that we could arbitrarily choose the way in which we insert the vectors. That this is not the case can be seen from (6–2) and (6–3). Only if the particular y_{rk} (or y_{ii} in terms of (6–15)) is different from zero can the insertion be carried out.

As we noted in the preceding section, the revised simplex method transforms B^{-1} at each stage rather than all the $\bar{y}_j$ as does the simplex method. We make use of the relationships such as (6–10) in order to do this. This will be seen in the next two sections where the revised simplex method is developed.

6–3 THE REVISED SIMPLEX METHOD: WITHOUT ARTIFICIAL VARIABLES

The linear programming problem that we shall begin with is assumed to have the form

$$\text{Max } z = \sum_{j=1}^{n} c_j x_j$$

subject to

$$\begin{aligned} \sum_{j=1}^{n} a_{ij} x_j &= b_i \qquad i = 1, 2, \ldots, m \\ x_j &\geq 0 \qquad j = 1, 2, \ldots, n \end{aligned} \tag{6–17}$$

where the slack or surplus variables are already included. Further, we assume that an identity submatrix can easily be found from the columns of the matrix A of the system of constraints of (6–17) which can then be used as an initial basis.

One of the changes we shall make in the revised simplex method, as compared with the simplex method, is to consider the objective function as a constraint. Since $z = \bar{c}'\bar{x}$ we can just as easily write $z - \bar{c}'\bar{x} = 0$. Hence we will write a set of $m + 1$ simultaneous linear equations as our constraints as follows:

$$\begin{array}{l} z - c_1 x_1 - c_2 x_2 - \ldots - c_n x_n = 0 \\ a_{11} x_1 + a_{12} x_2 + \ldots + a_{1n} x_n = b_1 \\ a_{21} x_1 + a_{22} x_2 + \ldots + a_{2n} x_n = b_2 \\ \quad \cdot \qquad \cdot \qquad\qquad \cdot \qquad \cdot \\ \quad \cdot \qquad \cdot \qquad\qquad \cdot \qquad \cdot \\ \quad \cdot \qquad \cdot \qquad\qquad \cdot \qquad \cdot \\ a_{m1} x_1 + a_{m2} x_2 + \ldots + a_{mn} x_n = b_m \end{array} \tag{6–18}$$

We wish to find a set of non-negative x_j which satisfy (6–18) and which make z as large as possible. It should be noted that z is not required to be non-negative. However, this will cause no difficulty as we shall see.

Since the system of equations (6–18) is of order $(m + 1) \times (n + 1)$ we require some additional notation. We then define the following:

$$A^* = \begin{bmatrix} 1 & -\bar{c}' \\ \bar{0} & A \end{bmatrix}; \qquad \bar{x}^* = \begin{bmatrix} z \\ \bar{x} \end{bmatrix}; \qquad \bar{b}^* = \begin{bmatrix} 0 \\ \bar{b} \end{bmatrix} \tag{6–19}$$

We can now write (6–18) as

$$\begin{bmatrix} 1 & -\bar{c}' \\ \bar{0} & A \end{bmatrix} \begin{bmatrix} z \\ \bar{x} \end{bmatrix} = \begin{bmatrix} \bar{0} \\ \bar{b} \end{bmatrix} \tag{6–20}$$

or more simply as $A^*\bar{x}^* = \bar{b}^*$. It is clear that a basis matrix for (6–20) will be of order $m + 1$. We will denote this matrix B^*. Therefore, a basis matrix for (6–20) will be

$$B^* = \begin{bmatrix} 1 & -\bar{c}_B' \\ \bar{0} & B \end{bmatrix} \tag{6–21}$$

It is a simple matter to compute the inverse of a partitioned matrix such as B^* if B^{-1} is known. It is easily verified that

$$B^{*-1} = \begin{bmatrix} 1 & \bar{c}_B'B^{-1} \\ \bar{0} & B^{-1} \end{bmatrix} \tag{6–22}$$

Therefore, we see that

$$\bar{x}_B^* = B^{*-1}\bar{b}^* = \begin{bmatrix} 1 & \bar{c}_B'B^{-1} \\ \bar{0} & B^{-1} \end{bmatrix} \begin{bmatrix} 0 \\ \bar{b} \end{bmatrix} = \begin{bmatrix} c_B'B^{-1}\bar{b} \\ B^{-1}\bar{b} \end{bmatrix} = \begin{bmatrix} \bar{c}_B'\bar{x}_B \\ \bar{x}_B \end{bmatrix} \tag{6–23}$$

Since $\bar{c}_B'\bar{x}_B = z$ we see that

$$\bar{x}_B^* = B^{*-1}\bar{b}^* = \begin{bmatrix} z \\ \bar{x}_B \end{bmatrix} \tag{6–24}$$

If we refer to (6–19) we see that any column of A^* except the first can be represented as $\bar{a}_j^*$ where $\bar{a}_j^* = [-c_j, \bar{a}_j]$. Therefore we can define $\bar{y}_j^*$ as

$$\bar{y}_j^* = B^{*-1}\bar{a}_j^* = \begin{bmatrix} 1 & \bar{c}_B'B^{-1} \\ \bar{0} & B^{-1} \end{bmatrix} \begin{bmatrix} -c_j \\ \bar{a}_j \end{bmatrix} = \begin{bmatrix} -c_j + \bar{c}_B'B^{-1}\bar{a}_j \\ B^{-1}\bar{a}_j \end{bmatrix} = \begin{bmatrix} z_j - c_j \\ \bar{y}_j \end{bmatrix} \tag{6–25}$$

for $j = 1, 2, \ldots, n$. Since we have added an additional column, we call the first column $\bar{y}_0^*$. It can be seen that $\bar{y}_0^*$ is simply the first column of B^* because

$$\bar{y}_0^* = B^{*-1}\bar{a}_0^* = \begin{bmatrix} 1 & \bar{c}_B' B^{-1} \\ \bar{0} & B^{-1} \end{bmatrix} \begin{bmatrix} 1 \\ \bar{0} \end{bmatrix} = \begin{bmatrix} 1 \\ \bar{0} \end{bmatrix} \tag{6–26}$$

Hence it never changes.

The motivation for including the objective function as one of the constraints in (6–18) can now be understood. We see from (6–25) that, with B^* defined as it is, the first component of $\bar{y}_j^*$ is simply $z_j - c_j$ and the remaining m components of $\bar{y}_j^*$ are $\bar{y}_j$. Another fact emerges from an examination of (6–25). We see that we can test for optimality, i.e., see whether $z_j - c_j \geq 0$ for all j, by multiplying the *first row* of B^{*-1} by the $\bar{a}_j^*$ which are not in the basis. It is not at all necessary to calculate the remaining components of all the $\bar{y}_j^*$.

With the preceding background, we are ready to consider the computational details of this form of the revised simplex method. We have assumed that the matrix A contained an identity submatrix. Hence, for our initial basis we will use the columns of $\bar{a}_j^*$ for which the corresponding $\bar{a}_j$ yielded an identity matrix in A. Hence, if

$$B^* = \begin{bmatrix} 1 & -\bar{c}_B' \\ \bar{0} & I \end{bmatrix} \tag{6–27}$$

then we have:

$$B^{*-1} = \begin{bmatrix} 1 & \bar{c}_B' I^{-1} \\ \bar{0} & I^{-1} \end{bmatrix} = \begin{bmatrix} 1 & \bar{c}_B' \\ \bar{0} & I \end{bmatrix} \tag{6–28}$$

Therefore our initial basic feasible solution is

$$\bar{x}_B^* = B^{*-1}\bar{b}^* = \begin{bmatrix} 1 & \bar{c}_B' \\ \bar{0} & I \end{bmatrix} \begin{bmatrix} 0 \\ \bar{b} \end{bmatrix} = \begin{bmatrix} \bar{c}_B' \bar{b} \\ \bar{b} \end{bmatrix} \tag{6–29}$$

We begin our computation as in any simplex method. We must compute the $z_j - c_j$ for each $\bar{a}_j^*$ not in the basis in order to determine whether or not the current solution is optimal. We do this, as has been mentioned, by multiplying the first row of B^{*-1} with each $\bar{a}_j^*$ not in the basis. As usual the vector to enter the basis is determined by

$$z_k - c_k = \min_j \{z_j - c_j \mid z_j - c_j < 0\} \tag{6–30}$$

Next we determine the vector to leave the basis. In this connection it can be seen from (6–21) that $[1, \bar{0}]$ and any subset of the columns of the matrix $\begin{bmatrix} -\bar{c}'_B \\ B \end{bmatrix}$ are linearly independent. Hence $[1, \bar{0}]$ must be in every basis B^* and is never a candidate for removal from the basis. Before we decide on which vector to remove from the basis we need to calculate $\bar{y}_k^*$ from

$$\bar{y}_k^* = B^{*-1}\bar{a}_k^* = \begin{bmatrix} z_k - c_k \\ \bar{y}_k \end{bmatrix} \tag{6–31}$$

Now we determine the vector to be removed (assuming the columns are now numbered 0, 1, 2, . . . , m) from

$$\frac{x_{Br}}{y_{rk}} = \min_i \left\{ \frac{x_{Bi}}{y_{ik}} \middle| y_{ik} > 0 \right\} \tag{6–32}$$

which tells us to remove column r. The zeroth column, which is $[1, \bar{0}]$, is never removed.

Having determined the vector to enter and to leave the basis, we may now proceed to transform the current basic feasible solution to a new basic feasible solution. We transform only the quantities $\bar{x}_B^*$ and B^{*-1}.

We can use the theory presented in the previous section to compute the new solution and inverse basis matrix. If the new solution is designated $\hat{\bar{x}}_B^*$ and the new basis inverse is designated $\hat{B}^{*-1}$ then we have, using (6–10),

$$\hat{B}^{*-1} = J^*B^{*-1}$$

$$\hat{\bar{x}}_B^* = J^*B^{*-1}\bar{b}^* = J^*\bar{x}_B^* \tag{6–33}$$

where

$$J^* = (\bar{e}_1, \bar{e}_2, \ldots, \bar{e}_r, \bar{\mu}_k^*, \bar{e}_{r+2}, \ldots, \bar{e}_{m+1}) \tag{6–34}$$

The numbering of the vectors in (6–34) is different from that in (6–8) because of the extra first column in B^* which is always present. Similarly $\bar{\mu}_k^*$ is given by

$$\bar{\mu}_k^* = \left[-\frac{z_k - c_k}{y_{rk}}, -\frac{y_{1k}}{y_{rk}}, -\frac{y_{2k}}{y_{rk}}, \ldots, \frac{1}{y_{rk}}, \ldots, -\frac{y_{mk}}{y_{rk}} \right] \tag{6–35}$$

In Chapter 5 we presented a unified vector transformation formula (see Equations (5–15) and (5–16)). The same can be done here to avoid actually using a J^* matrix. We need some additional notation to do this. Let us designate the last m column of B^{*-1} by $\bar{\beta}_j^*$, $j = 1, 2, \ldots, m$, where

$$\bar{\beta}_j^* = [\beta_{0j}, \beta_{1j}, \ldots, \beta_{mj}] \tag{6–36}$$

Using this notation we can write

$$\hat{B}^{*-1} = (\bar{e}_1, \hat{\bar{\beta}}_1^*, \hat{\bar{\beta}}_2^*, \ldots, \hat{\bar{\beta}}_m^*) = J^*B^{*-1}$$
$$= (\bar{e}_1, J^*\bar{\beta}_1^*, J^*\bar{\beta}_2^*, \ldots, J^*\bar{\beta}_m^*) \quad (6\text{–}37)$$

We see then from (6–37) that

$$\hat{\bar{\beta}}_j^* = J^*\bar{\beta}_j^* \qquad j = 1, 2, \ldots, m \qquad (6\text{–}38)$$

and also

$$\hat{\bar{x}}_B^* = J^*\bar{x}_B^* \qquad (6\text{–}39)$$

If we now define $\bar{\gamma}_k^* = \bar{\mu}_k^* - \bar{e}_{r+1}$, then we have

$$\begin{aligned} \hat{\bar{\beta}}_j^* &= \bar{\beta}_j^* + \beta_{rj}\bar{\gamma}_k^* \\ \hat{\bar{x}}_B^* &= \bar{x}_B^* + x_{Br}\bar{\gamma}_k^* \end{aligned} \qquad j = 1, 2, \ldots, m \qquad (6\text{–}40)$$

where

$$\bar{\gamma}_k^* = \left[-\frac{z_k - c_k}{y_{rk}}, -\frac{y_{1k}}{y_{rk}}, \ldots, \frac{1}{y_{rk}} - 1, \ldots, -\frac{y_{mk}}{y_{rk}} \right] \qquad (6\text{–}41)$$

We can also make use of one set of transformation formulae for the β_{ij} in component form, which is in the same form as we used to transform the y_{ij} in Chapter 5. To do this we use the convention that

$$\begin{aligned} \beta_{00} &= z \\ \beta_{i0} &= x_{Bi} \qquad i = 1, 2, \ldots, m \\ y_{0k} &= z_k - c_k \end{aligned} \qquad (6\text{–}42)$$

Then if we write equations (6–40) in component form we have

$$\begin{aligned} \hat{\beta}_{ij} &= \beta_{ij} - \frac{y_{ik}}{y_{rk}}\beta_{rj} \qquad i = 0, 1, \ldots, m; \quad i \neq r \\ & \qquad\qquad\qquad\qquad j = 0, 1, \ldots, m \\ \hat{\beta}_{rj} &= \frac{\beta_{rj}}{y_{rk}} \qquad j = 0, 1, \ldots, m \end{aligned} \qquad (6\text{–}43)$$

Note that equations (6–43) have the same form as the simplex transformation equations for the y_{ij} in the simplex method presented in Chapter 5.

It is a relatively simple matter to arrange the revised simplex calculations into a tableau format, just as we did for the simplex calculations. Such a tableau format is given in Figure 6–1. Since the first column of B^{*-1} is always $\bar{e}_1$, which corresponds to the variable z, it is not necessary to include this column. What is given in the tableau are columns corresponding to $\bar{\beta}_j^*$ $j = 1, 2, \ldots, m$ of B^{*-1}. We also include a column for $\bar{x}_B^*$ and $\bar{y}_k^*$. Unlike

Basis	$\bar{\beta}_1^*$	$\bar{\beta}_2^*$	$\cdots$	$\bar{\beta}_m^*$	$\bar{x}_B^*$	$\bar{y}_k^*$
$\bar{a}_0$	β_{01}	β_{02}	$\cdots$	β_{0m}	z	$z_k - c_k$
$\bar{b}_1$	β_{11}	β_{12}	$\cdots$	β_{1m}	x_{B1}	y_{1k}
$\bar{b}_2$	β_{21}	β_{22}	$\cdots$	β_{2m}	x_{B2}	y_{2k}
$\vdots$	$\vdots$	$\vdots$	$\vdots$	$\vdots$	$\vdots$	$\vdots$
$\bar{b}_m$	β_{m1}	β_{m2}	$\cdots$	β_{mm}	x_{Bm}	y_{mk}

Figure 6–1 Revised simplex tableau format.

the simplex tableau, the revised simplex tableau does not contain all the information we require to carry out the calculations. We need to have another table of the original vectors $\bar{a}_j^*$ of the matrix A in order to compute the $z_j - c_j$ and also $\bar{y}_k^*$.

It should be clear now how the revised simplex calculation is carried out using the tableau we have shown in Figure 6–1. First we enter the values of the β_{ij} and the x_{Bi} of the new tableau. We cannot fill in the values in the $\bar{y}_k^*$ column at this point. Next, we compute the values of $z_j - c_j$ by multiplying the first row of the tableau by each $\bar{a}_j^*$ not in the basis. It should be remembered, however, that the first component of the vector of the first row is "1," corresponding to the variable z, which is not shown in the tableau. When the vector to enter has been determined, we then compute $\bar{y}_k^*$ since we now know the $\bar{a}_k^*$ to enter the basis. Using the $\bar{y}_k^*$ we then determine the vector to leave the basis.

EXAMPLE 1: REVISED SIMPLEX METHOD WITHOUT ARTIFICIAL VARIABLES

Consider the following problem:

$$\begin{aligned} \text{Max } z &= x_1 + 2x_2 + 3x_3 + 3x_4 \\ 2x_1 + x_2 + 3x_3 + 5x_4 &\leq 30 \\ x_1 + 2x_2 + 4x_3 + 2x_4 &\leq 16 \\ 3x_1 + 2x_2 + 3x_3 + 4x_4 &\leq 24 \\ x_j \geq 0 \qquad j &= 1, 2, 3, 4 \end{aligned} \tag{6–44}$$

We convert the problem in (6–44) to the standard equality form as follows:

$$\begin{aligned} \text{Max } z &= x_1 + 2x_2 + 3x_3 + 3x_4 \\ 2x_1 + x_2 + 3x_3 + 5x_4 + x_5 &= 30 \\ x_1 + 2x_2 + 4x_3 + 2x_4 + x_6 &= 16 \\ 3x_1 + 2x_2 + 3x_3 + 4x_4 + x_7 &= 24 \\ x_j \geq 0, \qquad j &= 1, 2, \ldots, 7 \end{aligned} \tag{6–45}$$

The problem is now written in the form we shall use for revised simplex as

$$\begin{aligned} z - x_1 - 2x_2 - 3x_3 - 3x_4 \qquad\qquad\qquad\quad &= 0 \\ 2x_1 + x_2 + 3x_3 + 5x_4 + x_5 \qquad\qquad &= 30 \\ x_1 + 2x_2 + 4x_3 + 2x_4 \qquad + x_6 \qquad &= 16 \\ 3x_1 + 2x_2 + 3x_3 + 4x_4 \qquad\qquad + x_7 &= 24 \end{aligned} \tag{6–46}$$

It can be seen that the $\bar{a}_j^*$ and $\bar{b}^*$ vectors are

$$\begin{aligned} \bar{a}_1^* &= [-1, 2, 1, 3] & \bar{a}_5^* &= [0, 1, 0, 0] \\ \bar{a}_2^* &= [-2, 1, 2, 2] & \bar{a}_6^* &= [0, 0, 1, 0] \\ \bar{a}_3^* &= [-3, 3, 4, 3] & \bar{a}_7^* &= [0, 0, 0, 1] \\ \bar{a}_4^* &= [-3, 5, 2, 4] & \bar{b}^* &= [0, 30, 16, 24] \end{aligned}$$

Our initial basis is $B^* = (\bar{a}_0^*, \bar{a}_5^*, \bar{a}_6^*, \bar{a}_7^*)$. Hence, the initial basis inverse is calculated from equation (6–28) as

$$B^{*-1} = \begin{bmatrix} 1 & 0 & 0 & 0 \\ 0 & 1 & 0 & 0 \\ 0 & 0 & 1 & 0 \\ 0 & 0 & 0 & 1 \end{bmatrix}$$

The initial basic feasible solution is calculated from equation (6–29) as

$$\bar{x}_B^* = B^{*-1}\bar{b}^* = \begin{bmatrix} 1 & 0 & 0 & 0 \\ 0 & 1 & 0 & 0 \\ 0 & 0 & 1 & 0 \\ 0 & 0 & 0 & 1 \end{bmatrix} \begin{bmatrix} 0 \\ 30 \\ 16 \\ 24 \end{bmatrix} = \begin{bmatrix} 0 \\ 30 \\ 16 \\ 24 \end{bmatrix}$$

Hence, we have the initial tableau shown in Figure 6–2. Note that, although Tableau I shows the $\bar{y}_k^*$ column with entries, we do not know $\bar{y}_k^*$ until after we determine the vector to enter the basis. In order to do this we must examine the $z_j - c_j$. We compute the $z_j - c_j$ by multiplying the first row by the $\bar{a}_j^*$ for those vectors not in the basis. For example,

$$z_1 - c_1 = (1, 0, 0, 0) \begin{bmatrix} -1 \\ 2 \\ 1 \\ 3 \end{bmatrix} = -1$$

Basis	$\bar{\beta}_1^*$	$\bar{\beta}_2^*$	$\bar{\beta}_3^*$	$\bar{x}_B^*$	$\bar{y}_k^*$
$\bar{a}_0^*$	0	0	0	0	−3
a_5^*	1	0	0	30	3
a_6^*	0	1	0	16	4
$\bar{a}_7^*$	0	0	1	24	3

Figure 6–2 Tableau I.

Similarly,

$$z_2 - c_2 = -2; \qquad z_3 - c_3 = -3; \qquad z_4 - c_4 = -3$$

Therefore, we have

$$z_k - c_k = \min(-1, -2, -3, -3) = -3$$

There is a tie between $\bar{a}_3^*$ and $\bar{a}_4^*$ to enter the basis. We arbitrarily let $\bar{a}_3^*$ enter the basis. Next we compute

$$\bar{y}_3^* = B^{*-1}\bar{a}_3^* = \begin{bmatrix} 1 & 0 & 0 & 0 \\ 0 & 1 & 0 & 0 \\ 0 & 0 & 1 & 0 \\ 0 & 0 & 0 & 1 \end{bmatrix} \begin{bmatrix} -3 \\ 3 \\ 4 \\ 3 \end{bmatrix} = \begin{bmatrix} -3 \\ 3 \\ 4 \\ 3 \end{bmatrix}$$

This result now appears in the $\bar{y}_k^*$ column of Tableau I. The vector to leave the basis is

$$\frac{x_{Br}}{y_{rk}} = \min\left(\tfrac{30}{3}, \tfrac{16}{4}, \tfrac{24}{3}\right) = 4$$

Hence $\bar{a}_6^* = \bar{b}_2^*$ leaves the basis and $y_{rk} = 4$. Using the transformation equations (6–40) and (6–41) we transform the columns of the tableau. First we calculate

$$\bar{\gamma}_3^* = [\tfrac{3}{4}, -\tfrac{3}{4}, \tfrac{1}{4} - 1, -\tfrac{3}{4}] = [\tfrac{3}{4}, -\tfrac{3}{4}, -\tfrac{3}{4}, -\tfrac{3}{4}]$$

Then we proceed to calculate the new $\bar{\beta}_j^*$ and $\bar{x}_B^*$ from

$$\hat{\bar{\beta}}_j^* = \bar{\beta}_j^* + \beta_{rj}\bar{\gamma}_k^* \qquad j = 1, 2, \ldots, m$$

$$\hat{\bar{x}}_B^* = \bar{x}_B^* + x_{Br}\bar{\gamma}_k^*$$

Therefore

$$\hat{\beta}_1^* = \bar{\beta}_1^* + \beta_{r1}\bar{\gamma}_3^*$$

$$\hat{\bar{\beta}}_1^* = [0, 1, 0, 0] + 0[\tfrac{3}{4}, -\tfrac{3}{4}, -\tfrac{3}{4}, -\tfrac{3}{4}] = [0, 1, 0, 0]$$

$$\hat{\bar{\beta}}_2^* = [0, 0, 1, 0] + 1[\tfrac{3}{4}, -\tfrac{3}{4}, -\tfrac{3}{4}, -\tfrac{3}{4}] = [\tfrac{3}{4}, -\tfrac{3}{4}, \tfrac{1}{4}, -\tfrac{3}{4}]$$

$$\hat{\bar{\beta}}_3^* = [0, 0, 0, 1] + 0[\tfrac{3}{4}, -\tfrac{3}{4}, -\tfrac{3}{4}, -\tfrac{3}{4}] = [0, 0, 0, 1]$$

$$\hat{\bar{x}}_B^* = [0, 30, 16, 24] + 16[\tfrac{3}{4}, -\tfrac{3}{4}, -\tfrac{3}{4}, -\tfrac{3}{4}] = [12, 18, 4, 12]$$

The new tableau, Tableau II, is given in Figure 6–3.

We now compute the $z_j - c_j$ for the new basis given in Tableau II:

$$z_1 - c_1 = -\tfrac{1}{4}; \qquad z_2 - c_2 = -\tfrac{1}{2}; \qquad z_4 - c_4 = -\tfrac{3}{2}$$

$$z_6 - c_6 = \tfrac{3}{4}$$

Hence $z_k - c_k = \min(-\tfrac{1}{4}, -\tfrac{1}{2}, -\tfrac{3}{2}) = -\tfrac{3}{2}$ and $\bar{a}_4^*$ enters the basis. We next compute $\bar{y}_4^*$

$$\bar{y}_4^* = \begin{bmatrix} 1 & 0 & \tfrac{3}{4} & 0 \\ 0 & 1 & -\tfrac{3}{4} & 0 \\ 0 & 0 & \tfrac{1}{4} & 0 \\ 0 & 0 & -\tfrac{3}{4} & 1 \end{bmatrix} \begin{bmatrix} -3 \\ 5 \\ 2 \\ 4 \end{bmatrix} = \begin{bmatrix} -\tfrac{3}{2} \\ \tfrac{7}{2} \\ \tfrac{1}{2} \\ \tfrac{5}{2} \end{bmatrix}$$

The vector to leave the basis is

$$\frac{x_{Br}}{y_{rk}} = \min\left(\frac{18}{\tfrac{7}{2}}, \frac{4}{\tfrac{1}{2}}, \frac{12}{\tfrac{5}{2}}\right) = \frac{24}{5}$$

Hence $\bar{a}_7^* = \bar{b}_3^*$ leaves the basis and $y_{rk} = \tfrac{5}{2}$. We again transform Tableau II as follows:

$$\bar{\gamma}_4^* = \left[\frac{\tfrac{3}{2}}{\tfrac{5}{2}}, \frac{-\tfrac{7}{2}}{\tfrac{5}{2}}, \frac{-\tfrac{1}{2}}{\tfrac{5}{2}}, \frac{1}{\tfrac{5}{2}} - 1\right] = [\tfrac{3}{5}, -\tfrac{7}{5}, -\tfrac{1}{5}, -\tfrac{3}{5}]$$

Basis	$\bar{\beta}_1^*$	$\bar{\beta}_2^*$	$\bar{\beta}_3^*$	$\bar{x}_B^*$	$\bar{y}_k^*$
$\bar{a}_0^*$	0	$\tfrac{3}{4}$	0	12	$-\tfrac{3}{2}$
a_5^*	1	$-\tfrac{3}{4}$	0	18	$\tfrac{7}{2}$
$\bar{a}_3^*$	0	$\tfrac{1}{4}$	0	4	$\tfrac{1}{2}$
$\bar{a}_7^*$	0	$-\tfrac{3}{4}$	1	12	$\tfrac{5}{2}$

Figure 6–3 Tableau II.

Therefore,

$$\hat{\beta}_1^* = [0, 1, 0, 0] + 0[\tfrac{3}{5}, -\tfrac{7}{5}, -\tfrac{1}{5}, -\tfrac{3}{5}] = [0, 1, 0, 0]$$

$$\hat{\beta}_2^* = [\tfrac{3}{4}, -\tfrac{3}{4}, \tfrac{1}{4}, -\tfrac{3}{4}] - \tfrac{3}{4}[\tfrac{3}{5}, -\tfrac{7}{5}, -\tfrac{1}{5}, -\tfrac{3}{5}] = [\tfrac{3}{10}, \tfrac{3}{10}, \tfrac{4}{10}, -\tfrac{3}{10}]$$

$$\hat{\beta}_3^* = [0, 0, 0, 1] + 1[\tfrac{3}{5}, -\tfrac{7}{5}, -\tfrac{1}{5}, -\tfrac{3}{5}] = [\tfrac{3}{5}, -\tfrac{7}{5}, -\tfrac{1}{5}, \tfrac{2}{5}]$$

$$\hat{\bar{x}}_B^* = [12, 18, 4, 12] + 12[\tfrac{3}{5}, -\tfrac{7}{5}, -\tfrac{1}{5}, -\tfrac{3}{5}] = [\tfrac{96}{5}, \tfrac{6}{5}, \tfrac{8}{5}, \tfrac{24}{5}]$$

The new tableau, Tableau III, is given in Figure 6–4. Once more, we compute the $z_j - c_j$.

$$z_1 - c_1 = \tfrac{11}{10}; \qquad z_2 - c_2 = -\tfrac{1}{5}; \qquad z_6 - c_6 = \tfrac{3}{10}; \qquad z_7 - c_7 = \tfrac{3}{5}$$

Therefore $\bar{a}_2^*$ enters the basis. We compute $\bar{y}_2^*$:

$$\bar{y}_2^* = \begin{bmatrix} 1 & 0 & \frac{3}{10} & \frac{3}{5} \\ 0 & 1 & \frac{3}{10} & -\frac{7}{5} \\ 0 & 0 & \frac{4}{10} & -\frac{1}{5} \\ 0 & 0 & -\frac{3}{10} & \frac{2}{5} \end{bmatrix} \begin{bmatrix} -2 \\ 1 \\ 2 \\ 2 \end{bmatrix} = \begin{bmatrix} -\frac{1}{5} \\ -\frac{6}{5} \\ \frac{2}{5} \\ \frac{1}{5} \end{bmatrix}$$

The vector to leave the basis is

$$\frac{x_{Br}}{y_{rk}} = \text{Min}\left(\frac{\frac{8}{5}}{\frac{2}{5}}, \frac{\frac{24}{5}}{\frac{1}{5}}\right) = 4$$

Hence $\bar{a}_3^* = \bar{b}_2$ leaves the basis and $y_{rk} = \frac{2}{5}$. We transform Tableau III once more,

$$\bar{\gamma}_2^* = \left[\frac{\frac{1}{5}}{\frac{2}{5}}, \frac{\frac{6}{5}}{\frac{2}{5}}, \frac{1}{\frac{2}{5}} - 1, \frac{-\frac{1}{5}}{\frac{2}{5}}\right] = [\tfrac{1}{2}, 3, \tfrac{3}{2}, -\tfrac{1}{2}]$$

Basis	$\bar{\beta}_1^*$	$\bar{\beta}_2^*$	$\bar{\beta}_3^*$	$\bar{x}_B^*$	$\bar{y}_k^*$
$\bar{a}_0^*$	0	$\frac{3}{10}$	$\frac{3}{5}$	$\frac{96}{5}$	$-\frac{1}{5}$
$\bar{a}_5^*$	1	$\frac{3}{10}$	$-\frac{7}{5}$	$\frac{6}{5}$	$-\frac{6}{5}$
$\bar{a}_3^*$	0	$\frac{4}{10}$	$-\frac{1}{5}$	$\frac{8}{5}$	$\frac{2}{5}$
$\bar{a}_4^*$	0	$-\frac{3}{10}$	$\frac{2}{5}$	$\frac{24}{5}$	$\frac{1}{5}$

Figure 6–4 Tableau III.

Basis	$\bar{\beta}_1^*$	$\bar{\beta}_2^*$	$\bar{\beta}_3^*$	$\bar{x}_B^*$	$\bar{y}_k^*$
$\bar{a}_0^*$	0	$\frac{1}{2}$	$\frac{1}{2}$	20	
$\bar{a}_5^*$	1	$\frac{3}{2}$	-2	6	
$\bar{a}_2^*$	0	1	$-\frac{1}{2}$	4	
$\bar{a}_4^*$	0	$\frac{1}{10}$	$\frac{1}{2}$	4	

Figure 6-5 Tableau IV.

Therefore

$$\hat{\bar{\beta}}_1^* = [0, 1, 0, 0] + 0[\tfrac{1}{2}, 3, \tfrac{3}{2}, -\tfrac{1}{2}] = [0, 1, 0, 0]$$

$$\hat{\bar{\beta}}_2^* = [\tfrac{3}{10}, \tfrac{3}{10}, \tfrac{4}{10}, \tfrac{3}{10}] + \tfrac{2}{5}[\tfrac{1}{2}, 3, \tfrac{3}{2}, -\tfrac{1}{2}] = [\tfrac{1}{2}, \tfrac{3}{2}, 1, \tfrac{1}{10}]$$

$$\hat{\bar{\beta}}_3^* = [\tfrac{3}{5}, -\tfrac{7}{5}, -\tfrac{1}{5}, \tfrac{2}{5}] - \tfrac{1}{5}[\tfrac{1}{2}, 3, \tfrac{3}{5}, -\tfrac{1}{2}] = [\tfrac{1}{2}, -2, -\tfrac{1}{2}, \tfrac{1}{2}]$$

$$\hat{\bar{x}}_B^* = [\tfrac{96}{5}, \tfrac{6}{5}, \tfrac{8}{5}, \tfrac{24}{5}] + \tfrac{8}{5}[\tfrac{1}{2}, 3, \tfrac{3}{2}, -\tfrac{1}{2}] = [20, 6, 4, 4]$$

The new tableau, Tableau IV, is given in Figure 6–5.
We compute the $z_j - c_j$ for Tableau IV:

$$z_1 - c_1 = 1; \qquad z_3 - c_3 = \tfrac{1}{2}; \qquad z_6 - c_6 = \tfrac{1}{2}; \qquad z_7 - c_7 = \tfrac{1}{2}$$

We see then that the solution given in Tableau IV is the optimal solution, i.e., $x_2 = 4$, $x_4 = 4$, $x_5 = 6$ and $z = 20$.

6-4 THE REVISED SIMPLEX METHOD: WITH ARTIFICIAL VARIABLES

The linear programming problem we start with is assumed to be in the form

$$\text{Max } z = \sum_{j=1}^{n} c_j x_j$$

subject to

$$\sum_{j=1}^{n} a_{ij} x_j = b_i \qquad i = 1, 2, \ldots, m$$
$$x_j \geq 0 \qquad j = 1, 2, \ldots, n \tag{6-47}$$

where the slack and surplus variables are already included. However, we now assume that it is necessary to add artificial variables to the problem in order to obtain an initial basis matrix. Just as we did with the simplex method in Chapter 5 we shall use a two-phase calculation. In Phase I, the artificial variables will be driven to zero and only in Phase II will the optimal solution to the original problem be sought.

It will be recalled from Chapter 5 that it is necessary to prevent any artificial variables remaining in the basis at a zero level at the end of Phase I from ever becoming positive in Phase II. This will also be the case in the revised simplex algorithm. However, we shall employ a somewhat different method for handling this problem than any of the methods we described in Chapter 5.

Just as in the revised simplex method without artificial variables, we shall again treat the objective function as just another constraint. If we then add an artificial variable to each of the original constraints we have the following equations for Phase I:

$$\begin{array}{lllllll} & z_a & + x_{a1} & + x_{a2} & + \ldots & + x_{am} & = 0 \\ a_{11} x_1 + a_{12} x_2 + \ldots + a_{1n} x_n & & + x_{a1} & & & & = b_1 \\ a_{21} x_1 + a_{22} x_2 + \ldots + a_{2n} x_n & & & + x_{a2} & & & = b_2 \\ \cdot \quad \cdot \quad \cdot & & & & \cdot & & \cdot \\ \cdot \quad \cdot \quad \cdot & & & & \cdot & & \cdot \\ \cdot \quad \cdot \quad \cdot & & & & \cdot & & \cdot \\ a_{m1} x_1 + a_{m2} x_2 + \ldots + a_{mn} x_n & & & & & + x_{am} & = b_m \end{array} \tag{6–48}$$

It will be recalled that the Phase I objective function is

$$z_a = -\sum_{i=1}^{m} x_{ai} \tag{6–49}$$

This results in the first constraint of (6–48). We can operate on the equations (6–48) to seek a basic feasible solution which maximizes z_a to zero. Only if max $z_a = 0$ do we proceed to Phase II.

Since it is possible at the end of Phase I that one or more artificial variables are basic but have values of zero, we must prevent their becoming positive during Phase II. The device we shall employ is to simply add a constraint of the form:

$$\sum_{i=1}^{m} x_{ai} = 0 \tag{6–50}$$

Hence, in Phase II we add to the original constraint set of (6–47) two additional constraints: one of these from (6–50) and also the original objective function, i.e.,

$$z = \sum_{j=1}^{n} c_j x_j \tag{6–51}$$

Therefore, our set of equations for Phase II will be

$$\begin{array}{llllllll}
z - c_1x_1 - c_2x_2 - \ldots - c_nx_n & & & & & & = 0 \\
& x_{a1} & + x_{a2} & + \ldots & + x_{am} & & = 0 \\
a_{11}x_1 + a_{12}x_2 + \ldots + a_{1n}x_n & + x_{a1} & & & & & = b_1 \\
a_{21}x_1 + a_{22}x_2 + \ldots + a_{2n}x_n & & + x_{a2} & & & & = b_2 \\
\vdots & & & \ddots & & & \vdots \\
a_{m1}x_1 + a_{m2}x_2 + \ldots + a_{mn}x_n & & & & + x_{am} & & = b_m
\end{array} \qquad (6\text{–}52)$$

It would appear at first glance that in Phase I we use one set of equations, viz., equations (6–48) and that in Phase II we need to use another set of equations, viz., equations (6–52). However, if we note the ways in which the two sets of equations differ, we are led to a very useful conclusion. One of the ways they differ is quite simple. The Phase II equations contain one additional constraint, namely, the one corresponding to the original objective function. The other way is that in Phase I there is a z_a in the constraint involving the x_{ai} whereas in Phase II no such term occurs. However, when we begin Phase II $z_a = 0$. Hence, to include this term does not cause any difficulty. Once this is realized it becomes apparent that we really can make use of one set of equations for both phases. That set of equations would be

$$\begin{array}{llllllll}
z - c_1x_1 - c_2x_2 - \ldots - c_nx_n & & & & & & = 0 \\
& z_a + x_{a1} & + x_{a2} & + \ldots & + x_{am} & & = 0 \\
a_{11}x_1 + a_{12}x_2 + \ldots + a_{1n}x_n & + x_{a1} & & & & & = b_1 \\
a_{21}x_1 + a_{22}x_2 + \ldots + a_{2n}x_n & & + x_{a2} & & & & = b_2 \\
\vdots & & & \ddots & & & \vdots \\
a_{m1}x_1 + a_{m2}x_2 + \ldots + a_{mn}x_n & & & & + x_{am} & & = b_m
\end{array} \qquad (6\text{–}53)$$

It can be seen that the only difference between (6–52) and (6–53) is the addition of z_a. The way in which we use the equations (6–53) is quite simple.

In Phase I we maximize z_a and completely ignore the first constraint. In Phase I all variables except z_a and z are required to be non-negative and are automatically kept so by the simplex process. In Phase II, z_a is zero and we now maximize z. The Phase I objective function, i.e., the second constraint, serves during Phase II to keep any artificial variables remaining in the basis at values of zero. Hence we can use the unified set of equations (6–53) to carry out the revised simplex calculations.

The computational scheme for the revised simplex method with artificial variables is only slightly more complicated than the case of no artificial variables. One obvious difference is that we have $m + 2$ equations in (6–53). Hence the vectors and basis matrix will need to have the appropriate dimensions. In order to avoid cumbersome notation we will use the following notation for the activity vectors and basis matrix.

$$\begin{aligned} \bar{\alpha}_j &= [-c_j, 0, \bar{a}_j] \qquad j = 1, 2, \ldots, n \\ \bar{\alpha}_j &= [0, 1, \bar{e}_i] \qquad j = n + 1 + i, \quad i = 1, 2, \ldots, m \end{aligned} \tag{6–54}$$

It is clear that $\bar{\alpha}_0 = \bar{e}_1$, corresponding to the original objective function and $\bar{\alpha}_{n+1} = \bar{e}_2$, corresponding to the Phase I objective function. Since the rank of the set of equations (6–53) is $m + 2$, a basis matrix will be of order $m + 2$. We will designate the basis matrix S and it will have vectors

$$S = (\bar{e}_1, \bar{s}_1, \bar{s}_2, \ldots, \bar{s}_{m+1}) \tag{6–55}$$

The $\bar{s}_i$ are the counterparts of $\bar{b}_i$ in previous basis matrices. The first column of S will always be $\bar{e}_1$. During Phase I, the second column of S will be $\bar{e}_2$ since the Phase I objective function will not be zero until the termination of Phase I.

We begin the calculation with a basic feasible solution obtained by setting all legitimate variables equal to zero. The artificial variables x_{ai} and z_a are in general not equal to zero. Hence our initial basis is

$$S = \begin{bmatrix} 1 & 0 & 0 & \ldots & 0 \\ 0 & 1 & 1 & \ldots & 1 \\ 0 & 0 & 1 & \ldots & 0 \\ \cdot & \cdot & \cdot & & \cdot \\ \cdot & \cdot & \cdot & & \cdot \\ \cdot & \cdot & \cdot & & \cdot \\ 0 & 0 & 0 & \ldots & 1 \end{bmatrix} \tag{6–56}$$

Even though (6–56) is not an identity matrix an inverse is easily obtained.

It may be verified that S^{-1} is given by

$$S^{-1} = \begin{bmatrix} 1 & 0 & 0 & \cdots & 0 \\ 0 & 1 & -1 & \cdots & -1 \\ 0 & 0 & 1 & \cdots & 0 \\ \cdot & \cdot & \cdot & & \cdot \\ \cdot & \cdot & \cdot & & \cdot \\ \cdot & \cdot & \cdot & & \cdot \\ 0 & 0 & 0 & \cdots & 1 \end{bmatrix} \tag{6–57}$$

Let us designate the right hand side of (6–53) by $\bar{d} = [0, 0, b_1, \ldots, b_m]$. Then an initial basic feasible solution $\bar{x}_S$ is given by

$$\bar{x}_S = S^{-1}\bar{d} = \left[0, -\sum_{i=1}^{m} b_i, b_1, \ldots, b_m\right] \tag{6–58}$$

The quantities needed for the initial tableau are given by equations (6–57) and (6–58).

During Phase I, as legitimate vectors enter the basis and artificial vectors leave the basis, the general form of S is given by

$$S = \begin{bmatrix} 1 & 0 & -\bar{c}_B' \\ 0 & 1 & -c_I' \\ \bar{0} & \bar{0} & B \end{bmatrix} \tag{6–59}$$

where in (6–59) B is a basis for $A\bar{x} = \bar{b}$ and

$$\bar{c}_I = [c_{I1}, c_{I2}, \ldots, c_{Im}] \tag{6–60}$$

where

$$c_{Ii} = 0 \quad \text{corresponding to } \bar{a}_j, j \leq n$$

$$c_{Ii} = -1 \quad \text{corresponding to } \bar{a}_j, j \geq n + 1$$

The inverse of S in (6–59) is easily seen to be

$$S^{-1} = \begin{bmatrix} 1 & 0 & \bar{c}_B'B^{-1} \\ 0 & 1 & \bar{c}_I'B^{-1} \\ \bar{0} & \bar{0} & B^{-1} \end{bmatrix} \tag{6–61}$$

We can now see that

$$\bar{x}_S = S^{-1}\bar{d} = \begin{bmatrix} 1 & 0 & \bar{c}_B'B^{-1} \\ 0 & 1 & \bar{c}_I'B^{-1} \\ \bar{0} & \bar{0} & B^{-1} \end{bmatrix} \begin{bmatrix} 0 \\ 0 \\ \bar{b} \end{bmatrix} = \begin{bmatrix} c_B'B^{-1}\bar{b} \\ c_I'B^{-1}\bar{b} \\ B^{-1}\bar{b} \end{bmatrix} = \begin{bmatrix} z \\ z_a \\ \bar{x}_B \end{bmatrix} \qquad (6\text{–}62)$$

Similarly we can define $\bar{\eta}_j$ (a quantity analogous to $\bar{y}_j$) as

$$\bar{\eta}_j = S^{-1}\bar{\alpha}_j = \begin{bmatrix} 1 & 0 & \bar{c}_B'B^{-1} \\ 0 & 1 & \bar{c}_IB^{-1} \\ \bar{0} & \bar{0} & B^{-1} \end{bmatrix} \begin{bmatrix} -c_j \\ 0 \\ \bar{a}_j \end{bmatrix} = \begin{bmatrix} z_j - c_j \\ (z_j - c_j)_a \\ \bar{y}_j \end{bmatrix} \qquad j = 1, 2, \ldots, n \qquad (6\text{–}63)$$

In (6–63) the notation $(z_j - c_j)_a$ indicates that these are $z_j - c_j$ corresponding to the Phase I prices. It is apparent that in Phase I we find the $(z_j - c_j)_a$ by multiplying the *second* row S^{-1} with any legitimate $\bar{\alpha}_j$ not in the basis. We determine the vector to enter the basis in the usual way. We next compute $\bar{\eta}_k$ for the vector to enter and then compute the vector to leave in the usual way. It should be remembered that in Phase I the first two columns of S never leave the basis. The transformation equations are the same as they were in the case of no artificial variables, except that there is one additional component. If we designate the columns of S^{-1} as $\bar{\sigma}_j$, then we have

$$\begin{aligned} \hat{\bar{\sigma}}_j &= \bar{\sigma}_j + \sigma_{rj}\bar{\rho}_k \\ \hat{\bar{x}}_S &= \bar{x}_S + x_{Sr}\bar{\rho}_k \end{aligned} \qquad j = 1, 2, \ldots, m+1 \qquad (6\text{–}64)$$

where

$$\bar{\rho}_k = \left[\frac{-z_k - c_k}{\eta_{rk}}, \frac{-\eta_{1k}}{\eta_{rk}}, \ldots, \frac{1}{\eta_{rk}} - 1, \ldots, \frac{-\eta_{mk}}{\eta_{rk}}, \frac{-\eta_{m+1,k}}{\eta_{rk}}\right] \qquad (6\text{–}65)$$

Phase I terminates when $z_a = 0$ and when all the $(z_j - c_j)_a \geq 0$.

In Phase II we no longer care about z_a. Hence the second column can be removed from the basis. The first column, however, which corresponds to z, must always be in the basis. A consequence of this, since a basis must contain $m + 2$ columns, is that there will always be at least one artificial vector in the basis. Its corresponding variable value will, of course, be zero.

It is clear from (6–63) that the scalar product of the first row of S^{-1} and any legitimate vector $\bar{\alpha}_j$ not in the basis will give us the values of $z_j - c_j$. We then determine the vector to enter the basis in the usual way. $\bar{\eta}_k$ is calculated for this vector and then we compute the vector to leave the basis in the usual way. Equations (6–64) and (6–65) are used for transforming the tableaux. Phase II ends when all $z_j - c_j \geq 0$.

We can use a tableau format very similar to the one we used in the previous section. The only difference is the addition of one more row and column. Again the first column is always $\bar{e}_1$ and is omitted from the tableaux. However, the second column of S^{-1}, which is always $\bar{e}_2$ in Phase I, is included because it may change during Phase II.

One last matter should be mentioned before we turn to an example. The initial basis S given by (6–56) is not an identity matrix. Even so, it was not at all difficult to find S^{-1}. However, if one chooses to use the product form of the inverse, it is a great deal more convenient to start with an identity matrix. It is a simple matter to transform the original problem, which is stated in equations (6–53) into such a form. We proceed as follows. In all, we have $m + 2$ rows. Let us subtract the sum of rows 3 through $m + 2$ from the second row. This yields

$$z_a + \sum_{i=1}^{m} x_{ai} - \sum_{i=1}^{m} a_{i1}x_1 - \sum_{i=1}^{m} a_{i2}x_2 - \dots - \sum_{i=1}^{m} a_{in}x_n - \sum_{i=1}^{m} x_{ai} = 0 - \sum_{i=1}^{m} b_i \quad (6\text{–}66)$$

Equation (6–66) reduces to

$$z_a - \sum_{i=1}^{m} a_{i1}x_1 - \sum_{i=1}^{m} a_{i2}x_2 - \dots - \sum_{i=1}^{m} a_{in}x_n = -\sum_{i=1}^{m} b_i \quad (6\text{–}67)$$

If we now substitute (6–67) for the second equation in (6–53) we obtain the following new system of equations:

$$\begin{array}{llllllll}
z - c_1x_1 & - c_2x_2 & - \dots - c_nx_n & & & & & = 0 \\
t_1x_1 & + t_2x_2 & + \dots + t_nx_n & + z_a & & & & = \tilde{b} \\
a_{11}x_1 & + a_{12}x_2 & + \dots + a_{1n}x_n & + & x_{a1} & & & = b_1 \\
a_{21}x_1 & + a_{22}x_2 & + \dots + a_{2n}x_n & & + & x_{a2} & & = b_2 \\
\cdot & \cdot & \cdot & & & \cdot & & \cdot \\
\cdot & \cdot & \cdot & & & \cdot & & \cdot \\
\cdot & \cdot & \cdot & & & \cdot & & \cdot \\
a_{m1}x_1 & + a_{m2}x_2 & + \dots + a_{mn}x_n & & & & + x_{am} & = b_m
\end{array} \quad (6\text{–}68)$$

where

$$t_j = -\sum_{i=1}^{m} a_{ij}; \qquad \tilde{b} = -\sum_{i=1}^{m} b_i \quad (6\text{–}69)$$

It can be seen from (6–68) that an identity submatrix of order $m + 2$ is present using columns $1, n + 2, n + 3, \ldots, n + m + 2$.

It is easy to see that the substitution of the second equation in (6–68) still enables us to maximize z_a during Phase I with the same results. We see this as follows. If each x_{ai} is zero, then,

$$\sum_{i=1}^{m} a_{ij}x_j = b_i \qquad i = 1, 2, \ldots, m \tag{6–70}$$

If we substitute this into (6–67) we see that $z_a = 0$ which is the result we wish at the conclusion of Phase I. Because of the non-negativity of the x_{ai}, if one or more of the x_{ai} is positive then it is clear that:

$$-\sum_{i=1}^{m}\sum_{j=1}^{n} a_{ij}x_j \geq -\sum_{i=1}^{m} b_i \tag{6–71}$$

and if at least one $x_{ai} > 0$ then $z_a < 0$. Hence, we can use this new equation just as we did the second equation in (6–53).

EXAMPLE 2

$$\begin{aligned} \text{Max } z &= -x_1 - 2x_2 - 4x_3 - 3x_4 \\ 2x_1 + x_2 + 3x_3 + 4x_4 &\geq 10 \\ x_1 + 3x_2 + 2x_3 + 5x_4 &\geq 20 \\ 3x_1 + 2x_2 + x_3 + x_4 &\geq 15 \\ x_1, x_2, x_3, x_4 &\geq 0 \end{aligned} \tag{6–72}$$

First we convert (6–72) to the equality form and also add the artificial variables. This yields

$$\begin{array}{llllllll} z + x_1 + 2x_2 + 4x_3 + 3x_4 & & & & & & & = 0 \\ & & & & z_a + x_{a1} & + x_{a2} & + x_{a3} & = 0 \\ 2x_1 + x_2 + 3x_3 + 4x_4 & - x_5 & & & + x_{a1} & & & = 10 \\ x_1 + 3x_2 + 2x_3 + 5x_4 & & - x_6 & & & + x_{a2} & & = 20 \\ 3x_1 + 2x_2 + x_3 + x_4 & & & - x_7 & & & + x_{a3} & = 15 \end{array} \tag{6–73}$$

The vectors $\bar{\alpha}_j$ and $\bar{d}$ are as follows:

$$\begin{aligned} \bar{\alpha}_1 &= [1, 0, 2, 1, 3] & \bar{\alpha}_5 &= [0, 0, -1, 0, 0] \\ \bar{\alpha}_2 &= [2, 0, 1, 3, 2] & \bar{\alpha}_6 &= [0, 0, 0, -1, 0] \\ \bar{\alpha}_3 &= [4, 0, 3, 2, 1] & \bar{\alpha}_7 &= [0, 0, 0, 0, -1] \\ \bar{\alpha}_4 &= [3, 0, 4, 5, 1] & \bar{d} &= [0, 0, 10, 20, 15] \end{aligned}$$

Our initial tableau can be obtained from (6–57) and (6–58). Hence,

$$S^{-1} = \begin{bmatrix} 1 & 0 & 0 & 0 & 0 \\ 0 & 1 & -1 & -1 & -1 \\ 0 & 0 & 1 & 0 & 0 \\ 0 & 0 & 0 & 1 & 0 \\ 0 & 0 & 0 & 0 & 1 \end{bmatrix}$$

$$\bar{x}_S = S^{-1}\bar{d} = [0, -45, 10, 20, 15]$$

It should be noted that $\bar{\alpha}_0 = \bar{e}_1$, $\bar{\alpha}_8 = \bar{e}_2$, and the three artificial vectors which we shall call for convenience $\bar{\alpha}_9$, $\bar{\alpha}_{10}$, $\bar{\alpha}_{11}$ are in the basis. Therefore,

$$\bar{x}_S = [x_0, x_8, x_9, x_{10}, x_{11}]$$

The initial tableau, Tableau I, is given in Figure 6–6.

We begin in Phase I and we compute the $(z_j - c_j)_a$ by multiplying the second row of S^{-1} by the $\bar{\alpha}_j$ for non-basic j. For example,

$$(z_1 - c_1)_a = (0, 1, -1, -1, -1) \begin{bmatrix} 1 \\ 0 \\ 2 \\ 1 \\ 3 \end{bmatrix} = -6$$

Similarly,

$$(z_2 - c_2)_a = -6; \quad (z_3 - c_3)_a = -6; \quad (z_4 - c_4)_a = -10$$
$$(z_5 - c_5)_a = 1 \quad (z_6 - c_6)_a = 1; \quad (z_7 - c_7)_a = 1$$

Basis	$\bar{\sigma}_1$	$\bar{\sigma}_2$	$\bar{\sigma}_3$	$\bar{\sigma}_4$	$\bar{x}_s$	η_k
$\bar{\alpha}_0$	0	0	0	0	0	3
$\bar{\alpha}_8$	1	−1	−1	−1	−45	−10
$\bar{\alpha}_9$	0	1	0	0	10	4
$\bar{\alpha}_{10}$	0	0	1	0	20	5
$\bar{\alpha}_{11}$	0	0	0	1	15	1

Figure 6–6 Tableau I.

Therefore, $(z_k - c_k)_a = \min(-6, -6, -6, -10) = -10$ and $\bar{\alpha}_4$ enters the basis. Now we compute $\bar{\eta}_4$ from

$$\bar{\eta}_4 = S^{-1}\bar{\alpha}_4 = \begin{bmatrix} 1 & 0 & 0 & 0 & 0 \\ 0 & 1 & -1 & -1 & -1 \\ 0 & 0 & 1 & 0 & 0 \\ 0 & 0 & 0 & 1 & 0 \\ 0 & 0 & 0 & 0 & 1 \end{bmatrix} \begin{bmatrix} 3 \\ 0 \\ 4 \\ 5 \\ 1 \end{bmatrix} = \begin{bmatrix} 3 \\ -10 \\ 4 \\ 5 \\ 1 \end{bmatrix}$$

The vector to leave the basis is determined by

$$\frac{x_{Sr}}{\eta_{rk}} = \min\left(\tfrac{10}{4}, \tfrac{20}{5}, \tfrac{15}{1}\right) = \tfrac{5}{2}$$

Hence $\bar{\alpha}_9 = \bar{s}_2$ leaves the basis and $\eta_{rk} = 4$. Using the transformation equations given by (6–64) and (6–65) we have

$$\bar{\rho}_4 = [-\tfrac{3}{4}, \tfrac{10}{4}, \tfrac{1}{4} - 1, -\tfrac{5}{4}, -\tfrac{1}{4}]$$

and

$$\hat{\bar{\sigma}}_j = \bar{\sigma}_j + \sigma_{rj}\bar{\rho}_k$$
$$\hat{\bar{x}}_S = \bar{x}_S + x_{Sr}\bar{\rho}_k$$

For example,

$$\begin{aligned} \hat{\bar{\sigma}}_2 &= [0, -1, 1, 0, 0] + 1[-\tfrac{3}{4}, \tfrac{10}{4}, -\tfrac{3}{4}, -\tfrac{5}{4}, -\tfrac{1}{4}] \\ &= [-\tfrac{3}{4}, \tfrac{3}{2}, \tfrac{1}{4}, -\tfrac{5}{4}, -\tfrac{1}{4}] \\ \hat{\bar{x}}_S &= [0, -45, 10, 20, 15] + 10[-\tfrac{3}{4}, \tfrac{10}{4}, -\tfrac{3}{4}, -\tfrac{5}{4}, -\tfrac{1}{4}] \\ &= [-\tfrac{15}{2}, -20, \tfrac{5}{2}, \tfrac{15}{2}, \tfrac{25}{2}] \end{aligned}$$

The remaining $\bar{\sigma}_j$ are computed as above. The transformed tableau is given in Figure 6–7, Tableau II.

Basis	$\bar{\sigma}_1$	$\bar{\sigma}_2$	$\bar{\sigma}_3$	$\bar{\sigma}_4$	$\bar{x}_s$	$\bar{\eta}_k$
$\bar{\alpha}_0$	0	$-\frac{3}{4}$	0	0	$-\frac{15}{2}$	$\frac{5}{4}$
$\bar{\alpha}_8$	1	$\frac{3}{2}$	−1	−1	−20	$-\frac{7}{2}$
$\bar{\alpha}_4$	0	$\frac{1}{4}$	0	0	$\frac{5}{2}$	$\frac{1}{4}$
$\bar{\alpha}_{10}$	0	$-\frac{5}{4}$	1	0	$\frac{15}{2}$	$\frac{7}{4}$
$\bar{\alpha}_{11}$	0	$-\frac{1}{4}$	0	1	$\frac{25}{2}$	$\frac{7}{4}$

Figure 6–7 Tableau II.

Since, in Tableau II, $z_a < 0$, we continue in Phase I. We compute the $(z_j - c_j)_a$ from the second row in Tableau II and the $\bar{\alpha}_j$ and obtain

$$(z_1 - c_1)_a = -1; \quad (z_2 - c_2)_a = -\tfrac{7}{2}; \quad (z_3 - c_3)_a = \tfrac{3}{2}$$
$$(z_5 - c_5)_a = -\tfrac{3}{2}; \quad (z_6 - c_6)_a = 1; \quad (z_7 - c_7)_a = 1$$

$(z_k - c_k)_a = \min(-1, -\frac{7}{2}, -\frac{3}{2}) = -\frac{7}{2}$. Therefore $\bar{\alpha}_2$ enters the basis. We compute $\bar{\eta}_2 = S^{-1}\bar{\alpha}_2 = [\frac{5}{4}, -\frac{7}{2}, \frac{1}{4}, \frac{7}{4}, \frac{7}{4}]$. This is now entered in Tableau II. The vector to leave the basis is given by

$$\frac{x_{Sr}}{\eta_{rk}} = \min\left(\frac{\frac{5}{2}}{\frac{1}{4}}, \frac{\frac{15}{2}}{\frac{7}{4}}, \frac{\frac{25}{2}}{\frac{7}{4}}\right) = \frac{30}{7}$$

Hence $\bar{\alpha}_{10} = \bar{s}_3$ leaves the basis and $\eta_{rk} = \frac{7}{4}$. We transform again and obtain Tableau III in Figure 6–8.

We see in Tableau III that $z_a = -5 < 0$. Hence we continue the Phase I calculation. We again compute the $(z_j - c_j)_a$ as before and obtain

$$(z_1 - c_1)_a = -5; \quad (z_3 - c_3)_a = -2; \quad (z_5 - c_5)_a = 1$$
$$(z_6 - c_6)_a = -1; \quad (z_7 - c_7)_a = 1$$

$(z_k - c_k)_a = \min(-5, -2, -1) = -5$. Therefore $\bar{\alpha}_1$ enters the basis. We compute $\bar{\eta}_1 = S^{-1}\bar{\alpha}_1 = [\frac{4}{7}, -4, \frac{5}{7}, -\frac{6}{7}, 4]$. This is now entered in Tableau III. The vector to leave the basis is

$$\frac{x_{Sr}}{\eta_{rk}} = \min\left(\frac{\frac{10}{7}}{\frac{5}{7}}, \frac{5}{4}\right) = \frac{5}{4}$$

Hence $\bar{\alpha}_{11} = \bar{s}_4$ leaves the basis and $\eta_{rk} = 4$. We transform again and obtain Tableau IV in Figure 6–9.

In Tableau IV we see that $z_a = 0$ and all the artificial variables are out of the basis. Hence we have concluded Phase I. To initiate Phase II we take the

Basis	$\bar{\sigma}_1$	$\bar{\sigma}_2$	$\bar{\sigma}_3$	$\bar{\sigma}_4$	$\bar{x}_s$	$\bar{\eta}_k$
$\bar{\alpha}_0$	0	$\frac{1}{7}$	$-\frac{5}{7}$	0	$-\frac{90}{7}$	$\frac{4}{7}$
$\bar{\alpha}_8$	1	-1	1	-1	-5	-4
$\bar{\alpha}_4$	0	$\frac{3}{7}$	$-\frac{1}{7}$	0	$\frac{10}{7}$	$\frac{5}{7}$
$\bar{\alpha}_2$	0	$-\frac{5}{7}$	$\frac{4}{7}$	0	$\frac{30}{7}$	$-\frac{6}{7}$
$\bar{\alpha}_{11}$	0	1	-1	1	5	4

Figure 6–8 Tableau III.

Basis	$\bar{\sigma}_1$	$\bar{\sigma}_2$	$\bar{\sigma}_3$	$\bar{\sigma}_4$	$\bar{x}_s$	$\bar{\eta}_k$
$\bar{\alpha}_0$	0	0	$-\frac{4}{7}$	$-\frac{1}{7}$	$-\frac{95}{7}$	
$\bar{\alpha}_8$	1	0	0	0	0	
$\bar{\alpha}_4$	0	$\frac{1}{4}$	$\frac{1}{28}$	$-\frac{5}{28}$	$\frac{15}{28}$	
$\bar{\alpha}_2$	0	$-\frac{1}{2}$	$\frac{5}{14}$	$\frac{3}{14}$	$\frac{75}{14}$	
$\bar{\alpha}_1$	0	$\frac{1}{4}$	$-\frac{1}{4}$	$\frac{1}{4}$	$\frac{5}{4}$	

Figure 6–9 Tableau IV.

product of the first row of S^{-1} with each of the legitimate variables not in the basis. This gives us, for example,

$$z_3 - c_3 = (1, 0, 0, -\tfrac{4}{7}, -\tfrac{1}{7})\begin{bmatrix}4\\0\\3\\2\\1\end{bmatrix} = \tfrac{19}{7}$$

Similarly,

$$z_5 - c_5 = 0; \qquad z_6 - c_6 = \tfrac{4}{7}; \qquad z_7 - c_7 = \tfrac{1}{7}$$

Since all these $z_j - c_j \geq 0$, there is no need for further iterations in Phase II. We have obtained the optimal solution. It is

$$x_1^* = \tfrac{5}{4}, \qquad x_2^* = \tfrac{75}{14}, \qquad x_3^* = 0, \qquad x_4^* = \tfrac{15}{28}, \qquad z^* = -\tfrac{95}{7}$$

6–5 ADVANTAGES OF THE REVISED SIMPLEX METHOD

At each iteration of the simplex method one has to transform an $(m + 1) \times (n - m + 1)$ array of numbers, assuming that the m basis columns need not be transformed. For simplicity, we shall assume artificial variables were not used. In the revised simplex method one transforms an $(m + 1) \times (m + 1)$ array at each iteration. In addition one must compute $\bar{y}_k^*$, which requires a computation of $(m + 1)^2$ multiplications, and the calculation of $z_j - c_j$ requires $(n - m)(m + 1)$ multiplications. The number of arithmetic operations in the transformation formulae are basically the same for both methods. However, the revised simplex method has the additional calculations that we have mentioned for the $\bar{y}_k^*$ and $z_j - c_j$.

It is quite amazing how these simple facts have been interpreted in various texts and papers on linear programming. In several books a comparison is made between the $(m + 1) \times (n - m + 1)$ array of simplex and the $(m + 1) \times (m + 1)$ array of revised simplex. The inference is then drawn that revised simplex is superior! Simple claims of superiority of this form are usually quite erroneous. Indeed, since the whole matter is related to the use of computers and the way in which the internal structure of computers has changed, comparisons between the methods must be related to the fact that most linear programming calculations are done on a digital computer. If one were to restrict the comparison to hand calculations and the size problem that can be done by hand, there would be little to choose between the two methods. However, for large-scale computation, the matter is quite different.

While many books make comparisons between operation counts and conclude that the revised simplex method involves more *arithmetical* operations than simplex for a full or very dense matrix, even this is not an unassailable conclusion. At a time when multiplications and divisions took ten times as long to perform as addition, subtraction and logical operations, this kind of comparison was perhaps valid. However, in modern large-scale computers, all arithmetical operations are comparable, differing at most by a factor of two. More important for any operation on these machines is the core fetch time. Hence, simple comparisons of arithmetic operations counts are probably not valid.

The initial appeal of revised simplex was the obviously reduced requirement for in-core storage and the reduced requirement for writing the columns of the tableaux at each iteration when non-core storage such as magnetic tape was utilized. Another major advantage of revised simplex over simplex was the preservation of zeros in the original matrix A of the problem. It should be realized that most large linear programming models have non-zero matrix densities of no more than a few per cent. It is not at all uncommon for a matrix to have zeros for more than 90 per cent of its entries. In the simplex method, since the original $\bar{a}_j$ are not preserved in the tableaux, as changes are made from one basis to the next, before long most of the y_{ij} are different from zero and we have a very dense matrix. Therefore, at each iteration the full $(m + 1) \times (n - m + 1)$ number of entries is being transformed. In revised simplex, by contrast, the $z_j - c_j$ and $\bar{y}_k^*$ are computed from the *original* $\bar{a}_j$. Hence the zeros are preserved. If advantage is taken of this fact, the actual number of computations is far less than a simple analysis of multiplications would indicate.

A further advantage attending the use of the revised simplex method relates to the control of round-off errors. However, there is a certain amount of confusion in discussions of this subject also. There is some theoretical evidence which indicates that the simplex and revised simplex methods, as embodied in most computer codes, are inherently unstable with respect to accumulation of round-off error. Empirically, this has been known for quite some time. The revised simplex method, since it preserves zeros in the $\bar{a}_j$, tends to require fewer actual arithmetic operations. This results in a decreased round-off accumulation. A further advantage exists because of the use of the product form of the inverse and the fact that most computer codes allow for direct inversion of the basis matrix at any point. This allows better control of round-off error accumulation. It can be seen that this is probably more convenient in revised simplex than simplex since the revised simplex tableau makes direct use of B^{-1}. In the simplex method conversion to the usual tableau would require a good deal more work.

PROBLEMS

1. Obtain the inverse of

$$A = \begin{bmatrix} 10 & 2 & 3 & 4 \\ 5 & 1 & 6 & 2 \\ 3 & 0 & 4 & 3 \\ 1 & 2 & 3 & 4 \end{bmatrix}$$

using the product form of the inverse representation and starting with $B = B^{-1} = I$.

2. Solve the following problem using the revised simplex method:

$$\begin{aligned} \text{Max } z &= 3x_1 + 2x_2 + 4x_3 \\ x_1 + 2x_2 + 3x_3 &\leq 10 \\ 2x_1 + 3x_2 + x_3 &\leq 15 \\ x_1, x_2, x_3 &\geq 0 \end{aligned}$$

3. Solve the following problem using the revised simplex method:

$$\begin{aligned} \text{Max } z &= 4x_1 + 6x_2 + 8x_3 + 10x_4 \\ 3x_1 + 2x_2 + 3x_3 + 6x_4 &\leq 60 \\ x_1 + 3x_2 + 4x_3 + 2x_4 &\leq 40 \\ 4x_1 + x_2 + 2x_3 + 3x_4 &\leq 50 \\ x_1, x_2, x_3, x_4 &\geq 0 \end{aligned}$$

4. Solve the following problem using the revised simplex method:

$$\begin{aligned} \text{Min } z &= 2x_1 + 3x_2 + 4x_3 \\ 3x_1 + 4x_2 + 6x_3 &\geq 20 \\ x_1 + 5x_2 + 2x_3 &\geq 10 \\ x_1, x_2, x_3 &\geq 0 \end{aligned}$$

5. Solve the following problem using the revised simplex method:

$$\begin{aligned} \text{Max } z &= 3x_1 + 4x_2 + 2x_3 + x_4 \\ 3x_1 + 2x_2 + 6x_3 + 3x_4 &= 80 \\ 4x_1 + 3x_2 + 3x_3 + 5x_4 &= 70 \\ x_1, x_2, x_3, x_4 &\geq 0 \end{aligned}$$

7

Computer Implementation of Linear Programming

7–1 INTRODUCTION

It should be clear to anyone who has solved a linear programming problem with some variant of the simplex method using pencil, paper and perhaps desk calculator that even for problems of 10 variables and five constraints, this can be an onerous and unrewarding labor. It may require 10 or 15 iterations. A problem with 50 constraints and two or three times that number of variables may require about 100 iterations. This amount of computation is very great. Therefore only very small problems can be solved by manual means. Generally speaking, a linear programming problem with more than ten constraints is too large to solve by hand.

If we are dealing with problems of the real world, similar to those described in Chapter 2, we probably are considering linear programming models with hundreds and even thousands of constraints and with many thousands of variables. Problems, for example, with 3000 constraints and 15,000 variables are reasonably common, and larger problems are by no means rare. Indeed, they are routinely solved each day utilizing digital computers. It was extremely fortunate and quite fortuitous, that the large-scale development and use of digital computers (post-World War II) paralleled almost exactly, the development of the simplex method of George Dantzig. Indeed, if digital computers had not been available, large models of realistic problems would never have been formulated because they could not have been solved. The use of the simplex method would have been confined to very small examples in textbooks. Indeed, most of the textbooks would never have been written. The influence of the digital computer in the development of linear programming has been both decisive and crucial. The direction and aim of linear programming research to find new and/or improved algorithms has been influenced by the existence and nature of the computational tools of the day. Hence most of this chapter will be concerned with solving linear programming problems on medium to large-scale digital computers.

7-2 LINEAR PROGRAMMING SYSTEMS

A linear programming "system" is a large, complex computer code, usually supplied by a computer manufacturer or a commercial vendor of "software" (computer programs). These systems are intended to be much more than a computer implementation of the simplex method for the purpose of obtaining the solution to some individual problem. Besides doing this, the system usually includes programs for doing various kinds of post-optimal analyses (see Chapter 12), various options which allow the user to specify or not specify initial solutions (feasible or not feasible), and programs to modify or periodically revise data, as well as a host of other programs for maintaining the problem data for subsequent calculations. In addition, these systems have very complex subprograms for data handling. Indeed, the computational aspects of a linear programming system are minimal. The bulk of the system coding is concerned with input data manipulation and the many options relating to the reporting of the output of the calculation for the user. Because of the large number of sequences of the above options that are available to the user, the system must also include some sort of control or command language which enables the user to specify the sequence of options or operations he wishes to execute.

Each linear programming system is somewhat different from any other and, contrary to statements of the originators, they all require a significant investment of time in order to become even mimimally proficient in their use. At the present time, the large linear programming systems are able to handle problems of somewhere between 8000 to 16,000 rows, depending upon the system. Because of the ability of these systems to use direct access auxiliary memory devices, for all practical purposes, the number of columns of the input data matrix may be considered infinite. Hence, problem size limitation is most directly that of the number of rows of the matrix A. Most, but not all systems, rely upon machine or assembly language programming for the bulk of the system programming, especially for certain critical aspects of the computational subroutines.

It should be noted that this high degree of flexibility in computational and data handling options these systems exhibit is not acquired without some price. The price generally includes one or more of the following:

1. A long learning period to become familiar with how to use the system.
2. A system that is never completely free of "bugs" (errors).
3. A certain ponderousness and oversophistication for the casual or one-time user.
4. A remarkable lack of interchangeability between systems.

Nevertheless, for the serious, committed user who requires many and repeated production runs, these systems are important and must be used.

In the subsequent sections of this chapter, we shall consider the computational algorithms; how the simplex method is implemented; how the data are handled; and finally, how the control or command languages are used in linear programming systems. Finally, we shall briefly mention some of the extensions of linear programming (properly speaking, not linear programming) which are usually included in such systems.

The most important single reference with respect to linear programming systems is the book of Orchard-Hays.[1] Two survey articles have also appeared by Bonner[2] and Beale,[3] which can also be consulted.

7–3 COMPUTATIONAL ASPECTS OF LINEAR PROGRAMMING SYSTEMS

Most large-scale linear programming systems utilize the revised simplex method coupled with the product form of the inverse (see Chapter 6). Unlike what was described in Chapter 6, the Phase I uses a "composite algorithm," which has been described by Wolfe[4] and by Orchard-Hays.[1] The composite simplex method is described in Chapter 13. All systems make use of periodic reinversion of the basis matrix in order to minimize round-off error, and for other reasons to be explained, in order to reduce the number of iterations. They also have a variety of pivot choice options; i.e., criteria by which the vector to enter the basis is chosen. Often a subset of vectors to enter is considered at one time. Some of these options will be described subsequently.

Since there is such a bewildering array of options available to the user and since all systems are somewhat different, we will describe here one kind of general iterative process which the computational section of the linear programming system uses. Our description will be incomplete but will mention some of the most important details. For further details see Orchard-Hays.[1] The general iteration process is shown in Figure 7–1. We will discuss each step of this iteration process in turn. It should be noted that we assume when we enter this iterative process that the data are in a form to be used directly by our computational algorithm. This is *not* the form in which the problem data are submitted via an input routine and is also not the form in which they were originally stored. We shall not concern ourselves with that part of the problem now. It will be discussed briefly in the next section. However, this conversion process to a usable form for the computation can be exceedingly time consuming.

In the first stage of the process depicted in Figure 7–1 (designated I) we obtain some kind of initial basis. Note we have stated that it is also possible to have an infeasible basis. This will be explained subsequently. There are many different options on how to begin. Let us consider some of them. The most obvious and perhaps most time consuming is to append a

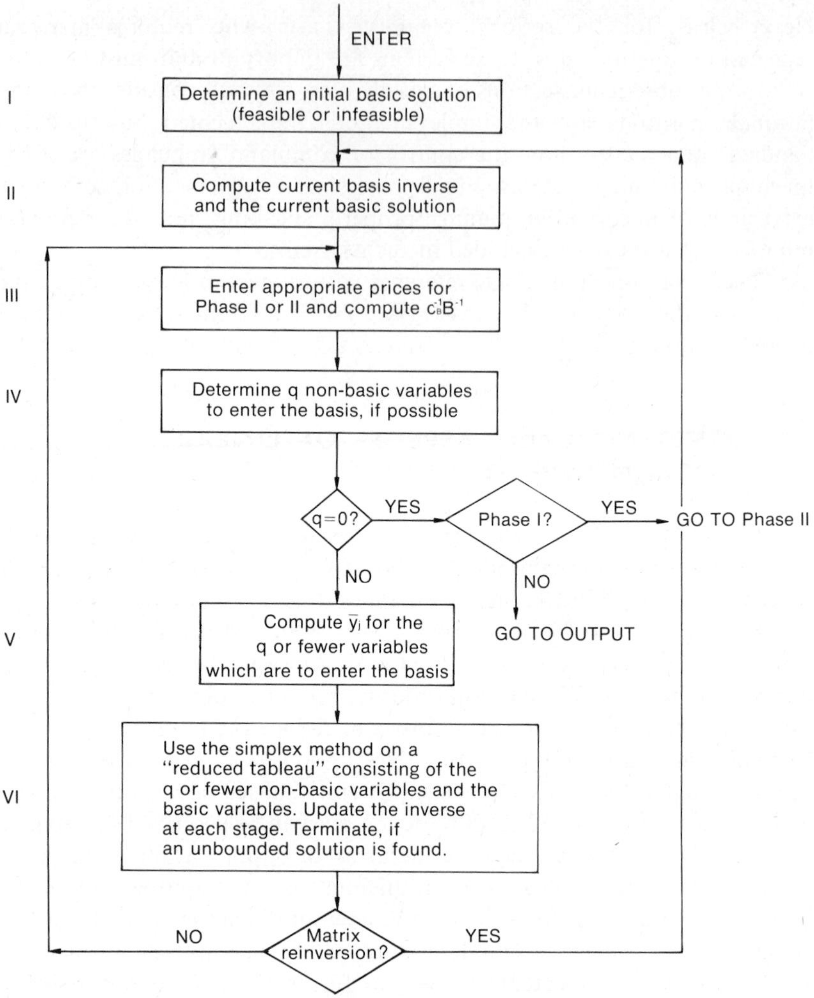

Figure 7–1

complete artificial basis as we have described previously, and go through a complete Phase I to eliminate the artificial variables. A second method is to use what are referred to as "crashing techniques."[1] This method is often used in conjunction with the use of a complete artificial basis or in other contexts. Crashing consists of choosing some number of legitimate non-basic variables to enter the basis and hence replace artificial variables. In some cases this is done with no reference to whether the variables which enter produce infeasibility or even an undesired change in the objective function. This is the origin of the term "crash." The only objective would seem to be to drive artificial variables out of the basis. In other versions of crash procedures some attention is paid to questions of infeasibility. There are many variations and options, and each system has to be considered different with respect to such procedures and options.

Still a third way of getting an initial basic solution is to allow the user to make use of empirical information he may have. Oftentimes one may have very good reasons for supposing that certain variables are likely to be in any feasible or indeed even in any optimal solution. It then makes sense to allow the user to specify these variables as being in an initial basis. This option is employed most frequently when the user has a series of problems to solve in which each problem differs from the previous problem in having slightly different coefficients (a_{ij}) or even a few additional or fewer variables. In this situation, it is advantageous to employ the previous optimal solution as the initial basic solution for the new problem. Hence, the ability of a linear programming system to allow user specification of an initial solution is quite important.

Let us now consider stage II of Figure 7–1. It will be recalled that we require B^{-1} for the revised simplex format. Hence, we must compute B^{-1}. Whether this is easy or difficult to do depends upon the nature of the initial solution. The method must be more general than this, however, since we reenter this block of the flow chart whenever we reinvert our basis. This is a complex problem and very careful attention has been paid to inversion of sparse matrices to take advantage of the very low density of linear programming matrices. The details need not concern us here. They can be found in references 1 or 5. It should be noted, however, that unless this is done extremely efficiently the inversion of very large matrices may be the most time-consuming step of the entire iteration algorithm of Figure 7–1. Once B^{-1} is calculated we then compute $\bar{x}_B = B^{-1}\bar{b}$, the current basic solution. This is straightforward.

In stage III of the process depicted in Figure 7–1 we simply enter the appropriate prices for either Phase I or Phase II. Actually, in a composite algorithm this is not quite correct (see Chapter 13). However, we still need to use the correct prices. Then we compute what are often called simplex multipliers,the elements of the vector $\bar{c}_B'B^{-1}$. They are actually dual variables (see Chapter 8). $\bar{c}_B'B^{-1}$ is sometimes called a "pricing vector" as well. It should be clear why we need this, since

$$z_j - c_j = \bar{c}_B'B^{-1}\bar{a}_j - c_j \tag{7–1}$$

Hence we need the dual variable vector $\bar{c}_B'B^{-1}$ to compute the $z_j - c_j$ in the next stage.

In stage IV we attempt to find some non-basic variables which are eligible to enter the basis. Generally we do this by examining the variables in their natural order and compute $z_j - c_j$. If no $z_j - c_j < 0$ are found we are optimal, of course. Assuming there are some, we wish to enter one or more of these into the basis. As we have noted previously in Chapter 4, there are many ways in which the vector to enter the basis can be chosen, and not much is *definitively* known about which method is "best" in terms of the number of iterations. A very common way to approach this in most linear programming systems is to simply find q (or fewer if there are not as many as q) non-basic vectors with $z_j - c_j < 0$ in the order in which they occur. No attempt is usually made to find the q vectors with the most negative $z_j - c_j$, and certainly no attempt has been made to apply the rigorous simplex criterion of choosing vectors with $(x_{Br}/y_{rk}\,(c_k - z_k))$ of greatest magnitude, even though there is some evidence that, despite commonly held beliefs, this may be most efficacious. Many unexamined and arbitrary decisions have been built into linear programming systems and codes.

In the previous stage we found q non-basic variables for entry into the basis. In stage V we now compute $\bar{y}_j$ for each of the vectors we have found. Hence, if L is the set of subscripts of the q vectors we have found then we compute

$$\bar{y}_l = B^{-1}\bar{a}_l, \qquad l \in L \tag{7–2}$$

We shall use these y_l, $l \in L$ in the next stage of the calculation.

We are now ready to perform in stage VI actual pivot operations or transformations, i.e., insertion of a vector into the basis and removal of the appropriate vector from the basis. We employ the simplex method but we restrict its application to the basic variables and *only* those q non-basic variables we have previously selected and for which we computed $\bar{y}_j$ vectors. This technique is referred to as "suboptimization" by Orchard-Hays.[1] The name refers to the fact that the subset of q variables is entered into the basis based on information that may not be completely correct in terms of which vectors are most eligible to enter the basis. In short, optimization of which vector is to enter the basis is made over a subset of the eligible non-basic vectors, viz., those q previously selected to enter the basis because they had $z_j - c_j < 0$ in stage V. The usual simplex method is employed: we select a vector to enter the basis, determine the vector to be removed, transform the revised simplex tableau, etc. However, all of this is done with respect to the variables $l \in L$, and the basic variables. The particular variable chosen to enter the basis at any stage may be determined by any of the usual criteria. The variable to leave the basis is rigidly determined in the usual way. If an unbounded solution is found at any stage, the computation is terminated. The inverse matrix B^{-1} is transformed and updated at each iteration so that we remain current when we return to stage II or III after stage VI.

Following stage VI we check to see if it is necessary to reinvert the basis matrix. There are several reasons why it may be necessary to reinvert the basis. The most important reason, and the one that probably accounts for most reinversions in actual linear programming runs, originates in the use of the product form of the inverse. Periodic reinversions help to keep the size of the product form of the inverse to manageable proportions. Every iteration adds a new matrix to this product. Hence, for large problems, which have large numbers of iterations, at each iteration the computation must deal with each of the product matrices which have been accumulated up to that point.† Reinversion enables one to begin again with a set of product matrices that does not exceed in number the size of the basis. This tends to reduce subsequent computational time and effort. A small number of product matrices tends to preserve more sparseness and hence also reduces computation.

The second major reason for reinverting the basis matrix is in order to preserve as much numerical accuracy as possible. As the simplex process proceeds through numerous iterations and the basis inverse is transformed at each iteration, round-off and other numerical errors combine and propagate through the calculation process. There is some evidence that the simplex process as usually implemented is inherently numerically unstable. In any case, close attention to the setting of tolerances with respect to many variables is of the utmost importance.[1,6] Periodic reinversion of the inverse basis matrix also helps to control numerical stability.

For all the above reasons reinversion is performed quite often in the course of a problem. Typically, a problem with a basis size of $m = 1000$ might be reinverted about every 100 iterations so that the number of product matrices does not increase by more than 10 per cent between reinversions.

7–4 FURTHER COMPUTATIONAL ASPECTS OF LINEAR PROGRAMMING SYSTEMS

While the major computational effort of a large linear programming system is devoted to the implementation of the simplex algorithm as described in the previous section, almost all such systems contain additional algorithms for subjecting the solution obtained by the simplex algorithm to further scrutiny of various kinds. The basic questions that these various algorithms consider relate to how sensitive the solution is to changes in prices, right hand sides, etc. In addition, they consider the effect on the solution in a continuous fashion to changes over a continuous range of values. These various algorithms are usually known as *ranging* algorithms and *parametric linear programming* algorithms. The theory of parametric linear programming and post-optimal analysis is discussed in Chapter 12. Orchard-Hays[1] also discusses these algorithms.

Ranging refers to a determination of the sensitivity of the optimal solution obtained by simplex to a change in one of the input constants of the problem. Almost all linear programming systems allow for "cost ranging," i.e., for changes in the c_j. Many also allow for ranging of the b_i. Most systems do not provide an analysis of sensitivity to changes in a_{ij}, although such an analysis can be made.

† Since each of the product matrices in the product form of the inverse differs from an identity matrix in only one column, only these columns need actually be stored, rather than the entire matrices.

As an example of what ranging does, consider the case of cost ranging. We determine for each variable x_j the range over which its corresponding c_j may vary without affecting which variables are in the optimal basic solution. If we should change c_j to some value outside this range, then the optimal basis will also change. Hence, what we are doing is a kind of sensitivity analysis. If we find that the range over which the optimal basis stays the same is very narrow, we know that our solution is very sensitive to small fluctuations or errors in this c_j. On the other hand, if we find that this range is fairly large and if also the current value of c_j is close to the middle of the range, then we can be reasonably certain of the relative insensitivity of our solution to errors in estimation of c_j or to small changes in c_j.

The other major computational aspect of post-optimal analysis is what is known as parametric linear programming. Here we are interested in examining in detail how the optimal solution changes as any of the parameters (constants) of the input are allowed to vary. These include the a_{ij}, c_j, and b_i. In contrast to ranging, parametric linear programming follows the changes or variation on the optimal solution as one or more of the problem parameters vary in a continuous fashion outside the range which would be examined by ranging. This topic is discussed briefly in Chapter 12.

Virtually all linear programming systems include a subroutine embodying the dual simplex algorithm (see Chapter 9). As is discussed in Chapter 9, there appear to be difficulties in getting started, i.e., in finding an initial solution to use with the dual simplex algorithm. Hence, normally there is no advantage to using it. However, under certain conditions it can be more efficient to use the dual simplex algorithm than to begin with the ordinary (primal) simplex algorithm. Such a situation arises when one has an optimal solution to a problem and then changes one or more of the coefficients of the $\bar{b}$ vector in such a way that the new $\bar{x}_B = B^{-1}\bar{b}$ becomes infeasible. However, even though this solution is primal infeasible, we are dual feasible and can proceed to solve our problem by employing the dual simplex iteration scheme as described in Chapter 9. Usually only a small number of iterations are required to reach a new optimal solution. By contrast, if the ordinary simplex algorithm was used, the regular composite or Phase I–Phase II procedure would have to be followed. This would result in as long a computation as the solution of the original problem. No effort is usually made to use any of the special techniques (e.g., multiple pricing) in the dual simplex algorithm. It is implemented almost exactly as described in Chapter 9, since it is used only for special purposes and not as a general purpose linear programming code.

Another feature which is included in linear programming systems is related to how upper bound constraints are handled. If some or all of the variables have constraints of the form: $x_j \leq d_j$ where d_j are upper bounds, then it is desirable to avoid having to enlarge the basis to handle such constraints. A computational procedure which is described in Chapter 13 allows one to implicitly use the knowledge of the existence of these upper bound constraints without having to enlarge the basis to include them. This is an important time-saving feature of any linear programming system since solution time is directly proportional to basis size.

There are many other computational features which some or all linear programming systems possess. These relate to such matters as numerical scaling of variables, negative variable ranges, and variables which are not constrained ("free variables"). For a discussion of these and still other features of these systems the reader should consult Orchard-Hays,[1] or examine some particular system in detail.

7-5 INPUT, OUTPUT AND DATA HANDLING

To a large extent the real complexity, as well as an overwhelming amount of the coding, of a linear programming system, is concerned with an imposing array of options for the handling and manipulation of large masses of data. That this should be so may not be obvious to the user who thinks of linear programming as a *computational* method with some minor attendant routines related to entering the data and printing the results of the calculation. In fact, just the reverse is the case. A large-scale linear programming system is, first and foremost, a data handling program with some additional computational routines for the simplex decision rules and matrix inverse manipulation. The system must obviously have the capability to enter the data for calculation and to print out the results. These functions cover a lot of ground, as the subsequent discussion will indicate. In addition, the system usually has the capability to store several different problems for future use, to update or change the stored data for any of the problems, to aggregate problems or select one or more portions of a larger problem for solution, as well as many other options. We will consider some of these capabilities now in more detail. The interested reader may also consult Orchard-Hays[1] for more details.

A linear programming problem is specified by its input data. This data has to enter a computer via punched cards, magnetic tape or some other external storage medium. The bulk of this data is the coefficient matrix A, of the linear constraints. Generally, this coefficient matrix is entered in a packed format. This means that those coefficients a_{ij} which are zero need not be entered but can be reconstituted later when they are required. The use of packed format necessitates identification of which particular coefficient is being entered. This is usually accomplished by having an identifying name for each row and each column of the matrix. Then the appropriate value of some element a_{ij} is entered, with its appropriate row name and column name. These names are usually mnemonics of some kind and allow the user easy reference, in terms of the model, to which variable is under consideration. It is also necessary to specify a $\bar{b}$ vector, i.e., a right hand side. Most codes allow one to enter several right hand sides for storage and for subsequent use when each individual problem is solved. There is also provision in the input routines for indicating whether each row constraint is an equality or an inequality, and which kind. As was mentioned earlier, it is also possible to provide information for starting solutions, or information on bounds on the variables.

The output routines of linear programming systems are quite complex and contain a large number of report generating capabilities which are of the greatest interest to the user. It is obvious that in order for the solution to a problem with hundreds or even thousands of both constraints and variables to be of use, a great deal of careful identification of information is required in a readily comprehended format. Therefore, the following information is almost always part of the main output report:

1. Selected input information.

2. Activity level (value of x_j) for each variable in the optimal solution, by a name in most cases, and the optimal value of the objective function.

3. The status or condition of each variable in the solution. This may include whether it is in the basis or not, whether it is feasible or not or whether it is at a lower or upper bound.

4. "Shadow prices" for each row and column. The column shadow prices ($z_j - c_j$), or dual variables, tell us how much a "small" change in x_j will change the objective function z. The row shadow prices indicate how great a penalty we pay for imposing the particular constraint. If a slack variable for a given row is positive, its shadow price is zero.

The foregoing information is the minimum that one would want. In addition, most users would also want an additional output report that provides the cost range of each variable in the final basis. By examining this report the user can easily determine how sensitive the solution is to changes in costs. Most systems also provide similar ranging of the right hand side. As previously mentioned, an additional output report could be requested to provide the result of a parametric linear programming study. It is also possible to obtain intermediate information during the course of problem solution. These reports provide such information as which variables are entering and leaving the basis, current value of the objective function, etc.

In addition to provisions for input handling and output reports, linear programming systems have other data handling capabilities. Some of these relate to modification and deletion of problems in the storage file and subsequent "editing" of the resultant problem storage file. Capability exists also for the merging of two or more problems in the storage file to construct a combined larger problem. In addition to deletion of problems one can delete individual columns from a problem. By the repeated use of these techniques one can effectively create new problems.

It is also possible to modify individual data items pertaining to a problem. These include changes in matrix coefficients, objective function costs or right hand side values, specification of new constraints or new variables (columns). As previously discussed, it is not always necessary to start from scratch in solving such a problem if one has an optimum solution to the unmodified problem at hand.

7-6 CONTROL LANGUAGES AND PROBLEM CONTROL

Present day large-scale computers generally operate within the framework of an "operating system" of some kind. These are very complex programs that generally handle the execution of programs, scheduling of various functions, allocation of memory, etc. Linear programming systems have to operate within the framework of some operating system, but they usually use the capabilities of an operating system less than some other programs, since linear programming systems have their own command and control language.

It is not difficult to see why a linear programming system would need a powerful control language. Since so many options exist to enable the user to decide precisely how he wants his data handled, problem solved and analyzed, etc., it is necessary to have a highly flexible and responsive language in which to specify the particular combination that is desired. The program that exists within the linear programming system, that enables the user to specify what he wants, is called the *control language* or *command language* of the system.

It is beyond the scope of this book to describe in detail the structure of some particular command language. Indeed it would not be productive or useful to do so. The interested reader can find some details in Orchard-Hays.[1] However, it is important to recognize what the control language enables the user to do. In brief, the user can tailor-make his own particular sequence of operations to be executed, including such particulars as determination of the initial operations (suppression of certain modes of operation), what input data are to be entered and how, the formulation and solution of his problem, what kinds of post-optimal analyses are to be performed, when to terminate a calculation, what data are to appear as results, etc. The specification of what the user wants is made by a sequential listing of the operations he desires. The operations are simply referred to by certain subroutine or other code names or designations.

Control languages are exceedingly complex and possess the ability to allow options to be specified under conditional transfers or other logical operations. For example, one might specify that a postoptimal analysis of some kind was to be undertaken only if the optimal solution possessed some characteristic. In addition, the control language allows the user to closely control the way in which a solution to a linear programming problem is sought—to specify the particular solution strategy that is employed. This includes such matters as the number of vectors involved in the multiple pricing option or the number of iterations between reinversions. The specification of such parameters as these can, of course, have a significant effect on the amount of computer time involved in obtaining a solution. The user is not forced to specify these parameters. In the absence of explicit specification, the program has certain values to which it automatically falls back. In a few systems there is dynamic adjustment of these parameters by the program during the process of solution. This tends to maximize the efficiency of the use of computer time.

7–7 AUXILIARY FEATURES OF LINEAR PROGRAMMING SYSTEMS

There are a number of additional features which are usually available to varying degrees in most linear programming systems. While they build on the basic primal simplex algorithm they are really extensions of the simplex algorithm to handle linear programming problems with special structural characteristics, or they are extensions to allow the handling of certain kinds of nonlinear programming problems. We will briefly discuss some of these extensions.

It is almost always the case that nonlinear programming problems can, with suitable transformations, be expressed in the form

$$\text{Max } z = \sum_{j=1}^{n} f_j(x_j) \tag{7–3}$$

subject to

$$\sum_{j=1}^{n} g_{ij}(x_j)\{\leq, =, \geq\}b_i, \qquad i = 1, 2, \ldots, m$$

The $f_j(x_j)$ and $g_{ij}(x_j)$ are all functions of a single variable. A problem that is in the form given by (7–3) is said to be in *separable form*. By means of a piecewise linear approximation to the $f_j(x_j)$ and $g_{ij}(x_j)$ such a problem can be reduced to a linear programming format and a modified simplex algorithm can be used to solve the problem (see Hadley,[7] Cooper and Steinberg[8] or Chapter 15 of this text for details). Most linear programming systems have subroutines which allow handling of such problems, once certain details of the linearization have been specified.

Another extension of the simplex method commonly found in linear programming systems is the ability to handle what are called *generalized upper bound problems*. These are problems in which there are upper bounds not on individual variables but on sums of sets of variables. These sums on the constraints, which are of the form

$$\sum_{j=p}^{s} x_j \leq u_j \tag{7–4}$$

are known as generalized upper bound constraints. Many distribution and scheduling problems possess such constraints. A very powerful technique has been developed (see Dantzig and Van Slyke[9] and Chapter 13) for solving such problems which allows these constraints to be treated implicitly, rather than as part of a larger basis matrix. This results in a considerable change in the simplex rules but nevertheless is a very worthwhile complication. In problems with a significant number of such constraints the use of generalized upper bounding considerably reduces the difficulty of solution. For some problems it is the only known practical method of solution. Hence, it is now being included in large-scale linear programming systems.

Another extension of great utility that is included in linear programming systems is a host of techniques that fall under the general name of *decomposition methods* (see Chapter 13). The most common of these is the Dantzig-Wolfe decomposition but others are important too. Generally speaking, they make it possible to solve problems of the requisite structure which otherwise could not be accommodated because the size of the problem is too great. Chapter 13 discusses these techniques in some detail. These methods are an increasingly important part of linear programming systems.

PROBLEM

Project

Examine a typical linear programming system such as MPS, etc., and describe its features in terms of the categorization given in this chapter. Try to learn to use the system on the computer for which it was developed.

REFERENCES

1. Orchard-Hays, W.: *Advanced Linear Programming Computing Techniques.* New York, McGraw-Hill, 1968.
2. Bonner, J. S.: *Mathematical Programming: Computer Systems.* In Aronofsky, J. S. (ed.): *Progress in Operations Research.* Vol. III, New York, Wiley, 1969.
3. Beale, E. M. L.: *Mathematical Programming: Algorithms.* In Aronofsky, J. S. (ed.): *Progress in Operations Research.* Vol. III, New York, Wiley, 1969.
4. Wolfe, P.: The Composite Simplex Method. *SIAM Review* **7**:42 (1965).
5. Willoughby, R. A. (ed.): Sparse Matrix Proceedings. Report RA 1 (#11707), Thomas J. Watson Research Center, IBM Corporation, Yorktown Heights, New York, March, 1969.
6. Clasen, R. J.: Techniques for Automatic Tolerance Control in Linear Programming. *Comm. of ACM* **9**:802, 1966.
7. Hadley, G.: *Nonlinear and Dynamic Programming.* Reading, Mass., Addison-Wesley, 1964.
8. Cooper, L., and Steinberg, D. I.: *Introduction to Methods of Optimization.* Philadelphia, W. B. Saunders, 1970.
9. Dantzig, G. B., and Van Slyke, R. M.: Generalized Upper Bounding Techniques for Linear Programming. J. *Syst., Comp. Sci.,* **1,** No. 3, 1967.

8

Duality

8–1 INTRODUCTION

With every linear programming problem there is a related linear programming problem called the *dual* problem of the original linear programming problem. The original problem, in this context, is called the *primal* problem.

The significant features of these two problems are that they are different mathematical problems in the sense that one is a maximization problem and the other is a minimization problem and that they are expressed in terms of different arrangements of the same basic data. Furthermore, if one has solved one problem, the optimal tableau contains all the information necessary for a simple calculation of the solution of the other problem. In addition, the optimal values of the objective functions of the two problems are equal. All of this will become clearer in the subsequent discussion.

In our initial discussion of duality we will assume that our primal problem is in the form

$$\begin{aligned} \text{Max } z &= \bar{c}'\bar{x} \\ A\bar{x} &\leq \bar{b} \\ \bar{x} &\geq \bar{0} \end{aligned} \tag{8–1}$$

That there is no loss in generality considering only a maximization problem can be seen from the following discussion.

Suppose our original problem was in the form

$$\text{Min } z = \bar{c}_1'\bar{x}$$

If we define $\bar{c} = -\bar{c}_1$, then Min $z = \bar{c}_1'\bar{x}$ is equivalent to

$$\text{Min } z_1 = -\bar{c}'\bar{x}$$

which is an equivalent problem to Max $z_2 = \bar{c}'\bar{x}$. That this is so follows from the fact that if $\bar{x}^*$ minimizes $z_1 = -\bar{c}'\bar{x}$ then

$$-\bar{c}'\bar{x}^* \leq -\bar{c}'\bar{x} \tag{8–2}$$

for all feasible $\bar{x}$. Therefore, if we multiply (8–2) by -1, we have

$$\bar{c}'\bar{x}^* \geq -\bar{c}'\bar{x}$$

which proves that $\bar{x}^*$ maximizes $z_2 = \bar{c}'\bar{x}$. Therefore, there is no loss in generality in considering the objective function as a maximization problem. The same $\bar{x}^*$ is the optimal solution to the two problems and $z_2 = -z_1$.

Now let us consider the constraints. We have already noted in Chapter 4 that any constraint which is an inequality can be converted into an equality by the inclusion of a non-negative slack or surplus variable in the constraint. Let us now consider the other possibilities. First, it should be noted that any inequality of the form

$$\sum_{j=1}^{n} \hat{a}_{ij}x_j \geq \hat{b}_i \tag{8–3}$$

can be multiplied by -1 and if we define

$$a_{ij} = -\hat{a}_{ij}$$
$$b_i = -\hat{b}_i$$

then (8–3) becomes

$$\sum_{j=1}^{n} a_{ij}x_j \leq b_i$$

and is now in the form given in (8–1).

Consider now the case of an equality constraint of the form

$$\sum_{j=1}^{n} a_{ij}x_j = b_i \tag{8–4}$$

It is clear that the equation (8–4) which is a hyperplane in E^n is the intersection of two closed half-spaces. Hence (8–4) is equivalent to:

$$\begin{aligned} \sum_{j=1}^{n} a_{ij}x_j &\leq b_i \\ \sum_{j=1}^{n} a_{ij}x_j &\geq b_i \end{aligned} \tag{8–5}$$

However, from the discussion in the preceding paragraph, we know that an inequality of the form of the second one in (8–5) can be converted to the form of the first.

What we have shown above then, is that any primal problem can be represented as

$$\begin{aligned} \text{Max } z &= \bar{c}'\bar{x} \\ A\bar{x} &\leq \bar{b} \\ \bar{x} &\geq \bar{0} \end{aligned} \tag{8–6}$$

8–2 DUAL LINEAR PROGRAMS

We are now ready to define a dual linear programming problem.
If we are given a linear programming problem in the form

$$\begin{aligned} \text{Max } z &= \bar{c}'\bar{x} \\ A\bar{x} &\leq \bar{b} \\ \bar{x} &\geq \bar{0} \end{aligned} \tag{8–7}$$

(where A is $m \times n$, $\bar{x}$ and $\bar{c}$ are n-vectors and $\bar{b}$ is an m-vector) which we shall call the primal problem, then the following problem is defined to be the dual problem of (8–7)

$$\begin{aligned} \text{Min } w &= \bar{b}'\bar{v} \\ A'\bar{v} &\geq \bar{c} \\ \bar{v} &\geq \bar{0} \end{aligned} \tag{8–8}$$

where $\bar{v}$ is an m–vector. All the other vectors and matrices have been defined. Note that since (8–7) has m constraints, (8–8) has n constraints and that while (8–7) has n variables, (8–8) has m variables. Hence, note that the number of variables in the primal problem is the same as the number of constraints in the dual problem. Similarly, the number of constraints in the primal problem is the same as the number of variables in the dual problem.

A convenient schematic representation of the relationship between the primal and dual problems is the following:

$$\begin{array}{c|cccc|c} & x_1 & x_2 & \cdots & x_n & \\ \hline v_1 & a_{11} & a_{12} & \cdots & a_{1n} & \leq b_1 \\ v_2 & a_{21} & a_{22} & \cdots & a_{2n} & \leq b_2 \\ \cdot & \cdot & \cdot & & \cdot & \cdot \\ \cdot & \cdot & \cdot & & \cdot & \cdot \\ \cdot & \cdot & \cdot & & \cdot & \cdot \\ v_m & a_{m1} & a_{m2} & \cdots & a_{mn} & \leq b_m \\ \hline & \geq c_1 & \geq c_2 & \cdots & \geq c_n & \end{array} \tag{8–9}$$

In the representation given in (8–9) the columns of the matrix A are the activity vectors $\bar{a}_j$ of the primal problem while rows of the matrix A (columns of the matrix A') are the activity vectors of the dual problem. Hence, we read the primal constraints across the rows and the dual constraints are read vertically down the columns.

To illustrate the connection between primal and dual problems consider the following example.

EXAMPLE 1

If the primal problem is

$$\text{Max } z = 2x_1 - 3x_2 + 4x_3 + 2x_4$$

subject to:

$$\begin{aligned} 3x_1 + 4x_2 - x_3 \phantom{{}- 2x_4} &\le 24 \\ x_1 \phantom{{}+ 4x_2} + 2x_3 - 2x_4 &\le 36 \\ x_1 + x_2 + x_3 + x_4 &\le 10 \\ x_1, x_2, x_3, x_4 &\ge 0 \end{aligned}$$

Then the dual problem is

$$\text{Min } w = 24v_1 + 36v_2 + 10v_3$$

subject to:

$$\begin{aligned} 3v_1 + v_2 + v_3 &\ge 2 \\ 4v_1 \phantom{{}+ 2v_2} + v_3 &\ge -3 \\ -v_1 + 2v_2 + v_3 &\ge 4 \\ - 2v_2 + v_3 &\ge 2 \\ v_1, v_2, v_3 \phantom{{}+ v_3} &\ge 0 \end{aligned}$$

It should be intuitively clear that it is a matter of indifference whether the original problem looks like the problem (8–7) or (8–8). We showed that any linear programming problem could be recast into the form (8–7). This should make us strongly suspect that every linear programming problem has a dual problem and that, therefore, the dual of the dual is the primal. We prove this result in the following theorem.

Theorem 8–1 The dual of the dual is the primal.

Proof: Consider the dual problem

$$\begin{aligned} \text{Min } w &= \bar{b}'\bar{v} \\ A'\bar{v} &\ge \bar{c} \\ \bar{v} &\ge \bar{0} \end{aligned} \tag{8–10}$$

It is clear that we can write the objective function as

$$\text{Min } w = \text{Max } (-w) = \text{Max } w_1 = -\bar{b}'\bar{v} \tag{8–11}$$

Also we can multiply $A'\bar{v} \geq \bar{c}$ by -1 to obtain

$$-A'\bar{v} \leq -\bar{c} \tag{8–12}$$

Therefore our original dual problem (8–10) is equivalent to

$$\begin{aligned} \text{Max } w_1 &= -\bar{b}'\bar{v} \\ -A'\bar{v} &\leq -\bar{c} \\ \bar{v} &\geq \bar{0} \end{aligned} \tag{8–13}$$

We see that (8–13) has the same form as (8–7). Hence its dual has the form of (8–8) and is given by

$$\begin{aligned} \text{Min } z_1 &= -\bar{c}'\bar{x} \\ -A\bar{x} &\geq -\bar{b} \\ \bar{x} &\geq \bar{0} \end{aligned} \tag{8–14}$$

We now note that

$$\text{Min } z_1 = \text{Max } (-z_1) = -\text{Max } z_1 = \bar{c}'\bar{x} = \text{Max } z \tag{8–15}$$

and that

$$-A\bar{x} \geq -\bar{b}$$

is equivalent to

$$A\bar{x} \leq \bar{b}.$$

Hence (8–14) can be written

$$\begin{aligned} \text{Max } z &= \bar{c}'\bar{x} \\ A\bar{x} &\leq \bar{b} \\ \bar{x} &\geq \bar{0} \end{aligned} \tag{8–16}$$

which is the primal problem.

From the foregoing results it should be clear that if we take the pair of problems

$$\begin{aligned} &\text{Max } z = \bar{c}'\bar{x}, \quad A\bar{x} \leq \bar{b}, \quad \bar{x} \geq \bar{0} \\ &\text{Min } w = \bar{b}'\bar{v}, \quad A'\bar{v} \geq c, \quad \bar{v} \geq \bar{0} \end{aligned}$$

as a primal-dual pair that we can equally well take

$$\begin{aligned} &\text{Min } z = \bar{c}'\bar{x}, \quad A\bar{x} \geq \bar{b}, \quad \bar{x} \geq \bar{0} \\ &\text{Max } w = \bar{b}'\bar{v}, \quad A'\bar{v} \leq \bar{c}, \quad \bar{v} \geq \bar{0} \end{aligned}$$

as a pair of primal and dual problems. There is complete symmetry in this mode of formulation.

There are a number of reasons why the relationship between the primal and dual problems is of interest. In the next section when some of the mathematical characteristics of the duality relationship are explored, it will be clear that there can be computational advantages to sometimes solving the dual of a problem rather than the original problem. In addition, the properties of duality have led to new and different algorithms for solving linear programming problems, i.e., different from the original simplex method. Also, theoretical considerations resulting from an awareness of duality have led to methods for solving certain kinds of nonlinear optimization problems. Finally, there are some economic interpretations of duality that are of interest and they will be discussed in a subsequent section of this chapter.

8–3 THE DUALITY THEOREM AND DUALITY PROPERTIES

We will now exhibit the fundamental relationships that exist between primal and dual problems. We standardize the notation at this point by reminding the reader that we will take as the primal problem

$$\begin{aligned} \text{Max } z &= \bar{c}'\bar{x} \\ A\bar{x} &\leq \bar{b} \\ \bar{x} &\geq \bar{0} \end{aligned} \tag{8–17}$$

and as the dual problem

$$\begin{aligned} \text{Min } w &= \bar{b}'\bar{v} \\ A'\bar{v} &\geq \bar{c} \\ \bar{v} &\geq \bar{0} \end{aligned} \tag{8–18}$$

We now prove the following important results.

Theorem 8–2 If $\bar{x}$ is any feasible solution to the primal problem and $\bar{v}$ is any feasible solution to the dual problem, then $\bar{c}'\bar{x} \leq \bar{b}'\bar{v}$

Proof: Since $\bar{x}$ is any feasible solution to the primal problem we know that

$$A\bar{x} \leq \bar{b} \tag{8–19}$$

$$\bar{x} \geq \bar{0} \tag{8–20}$$

Similarly since $\bar{v}$ is any feasible solution to the dual problem we know that

$$A'\bar{v} \geq \bar{c} \tag{8–21}$$

$$\bar{v} \geq \bar{0} \tag{8–22}$$

If we multiply (8–19) by $\bar{v} \geq \bar{0}$ (by (8–22)), we obtain

$$\bar{v}'A\bar{x} \leq \bar{v}'\bar{b} = \bar{b}'\bar{v} \tag{8–23}$$

Therefore, we have

$$\bar{v}'A\bar{x} \leq \bar{b}'\bar{v} \tag{8–24}$$

Now let us, using (8–20) and (8–21), multiply (8–21) by $\bar{x} \geq \bar{0}$. We obtain

$$\bar{x}'A'\bar{v} \geq \bar{x}'\bar{c} \tag{8–25}$$

Taking the transpose of (8–25) we have

$$\bar{v}'A\bar{x} \geq \bar{c}'\bar{x} \tag{8–26}$$

From (8–24) and (8–26) it follows that

$$\bar{c}'\bar{x} \leq \bar{b}'\bar{v} \tag{8–27}$$

Hence, we have shown that for the objective function z for the primal and the objective function w for the dual, it is always true that $z \leq w$.

Theorem 8–3 If $\bar{x}^*$ is a feasible solution to the primal problem and $\bar{v}^*$ is a feasible solution to the dual problem such that $\bar{c}'\bar{x}^* = \bar{b}'\bar{v}^*$, then both $\bar{x}^*$ and $\bar{v}^*$ are optimal solutions to their respective problems.

Proof: We are given that

$$\bar{c}'\bar{x}^* = \bar{b}'\bar{v}^* \tag{8–28}$$

However, by Theorem 8–2 for any feasible $\bar{x}$ and any feasible $\bar{v}$, such as $\bar{v}^*$ we know that

$$\bar{c}'\bar{x} \leq \bar{b}'\bar{v}^* \tag{8–29}$$

Combining (8–28) and (8–29) we see that

$$\bar{c}'\bar{x} \leq \bar{c}'\bar{x}^*$$

Therefore, $\bar{x}^*$ is an optimal solution to the primal. Similarly for a feasible $\bar{x}$ such as $\bar{x}^*$ and any feasible $\bar{v}$ we have that

$$\bar{c}'\bar{x}^* \leq \bar{b}'\bar{v} \tag{8–30}$$

Combining (8–28) and (8–30) we have

$$\bar{b}'\bar{v} \geq \bar{b}'\bar{v}^*$$

Therefore $\bar{v}^*$ is an optimal solution to the dual.

Finally, we prove the fundamental and most important theoretical result of duality in Theorem 8–4.

Theorem 8–4 A feasible solution $\bar{x}^*$ to the primal problem is optimal if and only if there exists a feasible solution $\bar{v}^*$ to the dual problem such that $\bar{c}'\bar{x}^* = \bar{b}'\bar{v}^*$.

Proof: We shall prove this theorem by starting with an optimal solution to the primal and constructing an optimal solution to the dual from the primal solution. This is sufficient because, by Theorem 8-1, the dual of the dual is the primal. Hence, we could just as well have started with the dual and converted it to the primal.

We start with the primal problem (8–17) and if we solved the primal problem by the simplex method we would have added a set of slack variables. Thus, we convert (8–17) to

$$\begin{aligned} \text{Max } z &= \bar{c}'x \\ A\bar{x} + I\bar{x}_s &= \bar{b} \\ \bar{x}, \bar{x}_s &\geq \bar{0} \end{aligned} \qquad (8\text{–}31)$$

where $\bar{x}_s$ is the vector of slack variables. Let $\bar{x}_B^*$ be the basic optimal feasible solution to (8–31) and B be its corresponding basis matrix. Since $\bar{x}_B^*$ is an optimal solution we know that

$$z_j - c_j \geq 0, \qquad \text{all } j \qquad (8\text{–}32)$$

Therefore, rearranging (8–32) in matrix form and using the definition of z, we have that

$$\bar{c}_B'B^{-1}(A, I) \geq (\bar{c}', \bar{0}') \qquad (8\text{–}33)$$

Let us now define $\bar{v}^{*\prime} = \bar{c}_B'B^{-1}$ and show that this is an optimal feasible solution to the dual. From $\bar{v}^{*\prime} = \bar{c}_B'B^{-1}$ and (8-33) we have that

$$\bar{v}^{*\prime}A \geq \bar{c}' \qquad (8\text{–}34)$$

$$\bar{v}^{*\prime} \geq \bar{0}' \qquad (8\text{–}35)$$

or rewriting it

$$\begin{aligned} A'\bar{v}^* &\geq \bar{c} \\ \bar{v}^* &\geq \bar{0} \end{aligned} \qquad (8\text{–}36)$$

Comparing (8–36) and (8–18) we see that $\bar{v}^*$ satisfies the constraints $A'\bar{v} \geq \bar{c}$, $\bar{v} \geq \bar{0}$.

Finally we must show that $\bar{v}^{*\prime} = \bar{c}_B'B^{-1}$ is an optimal solution to (8–18), the dual problem. By assumption $\bar{x}_B^*$ was optimal for the primal. Therefore

$$\text{Max } z = \bar{c}_B'\bar{x}_B^* = \bar{c}_B'B^{-1}\bar{b} = \bar{v}^{*\prime}\bar{b} = \bar{b}'\bar{v}^* = \text{Min } w \qquad (8\text{–}37)$$

Hence we have shown by an actual construction, that the optimal objective functions are equal if we use the definition of optimal dual solution we proposed, and the theorem is proved.

This last result is worthy of emphasis. We have shown by actual construction that if there is an optimal solution to the primal, then there is an optimal solution to the dual. Furthermore, we have shown how to calculate the optimal dual solution from the optimal primal basis and the prices corresponding to the vectors in the primal basis. We have also shown that the optimal values of the objective functions to the primal and dual problems are equal, i.e.,

$$\text{Max } z = z^* = w^* = \text{Min } w$$

The duality theory we have discussed thus far can sometimes be of assistance in actual computation. If a given linear programming problem contained many more constraints than variables, the basis size would be equal to the number of constraints. In such a case, it is desirable to solve the dual whose basis size would be equal to the number of variables in the primal. From the dual variables we could construct the solution to the primal. Let us consider an example:

EXAMPLE 2

Suppose we wish to solve

$$\text{Max } z = 3x_1 + 4x_2$$

subject to

$$\begin{aligned} x_1 + x_2 &\leq 10 \\ 2x_1 + 3x_2 &\leq 18 \\ x_1 &\leq 8 \\ x_2 &\leq 6 \\ x_1, x_2 &\geq 0 \end{aligned} \tag{8–38}$$

Since there are 4 constraints and 2 variables it is reasonable to solve the dual of (8–38) which will have only 2 constraints. It is given by

$$\text{Min } w = 10v_1 + 18v_2 + 8v_3 + 6v_4$$

subject to

$$\begin{aligned} v_1 + 2v_2 + v_3 &\geq 3 \\ v_1 + 3v_2 + v_4 &\geq 4 \\ v_1, v_2, v_3, v_4 &\geq 0 \end{aligned} \tag{8–39}$$

			10	18	8	6	0	0
$\bar{c}_B$	Basis	$\bar{v}_B$	$\bar{y}_1$	$\bar{y}_2$	$\bar{y}_3$	$\bar{y}_4$	$\bar{y}_5$	$\bar{y}_6$
8	$\bar{a}_3$	3	1	2	1	0	−1	0
6	$\bar{a}_4$	4	1	③	0	1	0	−1
		48	4	16	0	0	−8	−6

Figure 8–1 Tableau I.

If we add surplus variables to (8–39) we obtain

$$\begin{aligned} \text{Min } w &= 10v_1 + 18v_2 + 8v_3 + 6v_4 \\ v_1 + 2v_2 + v_3 - v_5 &= 3 \\ v_1 + 3v_2 + v_4 - v_6 &= 4 \\ v_j \geq 0, \quad j &= 1, 2, \ldots, 6 \end{aligned} \tag{8–40}$$

The initial tableau for this problem, utilizing the fact that it already contains an identity matrix with basis $B = (\bar{b}_1, \bar{b}_2) = (\bar{a}_3, \bar{a}_4)$, is given in Tableau I in Figure 8.1. From Tableau I we see that $\bar{a}_2$ will enter the basis and $\bar{a}_4$ will leave. The transformed tableau is given in Figure 8–2.

Since, in Tableau II, all $z_j - c_j \leq 0$ (this is a minimization problem), we have obtained the optimal solution. The optimal solution is

$$w^* = \tfrac{80}{3}, \qquad v_1^* = 0, \qquad v_2^* = \tfrac{4}{3}, \qquad v_3^* = \tfrac{1}{3}, \qquad v_4^* = v_5^* = v_6^* = 0$$

$$\bar{c}_B = [8, 18] \qquad B = \begin{bmatrix} 1 & 2 \\ 0 & 3 \end{bmatrix} \quad \text{and} \quad B^{-1} = \begin{bmatrix} 1 & -\frac{2}{3} \\ 0 & \frac{1}{3} \end{bmatrix}$$

Therefore,

$$\bar{x}^{*\prime} = \bar{c}'_B B^{-1} = [8, 18]\begin{bmatrix} 1 & -\frac{2}{3} \\ 0 & \frac{1}{3} \end{bmatrix} = [8, \tfrac{2}{3}]$$

We see that $x_1^* = 8$, $x_2^* = \frac{2}{3}$ is the optimal solution to the original problem

			10	18	8	6	0	0
$\bar{c}_B$	Basis	$\bar{v}_B$	$\bar{y}_1$	$\bar{y}_2$	$\bar{y}_3$	$\bar{y}_4$	$\bar{y}_5$	$\bar{y}_6$
8	$\bar{a}_3$	$\frac{1}{3}$	$\frac{1}{3}$	0	1	$-\frac{2}{3}$	−1	$\frac{2}{3}$
18	$\bar{a}_2$	$\frac{4}{3}$	$\frac{1}{3}$	1	0	$\frac{1}{3}$	0	$-\frac{1}{3}$
		$\frac{80}{3}$	$-\frac{4}{3}$	0	0	$-\frac{16}{3}$	−8	$-\frac{2}{3}$

Figure 8–2 Tableau II.

because

$$z^* = 3(8) + 4(\tfrac{2}{3}) = 24 + \tfrac{8}{3} = \tfrac{80}{3}$$

and

$$\begin{aligned} 8 + \tfrac{2}{3} &\leq 10 \\ 16 + 2 &\leq 18 \\ 8 &\leq 8 \\ \tfrac{2}{3} &\leq 6 \end{aligned}$$

By Theorem 8–4, a feasible solution such that

$$z = \text{Min } w$$

must be the optimal solution.

8–4 ADDITIONAL DUALITY RELATIONSHIPS

We showed earlier, in Section 8–1, that any formulation of a linear programming problem, regardless of the mixture of constraints, could be put in the form we assumed, namely (8–1), and hence its dual could be written down by inspection. However, it would be more convenient if we could write the dual of any linear programming problem without first having to put it into the "standard" symmetric form we have assumed. Let us consider how to generalize our duality results.

A natural question would concern what the dual of the standard linear programming statement would be, i.e., what is the dual of

$$\begin{aligned} \text{Max } z &= \bar{c}'\bar{x} \\ A\bar{x} &= \bar{b} \\ \bar{x} &\geq \bar{0} \end{aligned} \tag{8–41}$$

We show this in the following theorem.

Theorem 8–5 A linear programming problem in the form

$$\begin{aligned} \text{Max } z &= \bar{c}'\bar{x} \\ A\bar{x} &= \bar{b} \\ \bar{x} &\geq \bar{0} \end{aligned} \tag{8–42}$$

has a corresponding dual problem given by

$$\begin{aligned} \text{Min } w &= \bar{b}'\bar{v} \\ A'\bar{v} &\geq \bar{c} \end{aligned} \tag{8–43}$$

where $\bar{v}$ is a vector of variables unrestricted in sign.

Proof: We begin by transforming (8–42) into the form of our previous results. We do this in the usual way by representing each equality as two equivalent inequalities. Hence (8–42) is equivalent to

$$\begin{aligned} \text{Max } z &= \bar{c}'\bar{x} \\ A\bar{x} &\leq \bar{b} \\ -A\bar{x} &\leq -\bar{b} \\ \bar{x} &\geq \bar{0} \end{aligned} \tag{8–44}$$

If we associate dual variables $\bar{v}_+$ with constraints $A\bar{x} \leq \bar{b}$ and dual variables $\bar{v}_-$ with constraints $-A\bar{x} \leq -\bar{b}$ then the dual of (8–44) is clearly

$$\begin{aligned} \text{Min } w &= \bar{b}'\bar{v}_+ - \bar{b}'\bar{v}_- \\ A'\bar{v}_+ - A'\bar{v}_- &\geq \bar{c} \\ \bar{v}_+, \bar{v}_- &\geq \bar{0} \end{aligned} \tag{8–45}$$

where, if we imagine the matrix $\hat{A}$ of (8–44) was partitioned into $\begin{bmatrix} A \\ -A \end{bmatrix} = \hat{A}$, $\hat{A}' = [A', -A']$ and $\hat{\bar{v}} = [\bar{v}_+, \bar{v}_-]$ and we get the result (8–45). However, (8–45) can be simplified. Since both $\bar{v}_+$ and $\bar{v}_-$ are required to be non-negative, it is obvious that if we define a new vector of variables

$$\bar{v} = \bar{v}_+ - \bar{v}_- \tag{8–46}$$

then $\bar{v}$ is an unrestricted set of variables since the difference given by (8–46) can be positive, negative or zero. Applying (8–46) to (8–45) we obtain

$$\begin{aligned} \text{Min } w &= \bar{b}'\bar{v} \\ A'\bar{v} &\geq \bar{c} \end{aligned} \tag{8–47}$$

and we see that we have proved the theorem.

What we have shown in Theorem 8–5 is that the dual of the standard linear programming problem with equality constraints leads to a dual with unrestricted variables. Hence, an equality in the primal gives rise to unrestricted variables in the dual. Since the dual of the dual is the primal, this strongly suggests the converse, i.e., that if there were any unrestricted variables in the primal problem, this would lead to equality constraints in the dual problem. Let us now prove this.

Theorem 8–6 A linear programming problem in the form

$$\begin{aligned} \text{Max } z &= \bar{c}'\bar{x} \\ A\bar{x} &\leq \bar{b} \end{aligned} \tag{8–48}$$

has a corresponding dual problem given by

$$\begin{aligned} \text{Min } w &= \bar{b}'\bar{v} \\ A'\bar{v} &= \bar{c} \\ \bar{v} &\geq \bar{0} \end{aligned} \tag{8–49}$$

Proof: Let $\bar{x} = \bar{x}_+ - \bar{x}_-$ which is merely a transformation of variables such that $\bar{x}_+, \bar{x}_- \geq \bar{0}$. If we substitute for $\bar{x}$ into (8–48) we then obtain

$$\begin{aligned} \text{Max } z &= \bar{c}'\bar{x}_+ - \bar{c}'\bar{x}_- \\ A\bar{x}_+ - A\bar{x}_- &\leq \bar{b} \\ x_+, \bar{x}_- &\geq \bar{0} \end{aligned} \tag{8–50}$$

Since we could write $\hat{x} = [\bar{x}_+, \bar{x}_-]$ and $\hat{A} = [A, -A]$ it is clear that (8–50) is in the usual form of the primal, and its dual is, quite clearly

$$\begin{aligned} \text{Min } w &= \bar{b}'\bar{v} \\ A'\bar{v} &\geq \bar{c} \\ -A'\bar{v} &\geq -\bar{c} \\ \bar{v} &\geq \bar{0} \end{aligned} \tag{8–51}$$

Finally, we note that the constraints $A'\bar{v} \geq c$ and $-A'\bar{v} \geq -\bar{c}$ are equivalent to $A'\bar{v} = \bar{c}$. We now can write (8–51) as

$$\begin{aligned} \text{Min } w &= \bar{b}'\bar{v} \\ A'\bar{v} &= \bar{c} \\ \bar{v} &\geq \bar{0} \end{aligned} \tag{8–52}$$

and hence we have proved our result.

It should be clear that there is nothing in the proofs of Theorems 8–5 and 8–6 that could not be used if both equality and unrestricted variables were present in the same problem. We simply state the theorem. Its proof should be quite obvious by now.

Theorem 8–7 A linear programming problem in the form

$$\begin{aligned} \text{Max } z &= \bar{c}'\bar{x} \\ A\bar{x} &= \bar{b} \end{aligned}$$

has a corresponding dual problem given by

$$\begin{aligned} \text{Min } w &= \bar{b}'\bar{v} \\ A'\bar{v} &= \bar{c} \end{aligned}$$

One might suspect that if only some of the primal constraints were equalities, then only the dual variables corresponding to those equality constraints would be unrestricted but the remaining dual variables would be non-negative. This is indeed the case. Since the dual of the dual is the primal, all of these results are true in the other direction; that is, from the dual to the primal problem.

By reapplying in the appropriate places the same arguments used to prove Theorems 8–5 and 8–6 we can prove the following theorem, which combines all our duality relationships into its most general form. We leave the proof as an exercise for the reader.

Theorem 8–8 If the primal problem is given by

$$\text{Max } z = c_1x_1 + c_2x_2 + \ldots + c_nx_n$$

subject to:

$$\begin{array}{l} a_{11}x_1 + a_{12}x_2 + \ldots + a_{1n}x_n \leq b_1 \\ \quad\vdots \\ a_{s1}x_1 + a_{s2}x_2 + \ldots + a_{sn}x_n \leq b_s \\ a_{s+1,1}x_1 + a_{s+1,2}x_2 + \ldots + a_{s+1,n}x_n = b_{s+1} \\ \quad\vdots \\ a_{m1}x_1 + a_{m2}x_2 + \ldots + a_{mn}x_n = b_m \\ x_j \geq 0, \quad j = 1, \ldots, t \\ x_j \text{ unrestricted}, \quad j = t+1, \ldots, n \end{array} \tag{8–53}$$

then the corresponding dual problem is given by

$$\text{Min } w = b_1v_1 + b_2v_2 + \ldots + b_mv_m$$

subject to:

$$\begin{array}{l} a_{11}v_1 + a_{21}v_2 + \ldots + a_{m1}v_m \geq c_1 \\ \quad\vdots \\ a_{1t}v_1 + a_{2t}v_2 + \ldots + a_{mt}v_m \geq c_t \\ a_{1,t+1}v_1 + a_{2,t+1}v_2 + \ldots + a_{m,t+1}v_m = c_{t+1} \\ \quad\vdots \\ a_{1n}v_1 + a_{2n}v_2 + \ldots + a_{mn}v_m = c_n \\ v_i \geq 0, \quad i = 1, \ldots, s \\ v_i \text{ unrestricted}, \quad i = s+1, \ldots, m \end{array} \tag{8–54}$$

and in addition the dual of the dual is the primal, i.e., the dual of (8–54) is given by (8–53).

Let us emphasize what is shown in Theorem 8–8. If the primal is a maximization problem, each inequality ($\leq$)† has a non-negative dual variable associated with it, and each equality constraint has an unrestricted dual variable associated with it. On the other hand, if the primal is taken to be a minimization problem, each inequality ($\geq$) has a non-negative dual variable associated with it, and each equality constraint has an unrestricted dual variable associated with it.

To illustrate the direct use of Theorem 8–8 consider the following examples.

EXAMPLE 3

If the primal problem is

$$\begin{aligned} \text{Max } z &= 2x_1 + 3x_2 \\ x_1 + 3x_2 + 4x_3 &= 20 \\ 2x_1 - 4x_2 - x_3 &\leq 10 \\ x_2, x_3 &\geq 0 \end{aligned}$$

Then the dual problem is given by

$$\begin{aligned} \text{Min } w &= 20v_1 + 10v_2 \\ v_1 + 2v_2 &= 2 \\ 3v_1 + 4v_2 &\geq 3 \\ 4v_1 - v_2 &\geq 0 \\ v_2 &\geq 0 \end{aligned}$$

EXAMPLE 4

If the primal problem is

$$\begin{aligned} \text{Min } z &= 5x_1 - 10x_2 + 3x_3 \\ x_1 + 2x_2 - 4x_3 &\geq 25 \\ 2x_1 - x_2 + 3x_3 &= 30 \\ x_1, x_3 &\geq 0 \end{aligned}$$

† If a maximization problem contains a $\geq$ inequality it can be multiplied by -1. It can actually be shown that such a constraint, if it is not multiplied by -1, will give rise to a non-positive dual variable, but these are rarely of any significance in practical problems and will seldom occur in practice.

then the dual problem is given by

$$\begin{aligned} \text{Max } w &= 25v_1 + 30v_2 \\ v_1 + 2v_2 &\leq 5 \\ 2v_1 - \quad v_2 &= -10 \\ -4v_1 + 3v_2 &\leq 3 \\ v_1 &\geq 0 \end{aligned}$$

8–5 DUALITY AND UNBOUNDEDNESS

In the relationships we have discussed in the preceding sections of this chapter, we have discussed feasible and optimal solutions of primal-dual pairs. However, we have not yet considered the consequences to duality relationships of the failure of a feasible solution to exist or the existence of unboundedness in either of the problems. Neither of these situations should occur in practice. Nevertheless, they are of some theoretical significance and have implications in certain computational algorithms. The principal theoretical interest in these results is that the first of these might be considered a central existence theorem for linear programming. Its proof follows directly from the nature of duality and what we have already proved.

Theorem 8–9 A linear programming problem has a finite optimal solution if and only if there exist feasible solutions to both the given problem and its dual.

Proof: First, we note that if the primal has an optimal feasible solution, then by Theorem 8–4 we are guaranteed a feasible solution to the dual.

Now we consider the reverse case. If we assume that $\bar{x}^0$ and $\bar{v}^0$ are feasible solutions to the primal and dual, respectively, then it is clear that $\bar{c}'\bar{x}^0$ and $\bar{b}'\bar{v}^0$ are finite. However, we note further by Theorem 8–2 that for any pair of feasible solutions $\bar{x}$, $\bar{v}$ that $\bar{c}'\bar{x} \leq \bar{b}'\bar{v}$. In particular, since $\bar{b}'\bar{v}^0$ is finite, then the optimal solution to the primal $\bar{c}'\bar{x}^* \leq \bar{b}'\bar{v}^0$. Hence, the primal must have a finite optimum, since it has a finite bound.

Let us now consider what the consequence is of *unboundedness* in the primal. The result is easily shown in the following.

Theorem 8–10 If the primal problem has an unbounded objective function then the dual problem has no feasible solution.

Proof: If the primal is unbounded, it obviously has a feasible solution. Now, suppose the dual also had a feasible solution. Then, by Theorem 8–9, the primal would have a finite optimal solution, i.e., it would not be unbounded, which is a contradiction. Hence, the dual cannot have a feasible solution if the primal is unbounded.

If the dual has no feasible solution, the primal may be either unbounded or it may have no feasible solution. We consider this in the following theorem.

Theorem 8–11 If the dual problem has no feasible solution and the primal problem has a feasible solution, then the primal objective function is unbounded.

Proof: Consider any feasible solution to the primal problem, $\bar{x}^0$. It is clear that $\bar{x}^0$ cannot be an optimal solution to the primal because if it were, by Theorem 8–4, a feasible solution to the dual problem would have to exist which contradicts our assumption. From this it follows that no feasible solution to the primal can be optimal, since $\bar{x}^0$ is any feasible solution. If no feasible solution to the primal can be optimal, then the primal problem must have an unbounded objective function.

We can summarize the content of these theorems in the diagram shown in Figure 8–3.

We have previously seen examples of problems where both primal and dual problems have optimal solutions. Let us consider the other cases.

Primal problem	Dual problem: Feasible solution	Dual problem: No feasible solution
Feasible solution	Both optimal solutions exist and are finite	Primal objective function is unbounded
No feasible solution	Dual objective function is unbounded	No solutions exist

Figure 8–3 Duality and unboundedness.

EXAMPLE 5: PRIMAL UNBOUNDEDNESS AND DUAL INFEASIBILITY

Consider the primal problem

$$\begin{aligned} \text{Max } z &= 2x_1 + 3x_2 \\ 3x_1 - 2x_2 &\leq 5 \\ -x_1 - x_2 &\leq -1 \\ x_1, x_2 &\leq 0 \end{aligned}$$

It should be obvious that if $x_1 = 0$ then as x_2 increases without limit the constraints remain satisfied and the objective function is unbounded. Now let us consider the dual of this problem. It is

$$\begin{aligned} \text{Min } w &= 5v_1 - v_2 \\ 3v_1 - v_2 &\geq 2 \\ -2v_1 - v_2 &\geq 3 \\ v_1, v_2 &\geq 0 \end{aligned}$$

It is clear that no values of $v_1, v_2 \geq 0$ will satisfy the second constraint of the dual problem. Hence no feasible solution exists.

EXAMPLE 6: PRIMAL AND DUAL INFEASIBILITY

Consider the primal problem

$$\begin{aligned} \text{Max } z &= 5x_1 + 3x_2 \\ x_1 - x_2 &\leq -4 \\ -x_1 + x_2 &\leq -2 \\ x_1, x_2 &\geq 0 \end{aligned}$$

This problem has no feasible solution since the second constraint is equivalent to (multiply by -1) $x_1 - x_2 \geq 2$, which clearly contradicts the first constraint. Now consider the dual problem

$$\begin{aligned} \text{Min } w &= -4v_1 - 2v_2 \\ v_1 - v_2 &\geq 5 \\ -v_1 + v_2 &\geq 3 \\ v_1, v_2 &\geq 0 \end{aligned}$$

It is clear that this problem has no feasible solution either since multiplying the second constraint by -1 gives $v_1 - v_2 \leq -3$ which contradicts the first constraint. Hence neither problem has a feasible solution.

8–6 COMPLEMENTARY SLACKNESS

We will now consider some fundamental relationships which grow out of the theory of duality. These results, while interesting in their own right, also have important economic interpretations and also have a bearing on a different method for solving a linear programming problem which we shall discuss in the next chapter.

We shall state these results in the form of two theorems. The first is

Theorem 8–12 For any pair of optimal solutions to a linear programming problem and its associated dual,

1. For each $j = 1, 2, \ldots, n$, the product of the jth primal variable and the jth dual surplus variable is zero.

2. For each $i = 1, 2, \ldots, m$, the product of the ith dual variable and the ith primal slack variable is zero.

Proof: Let us consider the usual form of the primal problem in pure inequality form. This is given in (8–55) as

$$\begin{aligned} \text{Max } z &= \bar{c}'\bar{x} \\ A\bar{x} &\leq \bar{b} \\ \bar{x} &\geq \bar{0} \end{aligned} \tag{8–55}$$

If we add one slack variable to each inequality we obtain the following equivalent problem

$$\begin{aligned} \text{Max } z &= \bar{c}'\bar{x} \\ A\bar{x} + \bar{x}_s &= \bar{b} \\ \bar{x}, \bar{x}_s &\geq \bar{0} \end{aligned} \tag{8–56}$$

In a similar fashion the dual of (8–55)

$$\begin{aligned} \text{Min } w &= \bar{b}'\bar{v} \\ A'\bar{v} &\geq \bar{c} \\ \bar{v} &\geq \bar{0} \end{aligned} \tag{8–57}$$

can be converted, by subtracting a surplus variable from each inequality, to the equivalent form

$$\begin{aligned} \text{Min } w &= \bar{b}'\bar{v} \\ A'\bar{v} - \bar{v}_s &= \bar{c} \\ \bar{v}, \bar{v}_s &\geq \bar{0} \end{aligned} \tag{8–58}$$

We now consider the constraints of (8–56) and multiply them by $\bar{v}'$ to obtain

$$\bar{v}'A\bar{x} + \bar{v}'\bar{x}_s = \bar{v}'\bar{b} \tag{8–59}$$

Since each of the terms in (8–59) are scalars, we may rewrite (8–59) as

$$\bar{x}'A'\bar{v} + \bar{v}'\bar{x}_s = \bar{b}'\bar{v} \tag{8–60}$$

In a similar fashion we may multiply the constraints of (8–58) by $\bar{x}'$ to obtain

$$\bar{x}'A'\bar{v} - \bar{x}'\bar{v}_s = \bar{x}'\bar{c} \tag{8–61}$$

Again (8–61) may be conveniently rewritten as

$$\bar{x}'A'\bar{v} - \bar{x}'\bar{v}_s = \bar{c}'\bar{x} \tag{8–62}$$

Equations (8–60) and (8–62) are valid for any feasible solutions $(\bar{x}, \bar{x}_s)$ and $(\bar{v}, \bar{v}_s)$.

Let us now assume that problems (8–56) and (8–58), our primal and dual problems, have optimal solutions $(\bar{x}^*, \bar{x}_s^*)$ and $(\bar{v}^*, \bar{v}_s^*)$. Recalling that the cost coefficients of slack and surplus variables are zero, since $(\bar{x}^*, \bar{x}_s^*)$ and $(\bar{v}^*, \bar{v}_s^*)$ are optimal, we know by Theorem 8–4 that

$$\bar{c}'\bar{x}^* = \bar{b}'\bar{v}^* \tag{8–63}$$

In addition, since an optimal solution is feasible, equations (8–60) and (8–62) must also hold for them. Hence we may write

$$\bar{x}^{*\prime}A'\bar{v}^* + \bar{v}^{*\prime}\bar{x}_s^* = \bar{b}'\bar{v}^* \tag{8–64}$$

$$\bar{x}^{*\prime}A'\bar{v}^* - \bar{x}^{*\prime}\bar{v}_s^* = \bar{c}'\bar{x}^* \tag{8–65}$$

Since $\bar{c}'\bar{x}^* = \bar{b}'\bar{v}^*$ by (8–63) if we substitute (8–63) into (8–64) and subtract (8–65) from (8–64) we obtain

$$\bar{v}^{*\prime}\bar{x}_s^* + \bar{x}^{*\prime}\bar{v}_s^* = 0 \tag{8–66}$$

If we recall that $\bar{x}, \bar{v}, \bar{x}_s, \bar{v}_s \geq 0$, we can conclude from (8–66) that

$$\begin{aligned} v_i^* x_{si}^* &= 0 \qquad i = 1, 2, \ldots, m \\ x_j^* v_{sj}^* &= 0 \qquad j = 1, 2, \ldots, n \end{aligned} \tag{8–67}$$

which is what we wished to prove.

The results embodied in (8–66) and (8–67) are usually referred to as the principle of *complementary slackness.* The name follows from a simple economic interpretation which says that if a primal slack variable is positive (some resource not fully utilized), to use more of it will not improve the objective function. Hence its marginal value, which—as we will see in the next section—corresponds to its dual variable, must be zero.

It should be noted that if a primal slack variable is zero it is not *necessary* that the corresponding dual variable be positive. They may both be zero, because the principle of complementary slackness merely says that the product must be zero.

There is an interesting relationship between complementary slackness and optimal solutions to linear programming problems that is the basis for an algorithm, called the primal-dual algorithm, for solving linear programming problems. Let us now prove this simple but useful result.

Theorem 8–13 If $(\hat{\bar{x}}, \hat{\bar{x}}_s)$ and $(\hat{\bar{v}}, \hat{\bar{v}}_s)$ are feasible solutions to the primal and dual problems under conditions where complementary slackness holds, then $(\hat{\bar{x}}, \hat{\bar{x}}_s)$ and $(\hat{\bar{v}}, \hat{\bar{v}}_s)$ are also optimal solutions to the primal and dual problems.

Proof: Since we are given that complementary slackness holds, we know that

$$\hat{\bar{v}}'\hat{\bar{x}}_s + \hat{\bar{x}}'\hat{\bar{v}}_s = 0 \tag{8–68}$$

from which we obtain

$$\hat{\bar{v}}'\hat{\bar{x}}_s = -\hat{\bar{x}}'\hat{\bar{v}}_s = -\hat{\bar{v}}_s'\hat{\bar{x}} \tag{8–69}$$

If we add $\hat{\bar{v}}'A\hat{\bar{x}}$ to both sides of (8–69) we have

$$\hat{\bar{v}}'A\hat{\bar{x}} + \hat{\bar{v}}'\hat{\bar{x}}_s = \hat{\bar{v}}'A\hat{\bar{x}} - \hat{\bar{v}}_s'\hat{\bar{x}} = \hat{\bar{x}}'A'\hat{\bar{v}} - \hat{\bar{x}}'\hat{\bar{v}}_s \tag{8–70}$$

We can factor (8–70) to obtain

$$\hat{\bar{v}}'(A\hat{\bar{x}} + \hat{\bar{x}}_s) = \hat{\bar{x}}'(A'\hat{\bar{v}} - \hat{\bar{v}}_s) \tag{8–71}$$

Since $(\hat{\bar{x}}, \hat{\bar{x}}_s)$ is a feasible solution it must satisfy the constraints

$$A\hat{\bar{x}} + \hat{\bar{x}}_s = \bar{b} \tag{8–72}$$

Similarly $(\hat{\bar{v}}, \hat{\bar{v}}_s)$ must satisfy the dual constraints

$$A'\hat{\bar{v}} - \hat{\bar{v}}_s = \bar{c} \tag{8–73}$$

Substituting from (8–72) and (8–73) into (8–70) we obtain:

$$\hat{\bar{v}}'\bar{b} = \hat{\bar{x}}'c$$

or

$$\bar{c}'\hat{\bar{x}} = \bar{b}'\hat{\bar{v}} \tag{8–74}$$

By Theorem 8–3, a pair of feasible solutions such that the primal and dual objective functions are equal, are also optimal. Hence we have proved the theorem.

8–7 ECONOMIC INTERPRETATION OF DUALITY

Suppose we have a basic feasible solution, $\hat{\bar{x}}$, to the linear programming problem

$$\begin{aligned} \text{Max } z &= \bar{c}'\bar{x} \\ A\bar{x} &= \bar{b} \\ \bar{x} &\geq \bar{0} \end{aligned} \tag{8–75}$$

Associated with this solution is a basis matrix, $\hat{B} = (\hat{\bar{b}}_1, \hat{\bar{b}}_2, \ldots, \hat{\bar{b}}_m)$. As we know, any vector in E^m can be expressed in terms of $\hat{B}$. In particular, we could express a unit vector $\bar{e}_l$ (zeroes in all positions except a "1" in the lth position) in terms of the $\hat{\bar{b}}_i$, i.e.,

$$\bar{e}_l = \sum_{i=1}^{m} \lambda_{il}\hat{\bar{b}}_i = \hat{B}\bar{\lambda}_l \tag{8–76}$$

Since $\bar{e}_l$ is a linear combination of the basis vectors, the set of values λ_{il} represents the change in $\hat{x}_{Bi}$ which would occur if the total amount of the lth resource is changed by one unit. We see this since $B\hat{\bar{x}}_B = \bar{b}$. Hence, if we are given any feasible solution $\hat{\bar{x}}$ to the primal problem, the *marginal* unit of the lth resource is given by λ_{1l} units of $\hat{x}_{B1}$, λ_{2l} units of $\hat{x}_{B2}$, etc. Hence the vector $\bar{\lambda}_l$ represents the value of the lth resource expressed in terms of the resulting change in activity levels caused by the marginal unit of the lth resource. Let us now define a quantity which represents the revenue or income generated by the marginal unit of the lth resource. If we call this marginal revenue R_l, then

$$R_l = \sum_{i=1}^{m} c_{Bi}\lambda_{il} = \bar{c}_B'\bar{\lambda}_l \tag{8–77}$$

We will now show that R_l is simply the lth dual variable of our problem, i.e., the v_l of the dual of (8–75).

Let us multiply (8–76) by $\hat{B}^{-1}$. If we do so, we obtain

$$\bar{\lambda}_l = \hat{B}^{-1}\bar{e}_l = \hat{\bar{\beta}}_l \tag{8–78}$$

where $\hat{\beta}_l$ is the lth column of $\hat{B}^{-1}$. Therefore, from (8–77) and (8–78) we have that

$$R_l = \bar{c}_B'\bar{\lambda}_l = \bar{c}_B'\hat{\beta}_l \tag{8–79}$$

It will be recalled from the proof of Theorem 8–4 that $\bar{v}' = \bar{c}_B'B^{-1}$. Hence $R_l = \hat{v}_l$, the lth dual variable. Therefore, a dual variable (also called *shadow price*) v_l for a given primal feasible solution represents an implicit value (also called *opportunity cost*) of the marginal unit of the lth resource, if we actually implement the current solution.

As we noted in the proof of Theorem 8–4, there is a close connection between dual variables and the $z_j - c_j$ for the primal problem solution. This is also closely connected to complementary slackness. Let us recall that the value of $z_j - c_j$ corresponding to any basis variable in the primal solution is zero. Hence if the lth slack vector is in the basis, $z_{n+l} = \bar{c}_B'B^{-1}\bar{a}_{n+l} = 0$ (since $c_{n+l} = 0$ for slack variables). In economic terms we are saying that the economic value of a resource is zero if part of it remains unused. We see that $z_{n+l} = v_l$ for a slack vector in the basis as follows. If

$$z_{n+l} = \bar{c}_B'B^{-1}\bar{a}_{n+l} \tag{8–80}$$

and $\bar{a}_{n+l} = \bar{e}_l$, then

$$z_{n+l} = \bar{c}_B'B^{-1}\bar{e}_l = \bar{v}'\bar{e}_l = v_l \tag{8–81}$$

In simple words, we can now state the economic interpretation of primal and dual problems.

The Primal Problem: If c_j is the value of each unit of output, and b_i is the upper bound on the availability of each input resource, how much of each output (x_j) should be produced in order to maximize the total value of the output?

The Dual Problem: If b_i is the given availability of each input and c_j is the lower bound on the unit value of each output, what unit values should be assigned to each input (v_i) in order to minimize the total value of the input?

The primal problem is the problem a manufacturer or producer faces. He wishes to maximize the value of his production or his profit. The dual problem has a slightly different motivation. It is the problem faced, say, by an accountant who needs to determine a "value" to each input for replacement or insurance, for example. He wishes to determine his input valuations so as to minimize the total cost to him. Another example might be an investor who is contemplating purchasing the resources of our manufacturer. He needs to determine a total amount to offer the manufacturer. One way of doing this is to solve the dual problem.

In terms of our previous discussion the meaning and value of shadow prices and marginal values should now be clear. Suppose the value of the

i^{th} dual variable in the optimal solution is v_i^*. This means that if the right hand side b_i of the ith primal constraint were increased by some small amount ε, then the optimal value of the primal objective function would be increased by εv_i^*. What this means for our primal problem manufacturer is that, if a very small additional amount of the ith input were available, then an extra εv_i^* dollars of income could be achieved by the manufacturer. In other words, v_i^* is the *maximum price* that the manufacturer should be willing to pay for an additional unit of the ith input, say, if he had to buy this resource on the open market. This concept gives rise to the term *shadow price* or *marginal value* of the input.

The reader who is particularly interested in all the economic ramifications of the theory of duality should consult reference 1 at the end of this chapter. Very detailed analyses of these matters are discussed there.

PROBLEMS

1. Formulate the dual of the following problem.

$$\begin{aligned} \text{Max } z = 2x_1 + 3x_2 + 4x_3 \\ 2x_1 + 5x_2 + 4x_3 \leq 10 \\ 3x_1 + 7x_2 + x_3 \leq 15 \\ x_1 + 2x_2 + 5x_3 \leq 20 \\ x_1, x_2, x_3 \geq 0 \end{aligned}$$

2. Formulate the dual of the following problem.

$$\begin{aligned} \text{Min } z = 3x_1 - 2x_2 + 4x_3 \\ 2x_1 - 3x_2 + 4x_3 \leq 25 \\ x_1 + 4x_2 + 2x_3 \geq 3 \\ x_1, x_2, x_3 \geq 0 \end{aligned}$$

3. Solve the following linear programming problem by formulating and solving the dual.

$$\begin{aligned} \text{Min } z = 5x_1 + 3x_2 \\ 2x_1 + x_2 \geq 20 \\ x_1 + 3x_2 \geq 15 \\ x_1 \qquad \geq 2 \\ x_2 \geq 4 \\ x_1, x_2 \geq 0 \end{aligned}$$

4. Formulate the dual of the following problem.

$$\begin{aligned} \text{Max } z = 3x_1 + 4x_2 + 5x_3 \\ x_1 + 7x_2 + 8x_3 = 210 \\ 10x_1 + 2x_2 + 3x_3 \leq 150 \\ 3x_1 + 2x_2 \qquad \geq 40 \\ x_1, x_3 \geq 0 \end{aligned}$$

5. Formulate the dual of the following problem.

$$\begin{aligned} \text{Min } z &= 2x_1 + 3x_2 \\ 10x_1 - 10x_2 &\leq 31 \\ 2x_1 + x_2 &= 14 \\ 5x_1 + 7x_2 &\geq 8 \\ x_1 &\leq 10 \\ x_2 &\geq 0 \end{aligned}$$

6. Formulate the dual of the following problem and solve geometrically.

$$\begin{aligned} \text{Max } z &= 4x_1 + 3x_2 \\ 2x_1 + x_2 &\geq 1 \\ x_1 + 2x_2 &\leq 8 \\ 2x_1 + 4x_2 &\geq 11 \\ x_2 &\leq 5 \\ x_1, x_2 &\geq 0 \end{aligned}$$

7. Construct an example of a primal-dual pair neither of which has a solution.

8. Prove Theorem 8–8.

REFERENCE

1. Dorfman, R., P. A. Samuelson, and R. M. Solow: *Linear Programming and Economic Analysis.* New York, McGraw-Hill Book Company, 1958.

9

The Dual Simplex and Primal-Dual Algorithms

9–1 INTRODUCTION

In this chapter we shall consider two different algorithms for solving the general linear programming problem, i.e., different from the standard or revised simplex method, which we will now refer to as the *primal simplex method*.

The general nature of the primal simplex method, as well as the two algorithms we shall consider, is such that they fit into a basic framework or scheme which can be described briefly as *adjacent extreme point*, or sometimes as *pivoting*, *algorithms*. What this means is that the algorithm can be described in the following general terms:

1. Select a vector to enter (leave) the basis. The entering and leaving vectors correspond to adjacent extreme points of the convex set of feasible solutions.

2. Select a vector to leave (enter) the basis. The leaving and entering vectors correspond to adjacent extreme points of the convex set of feasible solutions.

3. The above steps are performed to insure both primal (dual) feasibility and a monotone increase (decrease) in the objective function. The new solution is then calculated by appropriate transformation formulae.

There are many conceivable variants of how the above sequence of steps can be carried out. The dual simplex method is a kind of mirror-image of the primal simplex method with respect to *feasibility*. The primal-dual algorithm uses features of both methods in a sense, in a way which will become apparent as the details are unfolded.

Neither the dual simplex method nor the primal-dual method is in general use for solving general linear programming problems. However, the dual simplex method is part of any large linear programming system (see Chapter

7) for it is extremely useful in conducting post optimality studies (see Chapter 12). The primal-dual algorithm has been used in a special form as a network flow algorithm for solving the transportation problem (see Chapter 14). Hence, both these algorithms have special and important practical applications. The matter of whether they should be used as general methods for solving linear programming problems has not received a great deal of attention. This will be discussed in subsequent sections.

9–2 THE DUAL SIMPLEX ALGORITHM

The dual simplex algorithm is entirely the work of C. E. Lemke.[1] It apparently originated in the course of applying the primal simplex method to the dual of a linear programming problem. While doing this, Lemke realized that an algorithm that was different in the sense of the sequence of feasible points leading to optimality could be devised, and that, further, it could be applied to either primal or dual problems.

Before developing the mathematical details of the dual simplex method, it is of value to contrast the basic conceptual differences of the primal and dual simplex algorithms. In the primal simplex method, we begin with and maintain at each iteration, a basic feasible solution, i.e., a primal feasible solution. However, not all the $z_j - c_j \geq 0$, else we would be optimal. Recall now that

$$z_j - c_j = \bar{c}_B' B^{-1} \bar{a}_j - c_j \quad \text{and} \quad \bar{v}' = \bar{c}_B' B^{-1} \tag{9–1}$$

Therefore, at a typical iteration of the primal simplex algorithm for $j \in J$, some subset of the j, $z_j - c_j \leq 0$ is equivalent to, by (9–1),

$$\bar{v}'\bar{a}_j \leq c_j \qquad j \in J \tag{9–2}$$

(9–2) would be a typical constraint of the dual if the inequality went the other way. Hence, the current solution $\bar{x}_B$ to the primal which is primal feasible is *dual infeasible*. Therefore, in primal simplex, at each iteration we are primal feasible and dual infeasible. When we are both, we have found the optimal solution. By contrast, in the dual simplex algorithm, at each iteration we have a solution for which all $z_j - c_j \geq 0$. Hence, we are dual feasible. However, one or more of the $x_{Bi} < 0$, i.e., we are not primal feasible at each iteration. When we become both primal and dual feasible for the first time, we have the optimal feasible solution to the primal and also the dual problems. To put it succinctly, in the primal simplex algorithm feasibility for the primal is maintained ($\bar{x}_B \geq \bar{0}$) at each iteration and we change bases until the optimality criterion ($z_j - c_j \geq 0$) is satisfied. In the dual simplex algorithm, at each iteration we maintain continual satisfaction of the optimality criterion ($z_j - c_j \geq 0$) but not primal feasibility ($\bar{x}_B \geq \bar{0}$) and we change bases until we obtain our first feasible solution, at which point we have the optimal solution.

Let us take our linear programming problem in the standard form:

$$\begin{aligned} \text{Max } z &= \bar{c}'\bar{x} \\ A\bar{x} &= \bar{b} \\ \bar{x} &\geq \bar{0} \end{aligned} \tag{9–3}$$

From Chapter 8, we know that the dual of (9–3) is given by:

$$\begin{aligned} \text{Min } w &= \bar{b}'\bar{v} \\ A'\bar{v} &\geq \bar{c} \\ \bar{v} &\text{ unrestricted} \end{aligned} \tag{9–4}$$

We assume that A is $m \times n$ and that $B = (\bar{b}_1, \bar{b}_2, \ldots, \bar{b}_m)$ is our basis matrix at any iteration. The constraints of (9–4) can also be written in the form as

$$\bar{v}'\bar{a}_j \geq c_j \qquad j = 1, 2, \ldots, n \tag{9–5}$$

Suppose that the basis B for the solution to the primal problem is such that

$$\bar{v}' = \bar{c}_B'B^{-1} \tag{9–6}$$

is also a solution to the dual problem, i.e.,

$$\begin{aligned} \bar{v}'\bar{b}_i &= c_{Bi} \qquad i = 1, 2, \ldots, m \\ \bar{v}'\bar{a}_j &\geq c_j \qquad j \text{ not in the basis} \end{aligned} \tag{9–7}$$

From (9–6) and (9–7) it follows that

$$v'a_j - c_j \geq 0 \qquad j = 1, 2, \ldots, n \tag{9–8}$$

and that:

$$\bar{c}_B'B^{-1}\bar{a}_j - c_j \geq 0 \qquad j = 1, 2, \ldots, n \tag{9–9}$$

or

$$z_j - c_j \geq 0 \qquad j = 1, 2, \ldots, n \tag{9–10}$$

Equation (9–10) indicates that a solution to the *dual* such as we have assumed has the property that the optimality criterion for the *primal* is satisfied. If it were also true that

$$\bar{x}_B = B^{-1}\bar{b} \geq \bar{0}$$

then we would also have the optimal solution to both the primal and dual problems, since

$$z = \bar{c}'\bar{x} = \bar{c}_B'B^{-1}\bar{b} = \bar{b}'\bar{v} = w \tag{9–11}$$

From the preceding discussion it is apparent that if we have a solution $\bar{v}' = \bar{c}_B'B^{-1}$ which satisfies the dual constraints (and hence the primal optimality criterion) and it is *not* optimal then it must not be primal feasible; i.e., one or more of the x_{Bi} must be negative. This, of course, never occurs in the primal simplex algorithm. However, we are developing a different approach. Let us then consider such a situation. In the following we designate the rows of B^{-1} as $\bar{\beta}^i$. Now we consider some $x_{Bi} < 0$. In particular, let $x_{Br} < 0$. The choice of the subscript r is deliberate as we shall see. We have, then,

$$x_{Br} = \bar{\beta}^r\bar{b} < 0 \tag{9–12}$$

Let us, seemingly arbitrarily, determine a new solution $\hat{\bar{v}}$ from $\bar{v}' = \bar{c}_B'B^{-1}$ as follows where θ is a scalar:

$$\hat{\bar{v}}' = \bar{v}' - \theta\bar{\beta}^r \tag{9–13}$$

Using (9–13), we see that the change in the dual objective function w is

$$\hat{w} = \bar{b}'\hat{\bar{v}} = \hat{\bar{v}}'\bar{b} = \bar{v}'\bar{b} - \theta\bar{\beta}^r\bar{b} = w - \theta x_{Br}$$

Therefore,

$$\hat{w} = w - \theta x_{Br} \tag{9–14}$$

Since $x_{Br} < 0$, if $\theta < 0$, then $\hat{w} < w$ and, therefore, we will have reduced the value of the dual objective function. If, in addition, the definition of the new solution $\hat{\bar{v}}$ could be made to satisfy the dual constraints, we would have a new solution to the dual with $\hat{w} < w$. Let us now consider this. We note that

$$\hat{\bar{v}}'\bar{b}_i = \bar{v}'\bar{b}_i - \theta\bar{\beta}^r\bar{b}_i \qquad i = 1, 2, \ldots, m \tag{9–15}$$

From (9–7), we know that

$$\bar{v}'\bar{b}_i = c_{Bi} \qquad i = 1, 2, \ldots, m \tag{9–16}$$

We also note that $B^{-1}B = I$, which implies that

$$\bar{\beta}^r\bar{b}_i = \delta_{ir} \tag{9–17}$$

where δ_{ir} is the Kronecker delta, $\delta_{ir} = \begin{cases} 1, i = r \\ 0, i \neq r \end{cases}$. Therefore, from (9–15), (9–16), and (9–17) we have

$$\hat{\bar{v}}'\bar{b}_i = c_{Bi} - \theta\delta_{ir} \qquad i = 1, 2, \ldots, m \tag{9–18}$$

An equivalent way of starting (9–18) is

$$\hat{\bar{v}}'\bar{b}_i = c_{Bi}, \qquad i \neq r \tag{9–19}$$
$$\hat{\bar{v}}'\hat{b}_r = c_{Br} - \theta \geq c_{Br} \quad \text{if} \quad \theta \leq 0$$

Referring again to (9–13), we see that for all $\bar{a}_j$ not in the primal basis we have

$$\hat{\bar{v}}'\bar{a}_j = \bar{v}'\bar{a}_j - \theta\bar{\beta}^r\bar{a}_j \tag{9–20}$$

Before we continue our general discussion let us dispose of the case of unboundedness in the dual problem, since, as we saw in the last chapter, this has implications for the primal problem. If $\bar{\beta}^r\bar{a}_j \geq 0$ for every $\bar{a}_j$ not in the primal basis, then for $\theta \leq 0$ (since from (9–7) $\bar{v}'\bar{a}_j \geq c_j$) we have that

$$\hat{\bar{v}}'\bar{a}_j \geq c_j \qquad j \text{ not in the basis} \tag{9–21}$$

If this is so, θ can be made arbitrarily small below zero and we will always have a solution to the dual. However, since

$$\hat{w} = w - \theta x_{Br} \tag{9–22}$$

then $\hat{w}$ can be made arbitrarily small and, therefore, the dual has an unbounded solution. However, from the previous chapter we know that if the dual is unbounded, the primal has no feasible solution. Hence, this should not be the case for a well posed problem.

Therefore, it is necessary for us to assume that there is at least one $\bar{a}_j$ for which $\bar{\beta}^r\bar{a}_j < 0$. If we recall that

$$\bar{y}_j = B^{-1}\bar{a}_j$$

then

$$y_{rj} = \bar{\beta}^r\bar{a}_j \tag{9–23}$$

If we now stipulate that $y_{rj} < 0$, then from (9–20) and (9–21) in order to satisfy

$$\hat{\bar{v}}'\bar{a}_j = \bar{v}'\bar{a}_j - \theta\bar{\beta}^r\bar{a}_j \geq c_j$$

it is necessary, using (9–23), that

$$\bar{v}'\bar{a}_j - c_j \geq \theta\, y_{rj} \tag{9–24}$$

Since $y_{rj} < 0$, this imposes the restriction on θ that

$$\theta \geq \frac{\bar{v}'\bar{a}_j - c_j}{y_{rj}} \tag{9–25}$$

Since we wish to decrease w as much as possible, the best we can possibly do is to choose θ so that

$$\theta = \max_j \left\{ \frac{\bar{v}'\bar{a}_j - c_j}{y_{rj}} \middle| y_{rj} < 0 \right\} \tag{9–26}$$

However, it is quite simple to express (9–26) in terms of the notation of the primal problem. We have already noted in equation (9–6) that $\bar{v}' = \bar{c}_B' B^{-1}$ and $z_j = \bar{c}_B' B^{-1} \bar{a}_j$ by definition. Therefore, $\bar{v}'\bar{a}_j - c_j = z_j - c_j \geq 0$. Hence, we can write (9–26) as

$$\theta = \frac{z_k - c_k}{y_{rk}} = \max_j \left\{ \frac{z_j - c_j}{y_{rj}} \middle| y_{rj} < 0 \right\} \tag{9–27}$$

Let us now summarize what we have been about in the preceding derivation. If one or more of the $x_{Bi} < 0$, say x_{Br}, then we have found a new solution to the dual which satisfies the dual constraints and also yields a reduced value of the dual objective function. If we take $\bar{a}_k$ as determined by (9–27) together with the $\bar{b}_i$, $i \neq r$, we have a new primal basis $\hat{B}$ in which column r of B has been replaced by $\bar{a}_k$. If $\hat{\bar{x}}_B = \hat{B}^{-1}\bar{b}$ still contains negative components, we can repeat the above process.

It should be noted that, while we had recourse to the dual problem in developing the basic argument, the dual simplex method never explicitly uses the dual of any stated linear programming problem. In fact, it solves the primal problem directly, using rules derived by considering how to minimize the dual objective function while remaining dual feasible. However, the solution obtained is to the primal problem. If one wanted the dual solution, it would be expressed in terms of the primal basis in the usual way.

Let us now summarize the dual simplex algorithm.

Dual Simplex Algorithm

0. We start with the linear programming problem as expressed by (9–3).

1. We must determine an initial basic solution to the primal problem such that $z_j - c_j \geq 0$ for all j and not necessarily feasible in the sense that one or more of the x_{Bi} may be < 0.

2. The vector to be removed from the basis is determined first. The rule is somewhat arbitrary but most commonly is given by

$$x_{Br} = \min_i \{x_{Bi} \mid x_{Bi} < 0\} \tag{9–28}$$

(9–28) tells us that $\bar{b}_r$, i.e., column r is removed from B and x_{Br} becomes zero.

3. The vector to enter the basis is determined from

$$\theta = \frac{z_k - c_k}{y_{rk}} = \max_j \left\{ \frac{z_j - c_j}{y_{rj}} \middle| y_{rj} < 0 \right\} \tag{9–29}$$

4. All quantities of interest are transformed by the usual simplex transformation formulae.

5. When a basic solution is obtained that is also feasible, i.e., $\bar{x}_B \geq \bar{0}$, we have obtained the optimal basic feasible solution.

There are a number of points worth noting about the algorithm we have just described. First, it will be seen that in the dual simplex method the order of determination of vectors to enter and leave the basis is reversed. Second, in dual simplex the rule to determine the vector to be *removed* is somewhat arbitrary, while the reverse is the case for the primal simplex algorithm. Third, it will be noted, and the subsequent example will show, that at each iteration we do *not* increase the primal objective function but rather decrease the dual objective function. Since the optimality criterion is satisfied, $z = w$ at each iteration, z will decrease or remain unchanged at each iteration. However, the final value of z will be optimal since all previous values correspond to feasible solutions.

The problem of cycling and dual degeneracy is present in the dual simplex algorithm, as one might expect. Its resolution involves precisely the same kinds of considerations and arguments that we used in the primal case. Details are given in Reference 3 for the interested reader.

9–3 THE DUAL SIMPLEX ALGORITHM—INITIAL BASIC SOLUTION

The dual simplex algorithm has never been used as a "general-purpose" linear programming algorithm because of the apparent difficulty attendant upon finding an initial basic solution with $z_j - c_j \geq 0$ for all j. However, neither has much effort been expended in investigating how to do so. We shall return to this point.

In parametric and post-optimal analyses there is a natural need for the dual simplex algorithm (see Chapter 12). This is also the case for certain integer programming algorithms. (See Reference 2 and Chapter 16.) However, let us now consider what methods do exist for finding an initial basic solution with $z_j - c_j \geq 0$ when one is not known *a priori*. We will briefly discuss several methods.

A. The Artificial Constraint Method

We assume a basis for the primal problem is known and for this basis, one or more of the $x_{Bi} < 0$ and unfortunately, one or more $z_j - c_j < 0$.

We then add to the problem the constraint

$$\sum_{j \in T} x_j \leq M \tag{9–30}$$

where $T = \{j \mid z_j - c_j < 0\}$ and M is a sufficiently large positive number so that it exceeds any value encountered in the computation. The constraint (9–30) can be converted to

$$\sum_{j \in T} x_j + x_M = M \tag{9–31}$$

We now select the variable x_p, $p \in T$ which has the largest *absolute* $z_p - c_p$ and from (9–31) we have

$$x_p = M - (x_M + \sum_{\substack{j \in T \\ j \notin p}} x_j) \tag{9–32}$$

We then substitute (9–32) into the original objective function and constraints and we obtain a modified but obviously equivalent problem. We calculate all the usual tableau quantities, and a check on these easily indicates that now all $z_j - c_j \geq 0$. This is left as an exercise for the reader. We can then proceed with the direct application of the dual simplex method. It should be noted that at each iteration a solution of the original problem is available by ignoring the value of x_M.

B. Surplus Variable—Negative Cost Case

Suppose one is fortunate enough to have a problem such that a slack or surplus variable is present in each constraint and $c_j \leq 0$ for all j. Then we have that $\bar{x}_B = B^{-1}\bar{b} = -b$ and $z_j - c_j \geq 0$ since $z_j = \bar{c}_B B^{-1}\bar{a}_j = 0$ because $\bar{c}_B = \bar{0}$. Hence the dual simplex method can be directly applied to such a problem.

C. Lemke's Method

We start with any m linearly independent vectors from A and let $\bar{b}_0$ be any linear combination of these vectors with positive coefficients. Now solve, instead of the original problem, the following:

$$\begin{aligned} \text{Max } z &= \bar{c}'\bar{x} \\ A\bar{x} &= \bar{b}_0 \\ \bar{x} &\geq \bar{0} \end{aligned}$$

using primal simplex. When the optimal solution is obtained, replace $\bar{b}_0$ with $\bar{b}$. This will give a basic but probably not feasible, in the sense of $\bar{x}_B \geq \bar{0}$, solution to the original problem with all $z_j - c_j \geq 0$, since the $z_j - c_j$ do not depend on $\bar{b}$. We can now apply the dual simplex method. Hasegawa[4] has investigated this method with some encouraging but not definitive computational results.

D. Dantzig's Method

We write the dual problem in standard form by adding n slack variables v_{sj}, which have n unit column vectors. A dual basis B_D will contain m column vectors transposed from rows of A. After suitable permutation and renumbering, the dual basis has the form

$$B_D = \begin{bmatrix} \bar{a}_1' & 0 & \dots & 0 \\ \cdot & & & \\ \cdot & & & \\ \cdot & & & \\ \bar{a}_m' & 0 & \dots & 0 \\ \bar{a}_{m+1}' & 1 & \dots & 0 \\ \cdot & \cdot & & \cdot \\ \cdot & \cdot & & \cdot \\ \cdot & \cdot & & \cdot \\ \bar{a}_n' & 0 & \dots & 1 \end{bmatrix} \tag{9–33}$$

By the construction of the dual basis B_D, the m vectors $\bar{a}_j$, $j = 1, \dots, m$ are independent and form a primal basis B such that the primal basic solution is dual feasible. If $c_j \leq 0$ for all j then we can let $v_{sj} = c_j$ be the initial solution and let the m column vectors which we transposed from A be the m dual slack vectors. We keep the slack variables of the basis non-negative and do not consider the optimality of the dual. This method requires a great deal of computation and is not practical. However, it does indicate the use of the dual problem to find a dual feasible solution to the primal.

9–4 DUAL SIMPLEX—COMPUTATIONAL EXAMPLES

EXAMPLE 1

$$\begin{aligned} \text{Max } z &= -2x_1 - 3x_2 - x_3 \\ 2x_1 + x_2 + 2x_3 &\geq 3 \\ 3x_1 + 2x_2 + z_3 &\geq 4 \\ x_1, x_2, x_3 &\geq 0 \end{aligned} \tag{9–34}$$

We convert (9–34) to the standard form:

$$\begin{aligned} \text{Max } z &= -2x_1 - 3x_2 - x_3 \\ 2x_1 + x_2 + 2x_3 - x_4 &= 3 \\ 3x_1 + 2x_2 + x_3 - x_5 &= 4 \\ x_j \geq 0, \quad j &= 1, 2, \ldots, 5 \end{aligned} \tag{9–35}$$

We take as our basis vectors $\bar{a}_4$ and $\bar{a}_5$ and, hence,

$$B = \begin{bmatrix} -1 & 0 \\ 0 & -1 \end{bmatrix} \quad \text{and} \quad B^{-1} = \begin{bmatrix} -1 & 0 \\ 0 & -1 \end{bmatrix}$$

Therefore,

$$\bar{x}_B = B^{-1}\bar{b} = \begin{bmatrix} -1 & 0 \\ 0 & -1 \end{bmatrix}\begin{bmatrix} 3 \\ 4 \end{bmatrix} = \begin{bmatrix} -3 \\ -4 \end{bmatrix}$$

Since

$$\bar{c}_B = \bar{0}, \qquad z_j - c_j = -c_j \geq 0, \qquad j = 1, 2, 3$$

Therefore, we are in a position to use the dual simplex method.

The initial tableau is shown in Figure 9–1. Note that $\bar{y}_j = B^{-1}\bar{a}_j = -\bar{a}_j$. The vector to leave the basis is determined from (9–28). Therefore, we have

$$x_{Br} = \min(-3, -4) = -4$$

and therefore, the second column of the basis is removed. The vector to enter is determined from (9–29) as

$$\frac{z_k - c_k}{y_{rk}} = \max\left\{\frac{2}{-3}, \frac{3}{-2}, \frac{1}{-1}\right\} = -\tfrac{2}{3}$$

Therefore, $\bar{a}_1$ enters the basis.

We transform the tableau in the usual way. The new tableau, Tableau 2, is given in Figure 9–2.

			−2	−3	−1	0	0
$\bar{c}_B$	Basis	$\bar{x}_B$	$\bar{y}_1$	$\bar{y}_2$	$\bar{y}_3$	$\bar{y}_4$	$\bar{y}_5$
0	$\bar{a}_4$	−3	−2	−1	−2	1	0
0	$\bar{a}_5$	−4	(−3)	−2	−1	0	1
	$z_j - c_j$	0	2	3	1	0	0

Figure 9–1 Example 1—Tableau 1

			-2	-3	-1	0	0
$\bar{c}_B$	Basis	$\bar{x}_B$	$\bar{y}_1$	$\bar{y}_2$	$\bar{y}_3$	$\bar{y}_4$	$\bar{y}_5$
0	$\bar{a}_4$	$-\frac{1}{3}$	0	$\frac{1}{3}$	$\boxed{-\frac{4}{3}}$	1	$-\frac{2}{3}$
-2	$\bar{a}_1$	$\frac{4}{3}$	1	$\frac{2}{3}$	$\frac{1}{3}$	0	$-\frac{1}{3}$
	$z_j - c_j$	$-\frac{8}{3}$	0	$\frac{5}{3}$	$\frac{1}{3}$	0	$\frac{2}{3}$

Figure 9–2 Example 1—Tableau 2

Since $x_4 < 0$ in Tableau 2 we are still not optimal. The vector to leave the basis then is $\bar{a}_4$, since only $x_{B1} = x_4 < 0$. Hence, the first column of the basis is removed. The vector to enter the basis is given by

$$\frac{z_k - c_k}{y_{rk}} = \frac{\frac{1}{3}}{-\frac{4}{3}} = -\tfrac{1}{4} \quad \text{since only } y_{13} < 0$$

Therefore $\bar{a}_3$ enters the basis.

We transform Tableau 2 and obtain Tableau 3 in Figure 9–3.

It can be seen that Tableau 3 is optimal. The optimal solution is $x_1^* = \frac{15}{12}$, $x_2^* = 0$, $x_3^* = \frac{1}{4}$, $z^* = -\frac{33}{12}$. Note that the value of z decreased at each iteration until the optimal solution was reached, since at each iteration we maintained $z = w$.

			-2	-3	-1	0	0
$\bar{c}_B$	Basis	$\bar{x}_B$	$\bar{y}_1$	$\bar{y}_2$	$\bar{y}_3$	$\bar{y}_4$	$\bar{y}_5$
-1	$\bar{a}_3$	$\frac{1}{4}$	0	$-\frac{1}{4}$	1	$-\frac{3}{4}$	$\frac{1}{2}$
-2	$\bar{a}_1$	$\frac{15}{12}$	1	0	0	$\frac{1}{4}$	$-\frac{1}{2}$
	$z_j - c_j$	$-\frac{33}{12}$	0	$\frac{21}{12}$	0	$\frac{1}{4}$	$\frac{1}{2}$

Figure 9–3 Example 1—Tableau 3

EXAMPLE 2

$$\begin{aligned} \text{Max } z &= 2x_1 - 3x_2 - 2x_3 \\ x_1 - 2x_2 - 3x_3 &= 8 \\ 2x_2 + x_3 &\le 10 \\ x_2 - 2x_3 &\ge 4 \\ x_1, x_2, x_3 &\ge 0 \end{aligned} \tag{9–36}$$

We convert (9–36) into the form

$$
\begin{aligned}
\text{Max } z &= 2x_1 - 3x_2 - 2x_3 \\
x_1 - 2x_2 - 3x_3 &= 8 \\
2x_2 + x_3 + x_4 &= 10 \\
-x_2 + 2x_3 + x_5 &= -4 \\
x_j \geq 0, \quad j &= 1, 2, \ldots, 5
\end{aligned}
\tag{9–37}
$$

We can have a basis with $\bar{a}_1$, $\bar{a}_4$, $\bar{a}_5$ which would yield

$$\bar{x}_B = B^{-1}\bar{b} = I\bar{b} = [6, 10, -4]$$

However, not all the $z_j - c_j$ are non-negative. For example,

$$
\begin{aligned}
z_2 - c_2 = \bar{c}_B' B^{-1}\bar{a}_2 - c_2 &= [2, 0, 0]'I[-2, 2, -1] - (-3) = -1 \\
z_3 - c_3 &= [2, 0, 0]'I[-3, 1, 2] - (-2) = -4
\end{aligned}
$$

We shall use the technique of adding an additional constraint to form an augmented problem on which we can use the dual simplex algorithm. Let

$$x_2 + x_3 + x_M = M \tag{9–38}$$

We substitute $x_3 = M - x_M - x_2$ into the objective function and the constraints and also add the constraint (9–38) to our problem. We then have the following augmented problem:

$$
\begin{aligned}
\text{Max } z &= 2x_M + 2x_1 - x_2 - 2M \\
3x_M + x_1 + x_2 &= 3M + 8 \\
-x_M + x_2 + x_4 &= -M + 10 \\
-2x_M - 3x_2 + x_5 &= -2M - 4 \\
x_M + x_2 + x_3 &= M \\
x_M, x_j \geq 0 \qquad & j = 1, 2, \ldots, 5
\end{aligned}
\tag{9–39}
$$

We have an obviously convenient basis in

$$B = [\bar{a}_1, \bar{a}_4, \bar{a}_5, \bar{a}_3] \tag{9–40}$$

Using the basis in (9–40) the initial tableau is shown in Figure 9–4.

We see in Figure 9–4 that we now have all $z_j - c_j \geq 0$ and we can now apply the dual simplex method to this problem. The vector to be removed

			2	2	−1	0	0	0
$\bar{c}_B$	Basis	$\bar{x}_B$	$\bar{y}_M$	$\bar{y}_1$	$\bar{y}_2$	$\bar{y}_3$	$\bar{y}_4$	$\bar{y}_5$
2	$\bar{a}_1$	$3M + 8$	3	1	1	0	0	0
0	$\bar{a}_4$	$-M + 10$	−1	0	1	0	1	0
0	$\bar{a}_5$	$-2M + 4$	−2	0	(−3)	0	0	1
0	$\bar{a}_3$	M	1	0	1	1	0	0
	$z_j - c_j$	$6M + 16$	4	0	3	0	0	0

Figure 9–4 Example 2—Tableau 1

from the basis is found from

$$\text{Min}\,(-M + 10, -2M - 4) = -2M - 4$$

Therefore, $\bar{a}_5$ leaves. The vector to enter the basis is found from

$$\frac{z_k - c_k}{y_{rk}} = \text{Max}\left(\frac{4}{-2}, \frac{3}{-3}\right) = -1 = \frac{z_2 - c_2}{y_{52}}$$

Therefore $\bar{a}_2$ enters the basis. The transformed tableau is given in Figure 9–5. The vector to be removed from Tableau 2 is found from

$$\text{Min}\left(\frac{-5M + 26}{3}, -M - 4\right) = \frac{-5M + 26}{3}$$

Therefore $\bar{a}_4$ leaves the basis. The vector to enter is $\bar{a}_M$, since it is the only one with $y_{2j} < 0$. The transformed tableau is given in Figure 9–6.

			2	2	−1	0	0	0
$\bar{c}_B$	Basis	$\bar{x}_B$	$\bar{y}_M$	$\bar{y}_1$	$\bar{y}_2$	$\bar{y}_3$	$\bar{y}_4$	$\bar{y}_5$
2	$\bar{a}_1$	$\frac{7M + 20}{3}$	$\frac{7}{3}$	1	0	0	0	$\frac{1}{3}$
0	$\bar{a}_4$	$\frac{-5M + 26}{3}$	($-\frac{5}{3}$)	0	0	0	1	$\frac{1}{3}$
−1	$\bar{a}_2$	$\frac{2M + 4}{3}$	$\frac{2}{3}$	0	1	0	0	$-\frac{1}{3}$
0	$\bar{a}_3$	$\frac{M - 4}{3}$	$\frac{1}{3}$	0	0	1	0	$\frac{1}{3}$
	$z_j - c_j$	$4M + 12$	2	0	0	0	0	1

Figure 9–5 Example 2—Tableau 2

			2	2	-1	0	0	0
$\bar{c}_B$	Basis	$\bar{x}_B$	$\bar{y}_M$	$\bar{y}_1$	$\bar{y}_2$	$\bar{y}_3$	$\bar{y}_4$	$\bar{y}_5$
2	$\bar{a}_1$	$\frac{94}{5}$	0	1	0	0	$\frac{7}{5}$	$\frac{4}{5}$
2	$\bar{a}_M$	$\dfrac{5M - 26}{5}$	1	0	0	0	$-\frac{3}{5}$	$-\frac{1}{5}$
-1	$\bar{a}_2$	$\frac{24}{5}$	0	0	1	0	$\frac{2}{5}$	$-\frac{1}{5}$
0	$\bar{a}_3$	$\frac{3}{5}$	0	0	0	1	$\frac{1}{5}$	$\frac{2}{5}$
	$z_j - c_j$	$\dfrac{10M + 112}{5}$	0	0	0	0	$\frac{6}{5}$	$\frac{7}{5}$

Figure 9–6 Example 2—Tableau 3

Since all $x_{Bi} > 0$, we have the optimal solution. Usually, we eliminate the extra row and column corresponding to $\bar{a}_M$. Often one or more of the $x_{Bi} < 0$ after the stage when $\bar{a}_M$ enters the basis is reached. We merely continue by disregarding or removing the extra row and column. Our optimal tableau is given in Figure 9–7.

It will be noted that in the final tableau the original prices are now restored.

			2	-3	-2	0	0
$\bar{c}_B$	Basis	$\bar{x}_B$	$\bar{y}_1$	$\bar{y}_2$	$\bar{y}_3$	$\bar{y}_4$	$\bar{y}_5$
2	$\bar{a}_1$	$\frac{94}{5}$	1	0	0	$\frac{7}{5}$	$\frac{4}{5}$
-3	$\bar{a}_2$	$\frac{24}{5}$	0	1	0	$\frac{2}{5}$	$-\frac{1}{5}$
-2	$\bar{a}_3$	$\frac{2}{5}$	0	0	1	$\frac{1}{5}$	$\frac{2}{5}$
	$z_j - c_j$	$\frac{112}{5}$	0	0	0	$\frac{6}{5}$	$\frac{7}{5}$

Figure 9–7 Example 2—Tableau 4

9–5 THE PRIMAL—DUAL ALGORITHM

The incentive or desire to develop an algorithm of the type we shall now discuss, apart from its inherent theoretical interest, stems from the fact that whether one uses a two-phase primal or revised simplex algorithm or a composite simplex algorithm, a substantial number of iterations and, hence, computing time, is used to drive artificial variables out of the basis. In the course of doing so, however, one has no assurance, nor has one any reason to expect, that the drive to feasibility has helped us, say at the end of Phase I or during certain stages of the composite algorithm, in getting closer to an optimal solution. Ideally, one might wish for some method which, while driving out artificial variables to become feasible, would also somehow take account of the desire to optimize the true objective function at one and the same time. It is this philosophy that the primal-dual algorithm, developed by Dantzig, Ford, and Fulkerson,[5] attempts to embody.

The general idea behind the primal-dual algorithm is easy to state. We introduce artificial variables into the primal problem and carry out a Phase I. However, the vectors to enter the basis during Phase I are chosen in a different way from that described in Chapter 5. We make use of considerations arising from an examination of the dual problem in order to decide which vectors may enter the primal basis. It is in this way that we keep in mind the goal of improving the original problem objective function. We execute this process in such a way that when we have driven the last artificial vector from the basis, i.e., concluded Phase I, and are feasible for the first time, we are also optimal. In effect, we have combined Phase I and Phase II in such a way that the goals of both phases are accomplished simultaneously. We accomplish this by using a solution to the dual of our original problem to give us information on which vectors may enter the primal basis. Then, after these vectors enter, we find a new solution to the dual and continue the process. Each new solution to the dual *decreases* the dual objective function. By maintaining complementary slackness at each stage, we know that when we have found a feasible solution, it will also be optimal (see Theorem 8–13, in Chapter 8). In this fashion, the iterative process enables us to pursue the goals of feasiblity and optimality at the same time. Let us now consider the derivation of the method.

We consider our primal problem to be:

$$\begin{aligned} \text{Max } z &= \bar{c}'\bar{x} \\ A\bar{x} &= \bar{b} \\ \bar{x} &\geq \bar{0} \end{aligned} \tag{9–41}$$

We need to find a solution to the dual of (9–41). However, as we have seen in the preceding sections of this chapter, this requires some effort. We shall use a device very similar to the one we used in the dual simplex method (see Section 9–3). We shall replace (9–41) by a modified primal problem which includes the addition of a constraint:

$$x_0 + \sum_{j=1}^{n} x_j = b_0 \tag{9–42}$$

where $x_0 \geq 0$, $c_0 = 0$ and b_0 is an arbitrarily large positive number that places no new restrictions on the $x_j, j = 1, 2, \ldots, n$. This modified primal is given by:

$$\text{Max } z = \bar{c}'\bar{x}$$
$$\begin{bmatrix} 1 & \bar{1}' \\ \bar{0} & A \end{bmatrix}\begin{bmatrix} x_0 \\ \bar{x} \end{bmatrix} = \begin{bmatrix} b_0 \\ \bar{b} \end{bmatrix} \tag{9–43}$$
$$[x_0, \bar{x}] \geq \bar{0}$$

We can readily write the dual of (9–43) as:

$$\text{Min } w = b_0 v_0 + \bar{b}'\bar{v}$$
$$\begin{bmatrix} 1 & \bar{0}' \\ \bar{1} & A' \end{bmatrix}\begin{bmatrix} v_0 \\ \bar{v} \end{bmatrix} \geq \begin{bmatrix} 0 \\ \bar{c} \end{bmatrix} \tag{9–44}$$

Since (9–43) has equality constraints the components of the dual variables $\bar{v}$ are unrestricted. However, the first constraint of (9–44) insures that $v_0 \geq 0$.

The main reason for modifying the primal was to give us a dual, viz., (9–44) for which a feasible solution could be written down immediately. We see that for (9–44) one obvious feasible solution is:

$$\bar{v} = \bar{0}, \qquad v_0 = \max c_j \tag{9–45}$$

$$j = 0, 1, \ldots, n$$

We can readily add surplus variables to (9–44) to obtain:

$$\text{Min } w = b_0 v_0 + \bar{b}'\bar{v}$$
$$\begin{bmatrix} 1 & \bar{0}' \\ \bar{1} & A' \end{bmatrix}\begin{bmatrix} v_0 \\ \bar{v} \end{bmatrix} - \begin{bmatrix} 1 & \bar{0}' \\ \bar{0} & I_n \end{bmatrix}\begin{bmatrix} v_{s0} \\ \bar{v}_s \end{bmatrix} = \begin{bmatrix} 0 \\ \bar{c} \end{bmatrix} \tag{9–46}$$
$$[v_{s0}, \bar{v}_s] \geq \bar{0}$$

If we now write the dual of our original primal (9–41) we have

$$\text{Min } w = \bar{b}'\bar{v}$$
$$A'\bar{v} \geq \bar{c} \tag{9–47}$$

We immediately see that if $v_0 = 0$ in (9–46), then (9–44) and (9–47) are identical. Since (9–46) is simply (9–44) in equality form, if $v_0 = 0$, any solution of (9–46) will be a solution to our dual (9–47).

In Chapter 8 we showed that if for a pair of feasible solutions to primal and dual problems, complementary slackness conditions held, then the solutions would be optimal. In terms of our problem, what this implies is that if $[x_0, \bar{x}]$ is a feasible solution and complementary slackness holds, i.e.,

$$x_0 v_{s0} + \bar{x}'\bar{v}_s = 0 \tag{9–48}$$

then $[x_0, \bar{x}]$ is an optimal solution to the modified primal problem (9–43) and $[v_0, \bar{v}]$ is an optimal solution to the modified dual problem (9–44). If it should also be true that $v_0 = 0$, then $z = \bar{c}'\bar{\bar{x}} = \bar{b}'\bar{v} = w$. Under these conditions, $\bar{x}$ would be the optimal solution to our original primal problem. On the other hand, if we had a solution to the modified dual (9–44) such that $v_0 \neq 0$ and still maintained (9–48), then

$$z = w = b_0 v_0 + \bar{b}'\bar{v} \tag{9–49}$$

We have developed (9–49) to consider and later dispose of the case of unboundedness. We have assumed that b_0 could be made arbitrarily large and still not affect the x_j. Since $z = \bar{c}'\bar{x}$ is the same for both (9–41) and (9–43) we see from (9–49) that z can be made arbitrarily large and, hence, the original problem has an unbounded solution. We shall use this result subsequently.

We now return to our modified primal problem (9–43) and add artificial variables just as we would for any Phase I calculation. As usual, the legitimate variables will have prices of zero and the artificial variables will have prices of -1. The resulting problem, which is the actual problem with which we do computation in our Phase I, is

$$\text{Max } z_a = -1x_{a0} - \bar{1}'\bar{x}_a$$
$$\begin{bmatrix} 1 & \bar{1}' \\ \bar{0} & A \end{bmatrix} \begin{bmatrix} x_0 \\ \bar{x} \end{bmatrix} + \begin{bmatrix} 1 & \bar{0} \\ \bar{0} & I_n \end{bmatrix} \begin{bmatrix} x_{a0} \\ \bar{x}_a \end{bmatrix} = \begin{bmatrix} b_0 \\ \bar{b} \end{bmatrix} \tag{9–50}$$
$$[x_0, x_{a0}, \bar{x}, \bar{x}_a] \geq \bar{0}$$

The activity vectors, i.e., the columns of the first matrix in (9–50), will be called $\bar{\alpha}_j$, $j = 0, 1, \ldots, n$. A basis matrix will be denoted B_0. The vector containing the prices in the basis (0 or -1, *not* c_j) will be designated $\tilde{\bar{c}}_{B_0}$ and $\tilde{z}_j - \tilde{c}_j = \tilde{\bar{c}}'_{B_0} B_0^{-1} \bar{\alpha}_j - \tilde{c}_j$.

Initially, our basis consists entirely of artificial vectors. In order to decide which vectors may enter the primal basis, we do *not* merely look at $\tilde{z}_j - \tilde{c}_j$, since this would just be an ordinary Phase I calculation. Instead, we turn to the dual problem (9–46) in order to find out which vectors may move us toward optimality. We have already noted that (9–45) provides an initial feasible solution to the dual and for this solution we see from (9–46) that:

$$\bar{v}_s = v_0 \bar{1} - \bar{c} \tag{9–51}$$

Because we selected $v_0 = \max\limits_j c_j$, at least one component of $[v_{s0}, \bar{v}_s]$ will vanish.

Recall that we wish to maintain complementary slackness so that when $z_a = 0$, we will have found both feasible and optimal solutions to our original problem and the $z = w$. Hence, our Phase I is all we will ever need, providing we use proper information from the dual problem to guide us in the selection of vectors to enter the basis B_0 at each stage.

We define

$$V = \{v_{sj} \mid v_{sj} = 0\} \tag{9–52}$$

and

$$W = \{\bar{\alpha}_j \mid v_{sj} \in V\} \tag{9–53}$$

Now we only allow $\bar{\alpha}_j \in W$ to enter the basis at any stage. We see in this way that complementary slackness is preserved, i.e., if $v_{sj} > 0$ the corresponding x_j cannot enter the basis and remains at zero. Because $v_0 = \max_j c_j$ we know that V is not empty, initially, and so at least one $\bar{\alpha}_j$ is eligible to enter the basis.

We now apply the simplex method in the usual way except that only vectors $\bar{\alpha}_j \in W$ may enter the basis. We continue this process until $\tilde{z}_j - \tilde{c}_j \geq 0$ for all $\bar{\alpha}_j \in W$. If B_0 is the basis at this point and $\tilde{\bar{c}}_{B0}$ contains the prices in the basis, then we define

$$\bar{r} = (r_0, r_1, \ldots, r_m) = \tilde{\bar{c}}'_{B_0} B_0^{-1} \tag{9–54}$$

We will now show that there exists a $\phi > 0$ such that

$$(\hat{v}_0, \hat{\bar{v}}') = (v_0, \bar{v}') + \phi\bar{r} \tag{9–55}$$

is a new solution to the dual with the property that $\hat{w} < w$. We do this as follows.

For all $\bar{\alpha}_j \in W$ we know that

$$\tilde{z}_j - \tilde{c}_j = \tilde{\bar{c}}'_{B_0} B_0^{-1} \bar{\alpha}_j = \bar{r}'\bar{\alpha}_j \geq 0 \tag{9–56}$$

(9–56) follows since $\tilde{c}_j = 0$ and we know that for $\bar{\alpha}_j \in W$ when we terminate, $\tilde{z}_j - \tilde{c}_j \geq 0$. We know further that if $\bar{\alpha}_j$ is in the basis then, clearly, $\bar{r}'\bar{\alpha}_j = 0$. Let us now multiply (9–55) by an $\bar{\alpha}_j \in W$, yielding

$$(\hat{v}_0, \hat{\bar{v}}')\bar{\alpha}_j = (v_0, \bar{v}')\bar{\alpha}_j + \phi\bar{r}'\bar{\alpha}_j \tag{9–57}$$

However, for $\bar{\alpha}_j \in W$, $(v_0, \bar{v}')\bar{\alpha}_j = c_j$, i.e., $v_{sj} = 0$. Since $(v_0, \bar{v}')\bar{\alpha}_j = c_j$ and $\bar{r}'\bar{\alpha}_j \geq 0$ by (9–56) we can then combine this with (9–57) to yield

$$\begin{aligned} (\hat{v}_0, \hat{\bar{v}}')\bar{\alpha}_j &= (v_0, \bar{v}')\bar{\alpha}_j + \phi\bar{r}'\bar{\alpha}_j \geq c_j \\ \bar{\alpha}_j &\in W \\ \phi &\geq 0 \end{aligned} \tag{9–58}$$

Let us now consider $\bar{\alpha}_j \notin W$. For these $\bar{\alpha}_j$, it is clear that:

$$(v_0, \bar{v}')\bar{\alpha}_j > c_j \tag{9–59}$$

Hence it will be equally true for $\bar{\alpha}_j \notin W$ that:

$$(\hat{v}_0, \hat{\bar{v}}')\bar{\alpha}_j = (v_0, \bar{v}')\bar{\alpha}_j + \phi\bar{r}'\bar{\alpha}_j \geq c_j \tag{9–60}$$

providing that ϕ satisfies the condition that:

$$0 \leq \phi \leq \phi_0 = \min_j \left\{ -\frac{[(v_0, \bar{v}')\bar{\alpha}_j - c_j]}{\bar{r}'\bar{\alpha}_j} \,\middle|\, \bar{r}'\bar{\alpha}_j < 0 \right\} \tag{9–61}$$

It is clear from (9–61) that if all $\bar{r}'\bar{\alpha}_j \geq 0$ then $(v_0, \bar{v}')$ will be a solution to the dual no matter how large ϕ becomes. Referring again to (9–61), if we recall that the numerator of the minimand:

$$(v_0, \bar{v}')\bar{\alpha}_j - c_j = v_{sj}$$

by (9–46) and that the denominator:

$$\bar{r}'\bar{\alpha}_j = \tilde{z}_j - \tilde{c}_j$$

we can then write that

$$\phi_0 = \min_j \left\{ \frac{v_{sj}}{\tilde{c}_j - \tilde{z}_j} \,\middle|\, \tilde{z}_j - \tilde{c}_j < 0 \right\} \tag{9–62}$$

or

$$\phi_0 = \infty$$

Let us now consider the value of the new dual objective function:

$$\begin{aligned} \hat{w} &= (b_0, \bar{b}')(\hat{v}_0, \hat{\bar{v}}')' \\ &= (v_0, \bar{v}')(b_0, \bar{b}')' + \phi\bar{r}'(b_0, \bar{b}')' \\ &= w + \phi\tilde{\bar{c}}'_{B_0}(b_0, \bar{b}')' = w + \phi z_a \end{aligned} \tag{9–63}$$

Since $z_a < 0$ and $\phi > 0$, it is clear that $\hat{w} < w$. In order to obtain the largest possible decrease in w, we use ϕ_0, if it is finite. If ϕ can be made arbitrarily large, the dual is unbounded, and by the results of Chapter 8, the primal has no feasible solution.

Therefore, using the value of ϕ_0 calculated from (9–62), we obtain a new solution to the dual which would be given by (9–55). What we actually do is equivalent to this, viz., calculate new values of the dual surplus variables, i.e.,

$$\hat{v}_{sj} = v_{sj} + \phi_0(\tilde{z}_j - \tilde{c}_j) \tag{9–64}$$

There will be at least one $\hat{v}_{sj} = 0$ and, therefore, one new $v_{sj} \in V$ than previously, since we determined ϕ_0 by (9–62). Therefore, we will also have at least one new vector $\bar{\alpha}_j \in W$. We now repeat the entire process until $\tilde{z}_j - \tilde{c}_j \geq 0$ for all the vectors eligible to enter the primal basis, i.e., $\bar{\alpha}_j \in W$. We then calculate a new ϕ_0 and a new solution to the dual, etc.

We know that when $z_a = 0$, we have an optimal solution to the primal and that max $z = w$ provided that $v_{s0} = v_0 = 0$. If $v_{s0} \neq 0$ then we see that the primal is unbounded since b_0 can be made arbitrarily large. Of course, if we reach a solution such that $z_a < 0$ and $\tilde{z}_j - \tilde{c}_j \geq 0, j = 1, 2, \ldots, n$, the original problem has no feasible solution.

The entire calculation can be done using the usual simplex tableau with the addition of an extra row for the v_{sj}, the dual surplus variables. Since complementary slackness is maintained throughout, $z = w$ at each iteration.

There are strong indications that the primal-dual algorithm is superior to the ordinary two-phase simplex method (see Reference 6) but no comparison has ever been made with the composite simplex method which is in general use.

9–6 THE PRIMAL-DUAL METHOD—AN EXAMPLE

Consider the following problem:

$$\begin{aligned} \text{Max } z &= 2x_1 + 5x_2 \\ 2x_1 + x_2 &\geq 3 \\ x_1 + x_2 &\leq 8 \\ x_1, x_2 &\geq 0 \end{aligned} \tag{9–65}$$

We add slack and surplus variables to (9–65) as well as the additional constraint (9–42) that we require and obtain

$$\begin{aligned} \text{Max } z &= 2x_1 + 5x_2 \\ x_0 + x_1 + x_2 + x_3 + x_4 &= b_0 \\ 2x_1 + x_2 - x_3 \quad &= 3 \\ x_1 + x_2 \quad + x_4 &= 8 \\ x_j \geq 0, \quad j &= 0, 1, 2, 3, 4, \end{aligned} \tag{9–66}$$

Our initial solution to the dual is given by

$$v_0 = \max c_j = \max(2, 5) = 5$$

Since $v_0 - v_{s0} = 0$ from (9–46) we see that $v_{s0} = 5$. We know that

$$\bar{v}_s = v_0\bar{1} - \bar{c}$$

from (9–51). Therefore,

$$v_{s1} = 5 - 2 = 3$$
$$v_{s2} = 5 - 5 = 0$$
$$v_{s3} = 5 - 0 = 5$$
$$v_{s4} = 5 - 0 = 5$$

We also have that $w = b_0 v_0 + \bar{b}'\bar{v} = 5b_0$ since $\bar{v} = \bar{0}$ initially. Hence, the initial tableau is as given in Figure 9-8. The last row gives the values of v_{sj}, the dual surplus variables, and under the $\bar{x}_B$ column the value of $w = z$. Initially, only artificial variables are in the basis. Since only $v_{s2} = 0$, $V = \{v_{s2}\}$ and $W = \{\bar{\alpha}_2\}$. The prices along the top of the tableau are the actual c_j since we use these to calculate v_{sj} from $\bar{v}_s = v_0\bar{1} - \bar{c}$. However, the prices $\tilde{c}_j = 0$ or -1 are used for the calculations in the tableau. Our basis is of order 3×3 and $\bar{e}_1$ never leaves because b_0 is assumed to be too large by comparison with other choices for removal. Hence, $\bar{e}_1$ is in every basis. However, $\bar{e}$ leaves in the last stage of the calculation. Since only $\bar{\alpha}_2 \in W$, i.e., $v_{s2} = 0$, $\bar{\alpha}_2$ enters the basis. The vector to be removed is determined from

$$\min\left(\tfrac{3}{1}, \tfrac{8}{1}\right) = 3$$

Therefore, $\bar{e}_2$ is removed from the basis. The new tableau is given in Figure 9–9.

We now need a new solution to the dual since the only vector in W is now in the primal basis. Therefore, we need to calculate new values of v_{sj} for the last row of Tableau 2 in Figure 9–9. The rest of the tableau is transformed in the usual way. We calculate v_{sj} from (9–62) and (9–64), using the new values of $\tilde{z}_j - \tilde{c}_j$:

$$\phi_0 = \min\left(\tfrac{5}{1}, \tfrac{5}{3}, \tfrac{5}{2}\right) = \tfrac{5}{3}$$

			0	2	5	0	0
$\tilde{c}_{B0}$	Basis	$\bar{x}_B$	$\bar{y}_0$	$\bar{y}_1$	$\bar{y}_2$	$\bar{y}_3$	$\bar{y}_4$
-1	$\bar{e}_1$	b_0	1	1	1	1	1
-1	$\bar{e}_2$	3	0	2	(1)	-1	0
-1	$\bar{e}_3$	8	0	1	1	0	1
	$\tilde{z}_j - \tilde{c}_j$	$-11 - b_0$	-1	-4	-3	0	-2
	v_{sj}	$5b_0$	5	3	0	5	5

Figure 9–8 Tableau 1

			0	2	5	0	0
c_{B0}	Basis	$\bar{x}_B$	$\bar{y}_0$	$\bar{y}_1$	$\bar{y}_2$	$\bar{y}_3$	$\bar{y}_4$
-1	$\bar{e}_1$	$b_0 - 3$	1	-1	0	2	1
0	$\bar{\alpha}_2$	3	0	2	1	-1	0
-1	$\bar{e}_3$	5	0	-1	0	(1)	1
	$z_j - c_j$	$-b_0 - 2$	-1	2	0	-3	-2
	v_{sj}	$\frac{10b_0 - 10}{3}$	$\frac{10}{3}$	$\frac{19}{3}$	0	0	$\frac{5}{3}$

Figure 9–9 Tableau 2

Therefore,

$$\hat{v}_{sj} = v_{sj} + \tfrac{5}{3}(\tilde{z}_j - \tilde{c}_j)$$
$$\hat{v}_{s0} = 5 + \tfrac{5}{3}(-1) = \tfrac{10}{3}$$
$$\hat{v}_{s1} = 3 + \tfrac{5}{3}(2) = \tfrac{19}{3}$$
$$\hat{v}_{s2} = 0 + \tfrac{5}{3}(0) = 0$$
$$\hat{v}_{s3} = 5 + \tfrac{5}{3}(-3) = 0$$
$$v_{s4} = 5 + \tfrac{5}{3}(-2) = \tfrac{5}{3}$$
$$w = 5b_0 + \tfrac{5}{3}(-b_0 - 2) = \frac{10b_0 - 10}{3}$$

Examining Tableau 2, we see that now $\bar{\alpha}_3$ may enter the basis since $v_{s3} = 0$ The vector to leave is clearly $\bar{e}_3$. The new tableau is given in Figure 9–10

			0	2	5	0	0
$\tilde{c}_{B0}$	Basis	$\bar{x}_B$	$\bar{y}_0$	$\bar{y}_1$	$\bar{y}_2$	$\bar{y}_3$	$\bar{y}_4$
-1	$\bar{e}_1$	$b_0 - 13$	(1)	1	0	0	-1
0	$\bar{\alpha}_2$	8	0	1	1	0	1
0	$\bar{\alpha}_3$	5	0	-1	0	1	0
	$\tilde{z}_j - \tilde{c}_j$	$-b_0 + 13$	-1	-1	0	0	1
	v_{sj}	40	0	3	0	0	5

Figure 9–10 Tableau 3

			0	2	5	0	0
$\tilde{\bar{c}}_{B0}$	Basis	$\bar{x}_B$	$\bar{y}_0$	$\bar{y}_1$	$\bar{y}_2$	$\bar{y}_3$	$\bar{y}_4$
0	$\bar{\alpha}_0$	$b_0 - 13$	1	1	0	0	-1
0	$\bar{\alpha}_2$	8	0	1	1	0	1
0	$\bar{\alpha}_3$	5	0	-1	0	1	0
	$\tilde{z}_j - \tilde{c}_j$	0	0	0	0	0	0
	v_{sj}	40	0	3	0	0	5

Figure 9–11 Tableau 4

The new values of v_{sj} are calculated and shown in Figure 9–10.

$$\phi_0 = \min\left(\frac{\frac{10}{3}}{1}, \frac{\frac{19}{3}}{1}\right) = \tfrac{10}{3}$$

$$\hat{v}_{s0} = \tfrac{10}{3} + \tfrac{10}{3}(-1) = 0$$

$$\hat{v}_{s1} = \tfrac{19}{3} + \tfrac{10}{3}(-1) = 3$$

$$\hat{v}_{s2} = 0 + \tfrac{10}{3}(0) = 0$$

$$\hat{v}_{s3} = 0 + \tfrac{10}{3}(0) = 0$$

$$v_{s4} = \tfrac{5}{3} + \tfrac{10}{3}(1) = 5$$

$$\hat{w} = \frac{10b_0 - 10}{3} + \tfrac{10}{3}(-b_0 + 13) = 40$$

We see from Tableau 3 in Figure 9–10 that $\bar{y}_0$ is the only nonbasic $\bar{\alpha}_j \in W$ and hence $\bar{\alpha}_0$ enters the basis. Now $\bar{e}_1$ must leave on this iteration and we will have obtained the optimal solution in the next tableau given in Figure 9–11. We note in Tableau 4 that $v_{s0} = 0$. Therefore, from (9–51), $v_0 = 0$. In addition, since $z_a = 0$, we have the optimal solution to the problem, which is

$$x_1^* = 0, \qquad x_2^* = 8, \qquad x_3^* = 5, \qquad x_4^* = 0, \qquad z^* = 40$$

PROBLEMS

1. Apply the perturbation method of Charnes that was used in Chapter 5 to resolve degeneracy in the primal simplex method, to the resolution of the degeneracy problem in the dual simplex algorithm.

2. Use Lemke's method to find an initial solution and then solve the following problem with the dual simplex method.

$$\text{Min } z = 4x_1 + 3x_2 + x_3 + 2x_4$$
$$3x_1 + 4x_2 + 5x_3 + 2x_4 \geq 10$$
$$2x_1 - x_2 + 6x_3 - x_4 \geq 3$$
$$6x_1 + 3x_2 + x_3 + 4x_4 \geq 18$$
$$x_1, x_2, x_3, x_4 \geq 0$$

3. Use the artificial constraint method to find an initial solution and then solve the following problem with the dual simplex method.

$$\text{Max } z = 3x_1 + 2x_2 + 4x_3$$
$$3x_1 + 2x_2 + 7x_3 \leq 29$$
$$2x_1 + 7x_2 + 5x_3 \leq 43$$
$$x_1, x_2, x_3 \geq 0$$

4. In Section 9–5 we proved that in the primal-dual algorithm the dual objective function w decreases strictly, i.e., $\hat{w} < w$ at each iteration. Does this mean that we do not need to resolve the degeneracy problem for this algorithm?

5. In the primal-dual algorithm, under what circumstances will more than one vector be released to enter the primal basis when a new solution to the dual is calculated?

6. Solve the following problem using the primal-dual algorithm.

$$\text{Max } z = 5x_1 - 2x_2 + 4x_3 + 2x_4 + 3x_5$$
$$x_1 + 2x_2 + 4x_3 + x_4 + 5x_5 = 40$$
$$4x_1 - 2x_2 + x_3 + 7x_4 + x_5 \geq 4$$
$$x_1 + 4x_2 + 3x_3 + 2x_4 - x_5 \geq 6$$
$$x_1, x_2, x_3, x_4, x_5 \geq 0$$

7. Show that the artificial constraint method of Section 9–3 does in fact lead to an initial solution in which all $z_j - c_j \geq 0$.

REFERENCES

1. Lemke, C. E.: The dual method of solving the linear programming problem. *Naval Res. Logistics Quarterly* **1**:48–54, 1954.
2. Hadley, G.: *Nonlinear and Dynamic Programming*. Addison-Wesley, Reading, Mass., 1964.
3. Simonnard, M.: *Linear Programming*. Prentice-Hall, Englewood Cliffs, N.J., 1966.
4. Hasegawa, H.: A Study of the Dual Simplex Algorithm. M.S. Thesis, Washington University, St. Louis, Mo., 1965.
5. Dantzig, G. B., L. R. Ford and D. R. Fulkerson: A Primal-dual algorithm for linear programming. In *Linear Inequalities and Related Systems*. Kuhn and Tucker, eds., Princeton University Press, Princeton, N.J., 1956.
6. Mueller, R. K. and L. Cooper: A comparison of the primal simplex and primal-dual algorithms for linear programming. *Comm. of the ACM* **8**:682–686, 1965.

10

The Transportation Problem

10–1 PROPERTIES OF TRANSPORTATION PROBLEMS

The phrase "transportation problem" is commonly used to describe a particular type of linear programming problem; specifically, one which has the following special structure:

$$\text{Minimize } z = \sum_{i=1}^{m} \sum_{j=1}^{n} c_{ij} x_{ij} \tag{10–1}$$

subject to

$$\sum_{j=1}^{n} x_{ij} = a_i, \qquad i = 1, 2, \ldots, m \tag{10–2}$$

$$\sum_{i=1}^{m} x_{ij} = b_j, \qquad j = 1, 2, \ldots, n \tag{10–3}$$

$$x_{ij} \geq 0, \qquad i = 1, 2, \ldots, m \quad \text{and} \quad j = 1, 2, \ldots, n \tag{10–4}$$

The above linear programming problem may be considered as one in which various amounts of a commodity are to be shipped from each of m "origins" to each of n "destinations." The amount available for shipment from the ith origin is a_i, $i = 1, 2, \ldots, m$; the amount required by the jth destination is b_j, $j = 1, 2, \ldots, n$. The cost of shipping each unit from origin i to destination j is c_{ij}. Obviously, x_{ij} denotes the quantity to be shipped from origin i to destination j. In order for this problem to have a feasible solution, the total supply must equal the total demand. Hence,

$$\sum_{i=1}^{m} a_i = \sum_{j=1}^{n} b_j \tag{10–5}$$

Frequently, it happens that total supply exceeds total demand, in which case the equations (10–2) would be replaced with inequalities:

$$\sum_{j=1}^{n} x_{ij} \leq a_i, \qquad i = 1, 2, \ldots, m$$

These inequalities allow solutions in which not all available units are shipped, in view of the excess supply. However, with the introduction of appropriate slack variables, the problem then becomes identical to the problem described by (10–1) through (10–4). This point, and several other variations of this problem, are discussed in the exercises. For the purpose of developing an algorithm for solving transportation problems, however, it is convenient to consider only the form given by (10–1) through (10–4), and to assume that (10–5) is satisfied. We shall refer to this problem as *the transportation problem.*

We observe that the transportation problem has $(m + n)$ constraints and mn variables. However, we shall soon learn that we can solve the problem with a tableau of size mn, instead of the $(m + n) \times mn$ tableau which would be required by the simplex method. Before doing so, however, we wish to make several other important observations about the transportation problem.

First, *the transportation problem always has a feasible solution:* If $T = \sum a_i = \sum b_j$, then $x_{ij} = a_i b_j / T$ is such a feasible solution, as the reader may easily verify. Secondly, *the transportation problem is never unbounded:* Each variable x_{ij} appears in exactly two constraints, both times with a coefficient of $+1$. Hence, it is easy to see that x_{ij} is bounded by

$$0 \leq x_{ij} \leq \min \{a_i, b_j\}$$

We shall conclude this section by demonstrating that a basis for the transportation problem consists of at most $(m + n - 1)$ variables, by showing that one of the $(m + n)$ constraints (10–2) and (10–3) is redundant and hence may be removed from the constraint set. If we sum the m constraints (10–2) we obtain, by (10–5),

$$\sum_{i=1}^{m} \sum_{j=1}^{n} x_{ij} = \sum_{i=1}^{m} a_i = \sum_{j=1}^{n} b_j \tag{10–6}$$

Summing the first $(n - 1)$ constraints of (10–3) yields

$$\sum_{j=1}^{n-1} \sum_{i=1}^{m} x_{ij} = \sum_{j=1}^{n-1} b_j$$

Upon subtracting the above from (10–6), we obtain

$$\sum_{i=1}^{m}\sum_{j=1}^{n} x_{ij} - \sum_{i=1}^{m}\sum_{j=1}^{n-1} x_{ij} = \sum_{j=1}^{n} b_j - \sum_{j=1}^{n-1} b_j$$

$$\sum_{i=1}^{m}\left[\sum_{j=1}^{n} x_{ij} - \sum_{j=1}^{n-1} x_{ij}\right] = b_n$$

$$\sum_{i=1}^{m} x_{nj} = b_n$$

which is the remaining (n^{th}) constraint of (10–3). Hence, this constraint is automatically satisfied whenever the other $(m + n - 1)$ constraints are satisfied. We have shown that a basis consists of at most $(m + n - 1)$ variables; in Section 10–3 we shall establish that it consists of exactly $(m + n - 1)$ variables.

10–2 A TABLEAU FOR THE TRANSPORTATION PROBLEM

We remarked at the beginning of this chapter that the transportation problem has a special structure. Let us examine this structure in more detail. We begin by rewriting the constraints (10–2) and (10–3) as follows:

$$\left.\begin{array}{lllllll}
x_{11} + x_{12} + x_{13} + \ldots + x_{1n} & & & & & & = a_1 \\
 & & x_{21} + x_{22} + \ldots + x_{2n} & & & & = a_2 \\
 & & & & & & \vdots \\
 & & & & x_{m1} + x_{m2} + \ldots + x_{mn} & & = a_m \\
x_{11} & & x_{21} + \ldots & & + x_{m1} & & = b_1 \\
 & x_{12} & & x_{22} + \ldots & & + x_{m2} & = b_2 \\
 & & & & & & \vdots \\
 & x_{1n} + & & x_{2n} + \ldots & & + x_{mn} & = b_n
\end{array}\right\} \tag{10–7}$$

Thus, if we were to write the system (10–7) in matrix notation, we could do so by defining $\bar{y} = [x_{11}, x_{12}, \ldots, x_{1n}, x_{21}, x_{22}, \ldots, x_{2n}, \ldots, x_{m1}, x_{m2}, \ldots, x_{mn}]$, $\bar{b} = [a_1, a_2, \ldots, a_m, b_1, b_2, \ldots, b_n]$,

$$A = [\bar{a}_{11}, \bar{a}_{12}, \ldots, \bar{a}_{1n}, \bar{a}_{21}, \bar{a}_{22}, \ldots, \bar{a}_{2n}, \ldots, \bar{a}_{m1}, \bar{a}_{m2}, \ldots, \bar{a}_{mn}]$$

where $\bar{a}_{ij}$ is the column of the coefficient matrix A corresponding to the variable x_{ij}. Then, (10–7) becomes $A\bar{y} = \bar{b}$.

If $\bar{e}_k$ denotes the kth column of the identity matrix $I_{(m+n)}$, then it is easy to see from (10–7) that

$$\bar{a}_{ij} = \bar{e}_i + \bar{e}_{m+j} \tag{10–8}$$

Equation (10–8) merely reflects the fact that variable x_{ij} appears only in the two constraints i and $(m + j)$, which correspond to the ith constraint of (10–2) and the jth constraint of (10–3), respectively.

The use of two subscripts to denote a column vector is a departure from our previous notational conventions, but it enables us to more readily identify a column of A with its corresponding variable.

The fact that the ith constraint of (10–7) (and of (10–2)) is the sum of the variables $x_{i1}, x_{i2}, \ldots, x_{im}$, while the $(m + j)^{\text{th}}$ constraint of (10–8) (the jth constraint of (10–3)) is the sum of the variables $x_{1j}, x_{2j}, \ldots, x_{mj}$, suggests a tableau format for recording the current values of the variables which at the same time provides an easy check on feasibility. This tableau has $(m + 1)$ rows and $(n + 1)$ columns, as shown in Figure 10–1.

A set of non-negative values for the variables x_{ij} is feasible if and only if the sum of the values in the ith row of the tableau is equal to a_i, for each $i = 1, 2, \ldots, m$, and the sum of the values in column j of the tableau is b_j, $j = 1, 2, \ldots, n$.

We have already seen that a *basic* feasible solution will contain no more than $(m + n - 1)$ positive variables, with the remaining variables being zero. However, not every feasible solution containing this number of positive variables is basic: the $(m + n - 1)$ columns of the coefficient matrix A corresponding to these variables must be linearly independent in order for such a solution to be basic. Hence, we must derive a method for determining

x_{11}	x_{12}	$\cdots$	x_{1j}	$\cdots$	x_{1n}	a_1
x_{21}	x_{22}	$\cdots$	x_{2j}	$\cdots$	x_{2n}	a_2
$\vdots$	$\vdots$		$\vdots$		$\vdots$	$\vdots$
x_{i1}	x_{i2}	$\cdots$	x_{ij}	$\cdots$	x_{in}	a_i
						$\vdots$
x_{m1}	x_{m2}	$\cdots$	x_{mj}	$\cdots$	x_{mn}	a_m
b_1	b_2	$\cdots$	b_j	$\cdots$	b_n	

Figure 10–1 A transportation problem tableau

whether a given set of $(m + n - 1)$ $\bar{a}_{ij}$ is linearly dependent. To do so, we first introduce the following terminology:

1. The "box" in row i and column j of the tableau is called the (i, j) *cell.*
2. A *loop* is a sequence of cells such that:
   a) each adjacent pair of cells lies in either the same row or the same column of the tableau;
   b) no group of three or more consecutive cells in the sequence lies in the same row or in the same column; and
   c) the first and last cells in the sequence are in either the same row or the same column.

In Figure 10–2 we have illustrated several examples of loops for a transportation problem in which $m = 4$ and $n = 6$. Loop (*a*) consists of four cells. Loops (*b*) and (*c*) contain six cells, and loop (*d*) contains ten cells. Note that the loops are closed; there is no "beginning" and no "end." Thus, the sequences

1. (1, 2), (1, 4), (2, 4), (2, 8), (4, 8), (4, 2)
2. (1, 4), (2, 4), (2, 8), (4, 8), (4, 2), (1, 2)
3. (2, 8), (4, 8), (4, 2), (1, 2), (1, 4), (2, 4)
4. (4, 2), (4, 8), (2, 8), (2, 4), (1, 4), (1, 2)

all yield the same loop; namely, loop (*b*).

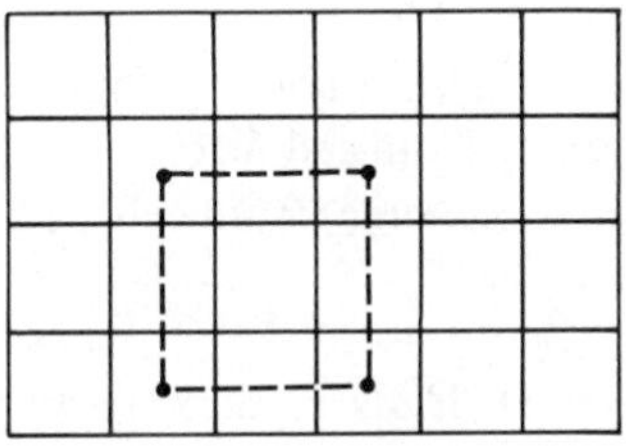

(a) (2,2), (2,4), (4,4), (4,2)

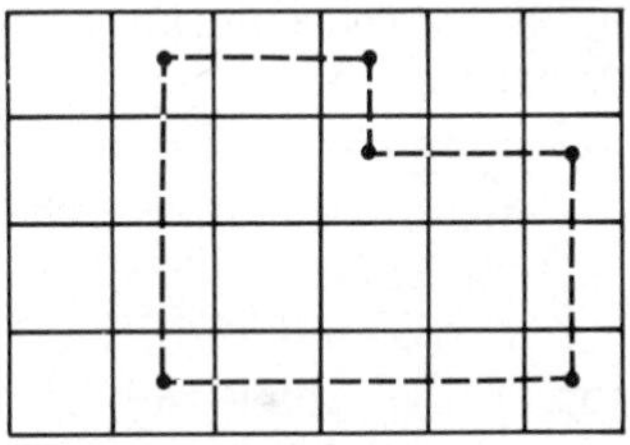

(b) (1,2), (1,4), (2,4), (2,6) (4,6), (4,2)

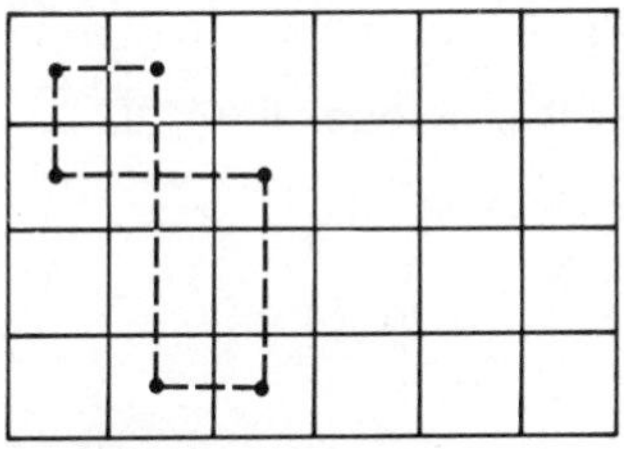

(c) (1,1), (2,1), (2,3) (4,3), (4,2), (1,2)

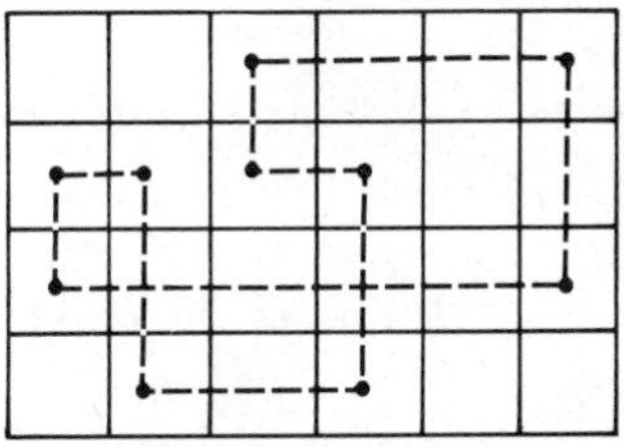

(d) (4,4), (2,4), (2,3), (1,3), (1,6), (3,6), (3,1), (2,1), (2,2), (4,2)

Figure 10–2 Examples of loops

Note that loop (*d*) contains four cells in row 2; however, no three of these are listed consecutively in the sequence describing the loop.

As Figure 10–2 indicates, given any loop, each row of the tableau contains either no cells in the loop or an even number of cells in the loop; similarly, every column of the tableau contains either no cells or an even number of cells in the loop. Furthermore, in going from one cell to an adjacent cell in a loop, we alternate between moving to a cell in the same row and between moving to a cell in the same column. Thus, every loop is described by a sequence of cells of the form

$$(k, p), (k, r), (s, r), (s, q), \ldots, (u, v), (u, p)$$

These facts suggest:

Theorem 10–1 Every loop contains an even number of cells.

Proof: Suppose we have a loop containing N cells. Choose any one of the cells as the starting point, and either of its two adjacent cells in the sequence as the second cell. Number these cells 1 and 2, and continue numbering the remaining cells in the loop, in order, from 3 to N. Now, cells 1 and 2 either lie in the same row or the same column. Assume they are in the same row. Then the step from cell 2 to cell 3 must be a column change, the step from cell 3 to cell 4 must be a row change, etc. Alternating in this manner, we see that the step from any cell t to cell $(t + 1)$ is a column change if t is even, and is a row change if t is odd. But, the step from cell N to cell 1 must be a column change, since the next change is a row change, from cell 1 to cell 2. Hence, N must be even.

Theorem 10–1, along with equation (10–8), leads us directly to a very important result, which we state as Theorem 10–2.

Theorem 10–2 Let T denote any subset of columns $\bar{a}_{ij}$ of coefficient matrix A of a transportation problem. Then the columns of T are linearly dependent if and only if the corresponding cells, or a subset of them, can be arranged into a sequence S which forms a loop.

Proof: a) Assume a sequence of cells forms a loop. Then, the sequence is of the form:

$$(k, p), (k, r), (s, r), (s, q), \ldots, (u, v), (u, p)$$

The corresponding columns of T are

$$\bar{a}_{kp}, \bar{a}_{kr}, \bar{a}_{sr}, \bar{a}_{sq}, \ldots, \bar{a}_{uv}, \bar{a}_{up}$$

Suppose we alternately add and subtract these columns, forming the linear combination

$$\bar{v} = \bar{a}_{kp} - \bar{a}_{kr} + \bar{a}_{sr} - \bar{a}_{sq} + \dots + \bar{a}_{uv} - \bar{a}_{up} \tag{10–9}$$

Substitution of (10–8) into (10–9) yields

$$\begin{aligned}\bar{v} &= (\bar{e}_k + \bar{e}_{m+p}) - (\bar{e}_k + \bar{e}_{m+r}) + (\bar{e}_s + \bar{e}_{m+r}) - (\bar{e}_s + \bar{e}_{m+q}) + \dots \\ &\quad + (\bar{e}_u + \bar{e}_{m+v}) - (\bar{e}_u + \bar{e}_{m+p}) \\ &= (\bar{e}_k - \bar{e}_k) + (\bar{e}_{m+p} - \bar{e}_{m+p}) + (\bar{e}_{m+r} - \bar{e}_{m+r}) + \dots + (\bar{e}_s - \bar{e}_s) \\ &= \bar{0},\end{aligned}$$

since there are an even number of terms in the right-hand side of (10–9). Hence, a nontrivial linear combination of the columns $\bar{a}_{kp}, \bar{a}_{kr}, \dots, \bar{a}_{up}$ equals the null vector, and the set is therefore linearly dependent.

b) We must now show that any linearly dependent set T of columns of A has a corresponding sequence of cells which forms a loop. For such a set T, there exist scalars α_{ij} not all zero such that

$$\sum_{\bar{a}_{ij} \in T} \alpha_{ij}\bar{a}_{ij} = \bar{0} \tag{10–10}$$

Henceforth, we shall say that a vector $\bar{a}_{pq}$ "appears" in the sum of (10–10) if and only if its corresponding scalar $\alpha_{pq} \neq 0$.

Suppose that $\bar{a}_{pq}$ appears in (10–10); hence, the term $\alpha_{pq}\bar{a}_{pq} = \alpha_{pq}(\bar{e}_p + \bar{e}_{m+q})$ appears. This implies that at least one other $\bar{a}_{ij}$ corresponding to row p of the tableau must also appear in (10–10); if this were not the case, then the pth component of the vector $\sum \alpha_{ij}\bar{a}_{ij}$ would be equal to $\alpha_{pq} \cdot 1 \neq 0$, which contradicts equation (10–10). Let $\bar{a}_{pk}$ be one such vector appearing in the sum of (10–10), $k \neq q$. Then, by the same reasoning, there must be at least one other vector appearing in the sum of (10–10) corresponding to column q and one corresponding to column k, of the tableau. These conditions can be met only in one of two ways:

(1) The vectors $\bar{a}_{rq}$ and $\bar{a}_{rk}$ both appear in the sum of (10–10), $r \neq p$. Then a loop is formed by the cells corresponding to the four vectors $\bar{a}_{pq}$, $\bar{a}_{pk}$, $\bar{a}_{rk}$, $\bar{a}_{rq}$.

(2) The vectors $\bar{a}_{rq}$ and $\bar{a}_{sk}$ appear in the sum of (10–10), $r \neq s$, and $r, s \neq p$. In this case, the sum of (10–10) must also contain at least one additional vector corresponding to row r and one corresponding to row s. These conditions can also be met only in one of two possible ways: either the vectors $\bar{a}_{rt}$ and $\bar{a}_{st}$ are in the sum of (10–10), in which case a loop is formed by the cells corresponding to the vectors $\bar{a}_{pq}$, $\bar{a}_{pk}$, $\bar{a}_{sk}$, $\bar{a}_{st}$, $\bar{a}_{rt}$, $\bar{a}_{rq}$, as illustrated schematically in Figure 10–3; otherwise, two vectors $\bar{a}_{rt}$ and $\bar{a}_{sw}$ must appear in the sum of (10–10), $t \neq w$. In this case, the same reasoning leads us to the

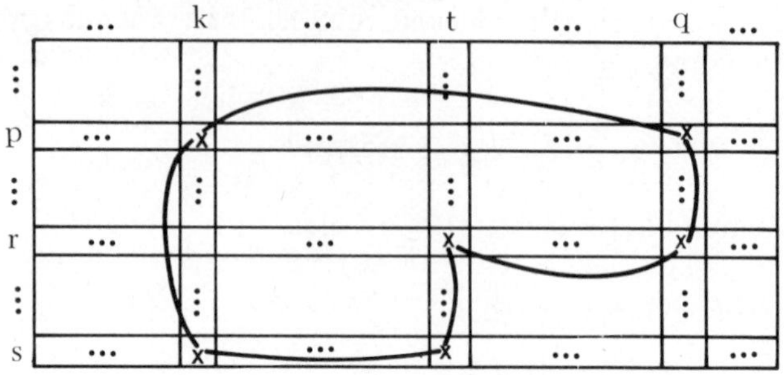

Figure 10–3 Schematic of a six cell loop

conclusion that either an eight cell loop is formed by the appearance of at least two more vectors corresponding to columns t and w of the tableau, $\bar{a}_{vt}$ and $\bar{a}_{vw}$; or that at least four more vectors must appear which, in turn, either leads to the conclusion that a ten cell loop is formed or else more vectors must appear in the sum of (10–10).

We can continue the above argument only a finite number of times, because there are only a finite number of vectors which can appear in the sum of (10–10). In each step of this argument we find that the required addition of two vectors either forms a loop or, in turn, requires the addition of at least two more vectors. Finally, we must reach a stage at which we have either formed a loop or must add the last two vectors. Clearly, when all vectors in the tableau are included, a loop is formed.

Hence, we have shown that any linearly dependent set of columns of A has a corresponding sequence of cells which forms a loop.

We are now ready to develop the formulas for the elements of the simplex tableau, which will demonstrate why it is unnecessary to actually construct a regular simplex tableau to solve the transportation problem.

In an ordinary simplex tableau, there are three sets of numbers to calculate:

1) The elements y_{kj}, where y_{kj} is the coefficient of the kth basis vector in the representation of the nonbasic vector $\bar{a}_j$ in terms of the current basis vectors;

2) The values of the current basis variables;

3) The values of the quantities $z_j - c_j$.

We shall now show how these quantities may be computed directly, given any set of basis vectors, without the use of a tableau:

First, let us consider the y_{kj} elements. Since we are now using two subscripts to denote a column, we shall define

$$\bar{b}_k = k^{th} \text{ current basis vector, } k = 1, 2, \ldots, (m + n - 1)$$

y_{ij}^k = coefficient of $\bar{b}_k$ in the representation of the nonbasic vector $\bar{a}_{ij}$ in terms of $\bar{b}_1, \bar{b}_2, \ldots, \bar{b}_{m+n-1}$.

Thus,

$$\bar{a}_{ij} = \sum_{k=1}^{m+n-1} y_{ij}^k \bar{b}_k \tag{10–11}$$

Now, the $m + n$ vectors $\bar{a}_{ij}, \bar{b}_1, \bar{b}_2, \ldots, \bar{b}_{m+n-1}$ are linearly dependent; hence, these vectors have a corresponding sequence of cells which forms a loop. Furthermore, the cell (i, j) must be in this loop, since the vectors $\bar{b}_1, \bar{b}_2, \ldots, \bar{b}_{m+n-1}$ are linearly independent, and their corresponding cells cannot form a loop.

Let this loop consist of the cells (i, j), (i, q), (p, q), (p, r), ..., (u, v), (u, j) (which implies that the vectors $\bar{a}_{iq}, \bar{a}_{pq}, \bar{a}_{pr}, \ldots, a_{uv}, \bar{a}_{uj}$ are basic vectors). Then, as we have seen in part a) of the proof of Theorem 10–1,

$$\bar{a}_{ij} - \bar{a}_{iq} + \bar{a}_{pq} - \bar{a}_{pr} + \ldots + \bar{a}_{uv} - \bar{a}_{uj} = \bar{0},$$

or

$$\bar{a}_{ij} = \bar{a}_{iq} - \bar{a}_{pq} + \bar{a}_{pr} - \ldots - \bar{a}_{uv} + \bar{a}_{uj} \tag{10–12}$$

Upon comparing equations (10–11) and (10–12) we see that

$$y_{ij}^k = \begin{Bmatrix} \pm 1, \text{ if } \bar{b}_k \text{ is one of the vectors appearing} \\ \text{in the right-hand side of (10–12)} \\ 0, \text{ if } \bar{b}_k \text{ is not one of the vectors in (10–12)} \end{Bmatrix} \tag{10–13}$$

Furthermore, if we number the vectors corresponding to the cells in the loop in sequence, starting with the nonbasic vector $\bar{a}_{ij}$ as number 1, so that basic vector $\bar{b}_k$ is the tth vector in this list, then

$$y_{ij}^k = \begin{Bmatrix} 1, & \text{if } t \text{ is even} \\ -1, & \text{if } t \text{ is odd} \end{Bmatrix} \tag{10–14}$$

Let us summarize these results:

Theorem 10–3 Given any set of basis vectors for the transportation problem, the elements y_{ij}^k are equal to either 0, +1, or − 1, and are given by equations (10–13) and (10–14).

Suppose now that we have somehow determined an initial basic feasible solution (we shall see how to do so in the next section). Specifically, suppose we know which $(m + n - 1)$ variables are basic and their current values. Let x_{Bk}, $k = 1, 2, \ldots, m + n - 1$, be these current values.

In the simplex method we first choose a nonbasic vector to enter the basis. Suppose we have chosen vector $\bar{a}_{ij}$. We then choose the vector to leave,

$\bar{b}_r$, from

$$\frac{x_{Br}}{y_{ij}^{r}} = \min_{k} \left\{ \frac{x_{Bk}}{y_{ij}^{k}} \,\middle|\, y_{ij}^{k} > 0 \right\} \tag{10–15}$$

Equation (10–15) is merely the counterpart of equation (4–44), using the notation of this section. Furthermore, Theorem 10–3 tells us that if $y_{ij}^k > 0$, then $y_{ij}^k = 1$. Hence, equation (10–15) becomes

$$x_{Br} = \min_{k} \{x_{Bk} \mid y_{ij}^{k} = 1\} \tag{10–16}$$

Once the vector to enter the basis, $\bar{a}_{ij}$, and the vector to leave the basis, $\bar{b}_r$, have been determined, the values of the new basic variables are computed by:

$$\left.\begin{aligned} \hat{x}_{Bk} &= x_{Bk} - y_{ij}^{k} x_{Br}, \qquad k \neq r \\ \hat{x}_{Br} &= x_{Br} \end{aligned}\right\} \tag{10–17}$$

The above discussion, along with equations (10–17) proves

Theorem 10–4 For any transportation problem, if an initial basic feasible solution is found in which all the variables are integer-valued, then all subsequent basic feasible solutions found by the simplex method will also be all integer-valued.

As we shall see shortly, if all the a_i and b_j in the transportation problem are integers, then an all integer initial basic feasible solution is very easy to find. Furthermore, in such a case, we can perform all computations in integer arithmetic, as opposed to real or floating point arithmetic. If we are performing these computations on a digital computer, the use of integer arithmetic means that there will be no round-off errors in our computations.

To complete our discussion of the quantities of interest in an ordinary simplex tableau, we need to derive an expression for the quantities $z_{ij} - c_{ij}$, which correspond to the column vectors $\bar{a}_{ij}$. The transportation problem definition of z_{ij}, analogous to the definition of z_j given in Chapter 4, $\left(z_j = \sum_{k=1}^{m} y_{kj}\, c_{Bk}\right)$ is

$$z_{ij} = \sum_{k=1}^{m+n-1} y_{ij}^{k} c_{Bk} \tag{10–18}$$

where c_{Bk} is the cost corresponding to the kth basic vector $\bar{b}_k$. Since all $y_{ij}^k = 0, \pm 1$, as given by equations (10–13) and (10–14), we see that each $z_{ij} - c_{ij}$ is easily computed once we know the loop formed by the basis vectors and $\bar{a}_{ij}$.

Hence, we now have all the quantities we need for the simplex method: choosing the vector whose $z_{ij} - c_{ij}$ is the most positive† as the vector to enter the basis, we find the loop formed by this vector and the basis vectors, and calculate the vector to leave the basis by (10–16). Next, we calculate the new values of the basic variables by (10–17) and the new $z_{ij} - c_{ij}$ by (10–18).

The reader should now be able to see that we can store all the necessary information in a tableau of the form given in Table 10–1, provided that we make the following modifications: In each cell corresponding to a basic variable, we enter the current value of that variable; in each nonbasic cell, we enter the corresponding value of $z_{ij} - c_{ij}$ (recall that $z_{ij} - c_{ij} = 0$ for basic variables). To help us distinguish between basic and nonbasic cells, we shall circle the values of the basic variables. Also, as another notational convenience, we shall insert the numbers c_{ij} in the upper left-hand corner of each cell.

Figure 10–4 illustrates our modified tableau for a transportation problem with three origins and five destinations, in which the current basis vectors are $\bar{a}_{11}, \bar{a}_{13}, \bar{a}_{22}, \bar{a}_{23}, \bar{a}_{25}, \bar{a}_{34}, \bar{a}_{35}$. Observe that these $m + n - 1 = 7$ vectors do not form a loop. The quantity $z_{12} - c_{12}$ is computed by finding the loop formed by cell (1, 2) and the basis cells; in this case the loop is (1, 2), (1, 3), (2, 3), (2, 2). Hence,

$$z_{12} - c_{12} = c_{13} - c_{23} + c_{22} - c_{12}$$

Similarly, we find that

$$\begin{aligned} z_{14} - c_{14} &= c_{13} - c_{23} + c_{25} - c_{35} + c_{34} - c_{14} \\ z_{15} - c_{15} &= c_{25} - c_{23} + c_{13} - c_{15} \\ z_{31} - c_{31} &= c_{11} - c_{13} + c_{23} - c_{25} + c_{35} - c_{31} \end{aligned}$$

And so forth.

c_{11} / (x_{B1})	c_{12} / $z_{12} - c_{12}$	c_{13} / (x_{B2})	c_{14} / $z_{14} - c_{14}$	c_{15} / $z_{15} - c_{15}$	a_1
c_{21} / $z_{21} - c_{21}$	c_{22} / (x_{B3})	c_{23} / (x_{B4})	c_{24} / $z_{24} - c_{24}$	c_{25} / (x_{B5})	a_2
c_{31} / $z_{31} - c_{31}$	c_{32} / $z_{32} - c_{32}$	c_{33} / $z_{33} - c_{33}$	c_{34} / (x_{B6})	c_{35} / (x_{B7})	a_3
b_1	b_2	b_3	b_4	b_5	

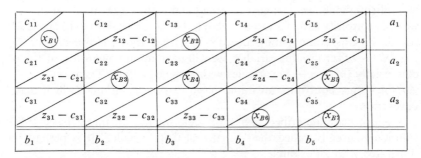

Figure 10–4 Example of transportation problem tableau

† We have formulated the transportation problem as a minimization problem. Hence, a basic feasible solution is optimal if all $z_{ij} - c_{ij} \leq 0$.

10–3 A TRANSPORTATION PROBLEM ALGORITHM

In the previous section we have developed all the necessary theory for specializing the simplex method for the transportation problem, with the exception of the determination of an initial basic feasible solution. Let us resolve this last point now, by describing a procedure for finding an initial basic feasible solution for any transportation problem.

We begin in row 1 by letting x_{11} be the first basic variable and by letting x_{11} be as large as possible. This means that either the supply at origin 1 will be exhausted or the demand at destination 1 will be completely satisfied. If the former, we proceed to row 2; if the latter, we continue in row 1 by allocating as much as possible to x_{12}, which will either exhaust the supply at origin 1 or completely satisfy the demand of destination 2. At each step of the process we either completely exhaust a supply or satisfy a demand. If, at a given step, we exhaust the supply, we proceed to the next row; if a demand is satisfied we continue in the same row by moving to the next column and allocating as much as possible. Thus, if we are assigning a value to variable x_{ij}, and in so doing exhaust the supply at resource i, then our next assignment will be to variable $x_{i+1,j}$. On the other hand, if the demand at destination j is satisfied by the assignment to x_{ij}, then our next assignment will be to variable $x_{i,j+1}$.

At each step of this process we either increase the first subscript or the second subscript of the variable for which we will make the next assignment. Hence, at most $(m + n)$ of the x_{ij}'s will be positive. Moreover, at the last step, we will simultaneously satisfy both the demand at destination n and exhaust the supply at resource m. This means that at most $(m + n - 1)$ variables will be positive. Furthermore, only one variable in either row m or column n will have been involved, and hence the cells corresponding to these variables do not form a loop.

In the above process, it is possible that at some step (other than the last) we will simultaneously exhaust a supply and satisfy a demand. In this case, we can arbitrarily assign a value of zero to the next variable in either the same row or the same column and proceed as described above. However, by assigning a value of zero to this variable, we will have upon conclusion of the process exactly $(m + n - 1)$ variables with assigned values. These $(m + n - 1)$ variables thus constitute an initial basic feasible solution† (which may be degenerate).

The process described above for determining an initial basic feasible solution is usually called the "Northwest Corner Rule." It is perhaps best illustrated with an example.

† The construction of a set of $m + n - 1$ cells which do not form a loop establishes that a basis does in fact consist of $m + n - 1$ linearly independent vectors.

EXAMPLE 1

Consider a transportation problem with four origins and six destinations, with origin supplies and destination demands as follows:

Origin	i	1	2	3	4		
Supply	a_i	10	15	12	9		
Destination	j	1	2	3	4	5	6
Demand	b_j	5	10	10	8	8	5

We first let $x_{11} = 5$, which satisfies the demand at destination 1. Next, we contine in row 1, and let $x_{12} = 5$, which exhausts the supply at origin 1. Moving to row 2, we let $x_{22} = 5$, which satisfies the demand at destination 2. We then let $x_{23} = 10$, which simultaneously exhausts the supply at origin 2 and satisfies the demand at destination 3. We arbitrarily assign $x_{24} = 0$. (We could just as easily have assigned $x_{33} = 0$, but we must designate one of these as having an assigned value of 0, so that the number of variables with assigned values is equal to $(m + n - 1) = 4 + 6 - 1 = 9$; hence, x_{24} is a basic variable equal to 0, and x_{33} is nonbasic). Next, we assign a value of 8 to x_{34}, satisfying the demand at destination 4. Continuing in this manner, we assign, in the following order, $x_{35} = 4$, $x_{45} = 4$, and $x_{46} = 5$. These calculations are summarized in Figure 10–5.

The Northwest Corner Rule is not a very efficient procedure for finding an initial basic feasible solution, because it does not take the objective function into account, and hence may yield a poor initial solution (i.e., one with a relatively large value of z). Several other methods for finding an initial solution are presented in the next section.

At this point, however, let us summarize the algorithm we have developed, and illustrate it by solving an example problem.

		Destinations 1	2	3	4	5	6	Remaining Supplies		
Origins	1	5	5					10	5	0
	2		5	10	0			15	10	0
	3				8	4		12	4	0
	4					4	5	9	5	0
Remaining Demands		5	10	10	8	8	5			
			5	0	0	4	0			
			0			0				

Figure 10–5 Initial basic feasible solution for Example 1

A Transportation Algorithm

1. Find an initial basic feasible solution.

2. Calculate the quantities $(z_{pq} - c_{pq})$ for each nonbasic variable. If $z_{pq} - c_{pq} \leq 0$ for all nonbasic variables the solution is optimal. If at least one $z_{pq} - c_{pq} > 0$, proceed to step 3.

3. Choose the variable x_{ij} to enter the basis, according to:

$$z_{ij} - c_{ij} = \underset{pq}{\text{maximum}} \{z_{pq} - c_{pq}\}$$

4. Determine the loop formed by the nonbasic cell (i, j) and the basic cells.† Number the cells in this loop consecutively, starting with (i, j); let $y_{ij}^k = 1$ for even-numbered cells, and $y_{ij}^k = -1$ for odd-numbered cells.

5. Choose the variable x_{Br} to leave the basis, according to: $x_{Br} = \min_k \{x_{Bk} \mid y_{ij}^k = 1\}$.

6. Compute the new basic feasible solution, by adding the quantity x_{Br} to each basic variable whose cell is odd-numbered in the loop determined in step 4, and by subtracting the quantity x_{Br} from each basic variable whose cell is even-numbered. Also, set $x_{ij} = x_{Br}$.

7. Return to Step 2.

Step 4 follows directly from equation (10–14), and Step 6 follows from equations (10–14) and (10–17).

EXAMPLE 2

Suppose the matrix of cost coefficients $C = (c_{ij})$ for the problem of Example 1 is as follows:

$$C = \begin{bmatrix} 10 & 15 & 10 & 12 & 20 & 15 \\ 5 & 10 & 8 & 15 & 10 & 12 \\ 15 & 10 & 12 & 12 & 10 & 20 \\ 20 & 15 & 15 & 10 & 12 & 12 \end{bmatrix}$$

Then, using the initial basic feasible solution found in Example 1,

† A nonbasic cell is one corresponding to a nonbasic variable; a basic cell is one corresponding to a basic variable.

we now must calculate

$$z_{13} - c_{13} = c_{23} - c_{22} + c_{12} - c_{13} = 3$$
$$z_{14} - c_{14} = c_{24} - c_{22} + c_{12} - c_{14} = 8$$
$$z_{15} - c_{15} = c_{35} - c_{34} + c_{24} - c_{22} + c_{12} - c_{15} = -2$$
$$z_{16} - c_{16} = c_{46} - c_{45} + c_{35} - c_{34} + c_{24} - c_{22} + c_{12} - c_{16} = 3$$
$$z_{21} - c_{21} = c_{11} - c_{12} + c_{22} - c_{21} = 0$$
$$z_{25} - c_{25} = c_{24} - c_{34} + c_{35} - c_{25} = 3$$
$$z_{26} - c_{26} = c_{46} - c_{45} + c_{35} - c_{34} + c_{24} - c_{26} = 1$$
$$z_{31} - c_{31} = c_{11} - c_{12} + c_{22} - c_{24} + c_{34} - c_{31} = -13$$
$$z_{32} - c_{32} = c_{22} - c_{24} + c_{34} - c_{32} = -3$$
$$z_{33} - c_{33} = c_{23} - c_{24} + c_{34} - c_{33} = -7$$
$$z_{36} - c_{36} = c_{46} - c_{45} + c_{35} - c_{36} = -10$$
$$z_{41} - c_{41} = c_{11} - c_{12} + c_{22} - c_{24} + c_{34} - c_{35} + c_{45} - c_{41} = -16$$
$$z_{42} - c_{42} = c_{22} - c_{24} + c_{34} - c_{35} + c_{45} - c_{42} = -6$$
$$z_{43} - c_{43} = c_{23} - c_{24} + c_{34} - c_{35} + c_{45} - c_{43} = -8$$
$$z_{44} - c_{44} = c_{34} - c_{35} + c_{45} - c_{44} = 4$$

Hence, the initial transportation tableau is as given in Figure 10–6.

Proceeding to step 3 of the algorithm, we find that $z_{14} - c_{14} = \max \{z_{pq} - c_{pq}\}$; hence, x_{14} enters the basis.

The loop formed by cell (1, 4) and the basic cells is: (1, 4), (1, 5), (2, 2), (1, 2). The variable to leave the basis is thus $x_{24} = \text{Min}\ \{x_{24}, x_{12}\}$. Since $x_{24} = 0$, the values of the remaining basis variables remain unchanged, but basic

Origins \ Destinations	1	2	3	4	5	6	Supply
1	10 / (5)	15 / (5)	10 / 3	12 / 8	20 / −2	15 / 3	10
2	5 / 0	10 / (5)	8 / (10)	15 / (0)	10 / 3	12 / 1	15
3	15 / −13	10 / −3	12 / −7	12 / (8)	10 / (4)	20 / −10	12
4	20 / −16	15 / −6	15 / −8	10 / 4	12 / (4)	12 / (5)	9
Demand	5	10	10	8	8	5	

Figure 10–6 First tableau for Example 2

variable x_{24} has been replaced in the basis by x_{14}. The $z_{pq} - c_{pq}$ must all be re-computed; these new values are computed below and are given in Figure 10–7.

$$z_{13} - c_{13} = c_{23} - c_{22} + c_{12} - c_{13} = 3$$
$$z_{15} - c_{15} = c_{35} - c_{34} + c_{14} - c_{15} = -10$$
$$z_{16} - c_{16} = c_{46} - c_{45} + c_{35} - c_{34} + c_{14} - c_{16} = -5$$
$$z_{21} - c_{21} = c_{22} - c_{12} + c_{11} - c_{21} = 0$$
$$z_{24} - c_{24} = c_{14} - c_{12} + c_{22} - c_{24} = -8$$
$$z_{25} - c_{25} = c_{35} - c_{34} + c_{14} - c_{12} + c_{22} - c_{25} = -5$$
$$z_{26} - c_{26} = c_{46} - c_{45} + c_{35} - c_{34} + c_{14} - c_{12} + c_{22} - c_{26} = -7$$
$$z_{31} - c_{31} = c_{11} - c_{14} + c_{34} - c_{31} = -5$$
$$z_{32} - c_{32} = c_{12} - c_{14} + c_{34} - c_{32} = 5$$
$$z_{33} - c_{33} = c_{23} - c_{22} + c_{12} - c_{14} + c_{34} - c_{33} = 1$$
$$z_{36} - c_{36} = c_{46} - c_{45} + c_{35} - c_{36} = -10$$
$$z_{41} - c_{41} = c_{11} - c_{14} + c_{34} - c_{35} + c_{45} - c_{41} = -8$$
$$z_{42} - c_{42} = c_{12} - c_{14} + c_{34} - c_{35} + c_{45} - c_{42} = 2$$
$$z_{43} - c_{43} = c_{23} - c_{22} + c_{12} - c_{14} + c_{34} - c_{35} + c_{45} - c_{43} = 0$$
$$z_{44} - c_{44} = c_{34} - c_{35} + c_{45} - c_{44} = 4$$

Since $z_{32} - c_{32} = \max\{z_{pq} - c_{pq}\} = 5$, we choose x_{32} as the entering variable. The loop formed by cell (3, 2) and the basic cells is (3, 2), (1, 2), (1, 4), (3, 4). The variable which leaves the basis is $\min\{x_{12}, x_{34}\} = x_{12} = 5$.

In the new basic solution,

$$\hat{x}_{14} = 0 + 5 = 5,$$
$$\hat{x}_{34} = 8 - 5 = 3,$$
$$\hat{x}_{32} = 5,$$

10 / (5)	15 / (5)	10 / 3	12 / (0)	20 / −10	15 / −5	10
5 / 0	10 / (5)	8 / (10)	15 / −8	10 / −5	12 / −7	15
15 / −5	10 / 5	12 / 1	12 / (8)	10 / (4)	20 / −10	12
20 / −8	15 / 2	15 / 0	10 / 4	12 / (4)	12 / (5)	9
5	10	10	8	8	5	

Figure 10–7 Second tableau for Example 2

10 / (5)	15 / −5	10 / −2	12 / (5)	20 / −10	15 / −5	10
5 / 5	10 / (5)	8 / (10)	15 / −3	10 / 0	12 / −2	15
15 / −5	10 / (5)	12 / −4	12 / (3)	10 / (4)	20 / −10	12
20 / −8	15 / −3	15 / −5	10 / 4	12 / (4)	12 / (5)	9
5	10	10	8	8	5	

Figure 10–8 Third tableau for Example 2

and all other basic variables remain unchanged. The new tableau is shown in Figure 10–8.

At the next iteration, x_{21} enters the basis; the loop formed by cell (2, 1) with the basic cells is (2, 1), (1, 1), (1, 4), (3, 4), (3, 2), (2, 2). The variable which leaves the basis is $x_{34} = \min\{x_{11}, x_{34}, x_{22}\} = 2$. The new values of the basic variables are $\hat{x}_{11} = 5 - 3 = 2$, $\hat{x}_{14} = 5 + 3 = 8$, $\hat{x}_{24} = 5 + 3 = 8$, $\hat{x}_{21} = 3$. The other basic variables are unchanged, and the new tableau is given in Figure 10–9.

In the following two iterations, given in Figures 10–10 and 10–11, we find that x_{13} enters the basis and x_{11} leaves the basis, and then that x_{44} enters the basis, with x_{45} leaving the basis. In Figure 10–11, we see that the current solution is optimal, as all $z_{pq} - c_{pq} \leq 0$. Hence, the optimal solution† is $x_{13} = 6$, $x_{14} = 4$, $x_{21} = 5$, $x_{22} = 6$, $x_{23} = 4$, $x_{32} = 4$, $x_{35} = 8$, $x_{44} = 4$, $x_{46} = 5$, and the optimal value of the objective function is $z = 6(10) + 4(12) + 5(5) + 6(10) + 4(8) + 4(10) + 8(10) + 4(10) + 5(12) = 445$.

10 / (2)	15 / 0	10 / 3	12 / (8)	20 / −5	15 / 0	10
5 / (3)	10 / (2)	8 / (10)	15 / −8	10 / 0	12 / −2	15
15 / 10	10 / (8)	12 / −4	12 / −5	10 / (4)	20 / −10	12
20 / −12	15 / −3	15 / −5	10 / −1	12 / (4)	12 / (5)	9
5	10	10	8	8	5	

Figure 10–9 Fourth tableau for Example 2

† The fact that $z_{25} - c_{25} = 0$ and $z_{26} - c_{26} = 0$ means that there are alternate optima to this problem.

10 / −3	15 / −3	10 / ②	12 / ⑧	20 / −8	15 / −3	10
5 / ⑤	10 / ②	8 / ⑧	15 / −5	10 / 0	12 / −2	15
15 / −10	10 / ⑧	12 / −4	12 / −2	10 / ④	20 / −10	12
20 / −13	15 / −3	15 / −5	10 / 2	12 / ④	12 / ⑤	9
5	10	10	8	8	5	

Figure 10–10 Fifth tableau for Example 2

An alternate—and computationally more efficient—method for calculating the values of $(z_{pq} - c_{pq})$ for each nonbasic variable is developed below. This approach is often called the "*uv* method."

Consider the following: The original objective function for the transportation problem may be written as

$$z - \sum_{i=1}^{m} \sum_{j=1}^{n} c_{ij} x_{ij} = 0 \tag{10–19}$$

Suppose we eliminate the basic variables from this expression, by adding a multiple of each of the constraint equations to (10–19); let u_i be the multiple of the constraint $\sum_{j=1}^{n} x_{ij} = a_i$ added to (10–19), and let v_j be the multiple of the constraint $\sum_{i=1}^{m} x_{ij} = b_j$ added to (10–19). Hence, we add in total the quantity

$$\sum_{i=1}^{m} u_i \sum_{j=1}^{n} x_{ij} + \sum_{j=1}^{n} v_j \sum_{i=1}^{m} x_{ij} = \sum_{i=1}^{m} u_i a_i + \sum_{j=1}^{n} v_j b_j$$

10 / −3	15 / −3	10 / ⑥	12 / ④	20 / −8	15 / −1	10
5 / ⑤	10 / ⑥	8 / ④	15 / −5	10 / 0	12 / 0	15
15 / −10	10 / ④	12 / −4	12 / −2	10 / ⑧	20 / −8	12
20 / −15	15 / −5	15 / −7	10 / ④	12 / −2	12 / ⑤	9
5	10	10	8	8	5	

Figure 10–11 Sixth tableau for Example 2

to equation (10–19). The result is:

$$z + \sum_{i=1}^{m} \sum_{j=1}^{n} \{u_i + v_j - c_{ij}\}x_{ij} = \sum_{i=1}^{m} a_i u_i + \sum_{j=1}^{n} b_j v_j \qquad (10\text{–}20)$$

The coefficient of each basic variable x_{ij} in equation (10–20) is zero, by our choice of values for the u_i and v_j. Moreover, by definition of $z_{ij} - c_{ij}$,

$$z + \sum_{i=1}^{m} \sum_{j=1}^{n} \{z_{ij} - c_{ij}\}x_{ij} = f \qquad (10\text{–}21)$$

where $z_{ij} - c_{ij} = 0$ for each basic x_{ij}, and where f is the current value of the objective function. Hence, upon comparing equations (10–20) and (10–21) we find that

$$z_{ij} - c_{ij} = u_i + v_j - c_{ij} \qquad (10\text{–}22)$$

$$f = \sum_{i=1}^{m} a_i u_i + \sum_{j=1}^{n} b_j v_j \qquad (10\text{–}23)$$

Thus, if we know the values of the m u_i and the n v_j, we can calculate $z_{ij} - c_{ij}$ for all nonbasic variables x_{ij}, from equation (10–22). But, these values may be easily computed by solving the system of $(m + n - 1)$ equations

$$u_p + v_q = c_{pq}, \quad \text{for each basic variable } x_{pq}$$

The system of equations has $(m + n)$ variables; hence, one of these variables may be set arbitrarily to zero, and the remaining variables may be solved for uniquely, by inspection.

Since each equation contains exactly two variables, we can solve the equations in an order such that one of the variables in each equation is already known (e.g., after arbitrarily setting one variable to zero, at least one other variable may also be solved for directly; these two values in turn may be substituted into other equations, which now contain only one unknown variable, etc.). If the equations are solved in this manner, there will be a total of $(m + n - 1)$ additions involved (counting additions and subtractions as the same type of arithmetic operation). Then, to compute the $[mn - (m + n - 1)]$ nonbasic $z_{ij} - c_{ij}$, requires $2[mn - (m + n - 1)]$ more additions. Therefore, the "*uv* method" requires a total of less than $2mn$ additions. However, the "loop method" requires a *minimum* of three additions per loop (since the minimum number of cells in a loop is four), and hence requires a minimum of $3[mn - m + n - 1)]$ additions. Since for sufficiently large m and n, $mn > (m + n - 1)$, this method is obviously inferior to the "*uv* method." Moreover, we did not consider the determination of the $[mn - (m + n - 1)]$ loops required for the loop method; if the "*uv* method" is used for the computation of the $(z_{ij} - c_{ij})$, then only one loop need be found at each iteration; namely, the loop involving the entering variable.

EXAMPLE 3

Let us re-compute the $z_{pq} - c_{pq}$ for the initial solution in the problem of Example 2, given in Figure 10–6. The basic variables in this solution are x_{11}, x_{12}, x_{22}, x_{23}, x_{24}, x_{34}, x_{35}, x_{45}, and x_{46}. Hence, we have the following $m + n - 1 = 9$ equations:

$$\begin{aligned}
u_1 + v_1 &= c_{11} = 10\\
u_1 + v_2 &= c_{12} = 15\\
u_2 + v_2 &= c_{22} = 10\\
u_2 + v_3 &= c_{23} = 8\\
u_2 + v_4 &= c_{24} = 15\\
u_3 + v_4 &= c_{34} = 12\\
u_3 + v_5 &= c_{35} = 10\\
u_4 + v_5 &= c_{45} = 12\\
u_4 + v_6 &= c_{46} = 12
\end{aligned}$$

As we have noted earlier, any one of the $m + n$ variables may be set to zero, and the resulting system may be solved (uniquely) by inspection. Suppose we set $u_1 = 0$. Then the first two equations above yield, respectively, $v_1 = 10$ and $v_2 = 15$. Substitution of $v_2 = 15$ into the third equation above yields $u_2 = -5$. Then, upon substituting $u_2 = -5$ into the fourth and fifth equations, we find that $v_3 = 13$ and $v_4 = 20$. Substitution of the latter into the sixth equation gives us $u_3 = -8$; substitution of this value into the seventh equation yields $v_5 = 18$, which in turn yields $u_4 = -6$ when substituted into the eighth equation. Finally, we obtain $v_6 = 18$ from the last equation.

Now, for each nonbasic cell, we must calculate $z_{pq} - c_{pq} = u_p + v_q - c_{pq}$:

$$\begin{aligned}
z_{13} - c_{13} &= u_1 + v_3 - c_{13} = 0 + 13 - 10 = 3\\
z_{14} - c_{14} &= u_1 + v_4 - c_{14} = 0 + 20 - 12 = 8\\
z_{15} - c_{15} &= u_1 + v_5 - c_{15} = 0 + 18 - 20 = -2\\
z_{16} - c_{16} &= u_1 + v_6 - c_{16} = 0 + 18 - 15 = 3\\
z_{21} - c_{21} &= u_2 + v_1 - c_{21} = -5 + 10 - 5 = 0
\end{aligned}$$

and so forth.

The reader should verify that this process produces the same results as were obtained by the "loop method" for this solution.

We conclude this section by observing that the u_i, v_j variables described above turn out to be the variables for the linear programming problem which is the dual of the transportation problem. The reader is asked to investigate this aspect more fully in the exercises.

10–4 FINDING AN INITIAL BASIC FEASIBLE SOLUTION

As we have already observed, the Northwest Corner Rule does not consider the objective function in its determination of an initial basic feasible solution, and hence may yield one which is far from optimal (in the sense of having a relatively high value of z). Using such an initial solution may result in a larger number of iterations to achieve optimality than might be the case if a better initial solution were used. Below are four methods for finding an initial solution, all of which tend to find reasonably good basic feasible solutions.

Method 1: Row Minima

Begin in row 1, and find the smallest c_{ij}. Allocate as much as possible to the corresponding x_{ij}. If this allocation exhausts a destination demand, continue in row 1, finding the smallest remaining c_{ij} and allocating as much as possible for this corresponding variable. Continue in this manner until the supply at origin 1 has been exhausted. Then, proceed to row 2 and repeat the process, each time allocating as much as possible to the variable whose cost is the smallest, and moving to the next row whenever a supply is exhausted. If a supply is exhausted simultaneously with a demand being satisfied, arbitrarily assign 0 to the variable in that row whose cost is the smallest of the remaining unallocated variables in the row.

EXAMPLE 4

Let us use the row minima method to find an initial basic feasible solution for the problem of Example 2.

The minimum cost in row 1 is $c_{11} = c_{13} = 10$. Let $x_{11} = 5$; the remaining supply in row 1 is now 5. Hence, we let $x_{13} = 5$, and proceed to row 2. In row 2, the minimum remaining cost (column 1 is not considered, since the demand at destination 1 has previously been satisfied) is $c_{23} = 8$. The remaining demand at destination 3 is 5 units; hence, we let $x_{23} = 5$, and find the smallest remaining cost in row 2, which is $c_{22} = c_{25} = 10$. Letting $x_{22} = 10$ simultaneously exhausts supply 2 and fulfills demand 2. Hence, we assign x_{25} a value of 0, and proceed to row 3. Now, since the demands at destinations 1, 2, and 3 are satisfied, we find $\min\{c_{34}, c_{35}, c_{36}\} = c_{35} = 10$, and let $x_{35} = 8$. Next, we find $\min\{c_{34}, c_{36}\} = c_{34} = 12$, and let $x_{34} = 4$, which exhausts the supply at origin 3. Proceeding to row 4, we find that the demands are satisfied

Origins \ Destinations	1	2	3	4	5	6	Supply
1	10 (5)	15	10 (5)	12	20	15	10
2	5	10 (10)	8 (5)	15	10 (0)	12	15
3	15	10	12	12 (4)	10 (8)	20	12
4	20	15	15	10 (4)	12	12 (5)	9
Demand	5	10	10	8	8	5	

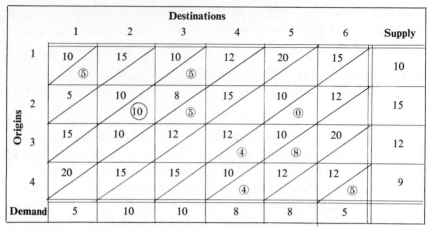

Figure 10–12 Initial solution by row minima for Example 2

at destinations 1, 2, 3, and 5. Hence, we find min $\{c_{44}, c_{46}\} = c_{44} = 10$, and let $x_{44} = 4$, which then requires that $x_{46} = 5$. These calculations are summarized in Figure 10–12. The value of the objective function corresponding to this solution is $z = 468$, compared with a value of $z = 494$ for the initial solution found by the Northwest Corner Rule, and with the optimal solution of $z = 445$.

Method 2: Column Minima

This method is essentially the "transpose" of the row minima method. It is illustrated in Example 5 below.

EXAMPLE 5

Again using the problem of Example 2, let us find an initial basic feasible solution by the method of column minima. We begin by finding the smallest cost in column 1, which is c_{21}, and let $x_{21} = 5$; this satisfies the demand at destination 1, so we proceed to column 2, and find min $\{c_{i2}\} = c_{22} = c_{32} = 10$. Arbitrarily letting $x_{22} = 10$ simultaneously exhausts the supply at origin 2 and satisfies the demand at destination 2; we assign $x_{32} = 0$ and proceed to column 3. Now we find min $\{c_{13}, c_{33}, c_{43}\} = c_{13} = 10$.

Hence, we let $x_{13} = 10$, which again simultaneously satisfies a demand and exhausts a supply. Since min $\{c_{23}, c_{43}\} = c_{33} = 10$, we assign $x_{33} = 0$, and proceed to column 4. Now, we find min $\{c_{34}, c_{44}\} = c_{44}$, and assign $x_{44} = 8$, which satisfies the demand at destination 4, so we proceed to column 5, where we find min $\{c_{35}, c_{45}\} = c_{35} = 10$. Letting $x_{35} = 8$ satisfies the demand at destination 5. In column 6, we find min $\{c_{36}, c_{46}\} = c_{46} = 12$. Hence, we let $x_{46} = 1$, which then forces $x_{36} = 4$, to complete the initial

Origins \ Destinations	1	2	3	4	5	6	Supply
1	10	15	10 (10)	12	20	15	10
2	5 (5)	10 (10)	8	15	10	12	15
3	15	10 (0)	12 (0)	12	10 (8)	20 (4)	12
4	20	15	15	10 (8)	12	12 (1)	9
Demand	5	10	10	8	8	5	

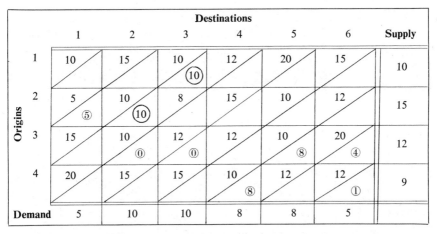

Figure 10–13 Initial solution by column minima for Example 2

basic feasible solution, which is summarized in Figure 10–13. The value of the objective function corresponding to this solution is $z = 477$.

Method 3: Matrix Minima

An obvious generalization of the row minima and column minima methods is to find the smallest c_{ij} in the entire cost matrix, from among those rows i and columns j for which the supplies have not been exhausted or the demands satisfied, respectively. This is precisely what the method of matrix minima does. Again, however, one must be careful to assign one variable in either row i or column j a value of zero each time an allocation causes a supply i to be exhausted simultaneously with a demand j being satisfied.

EXAMPLE 6

Let us find an initial basic feasible solution for the problem of Example 2 by the matrix minima method. We begin by finding $\min_{i,j} \{c_{ij}\} = c_{21} = 5$. Hence, we let $x_{21} = 5$, delete column 1 from further consideration, and find $\min_{\substack{i,j \\ j \neq 1}} \{c_{ij}\} = c_{23} = 8$, and let $x_{23} = 10$. This causes supply 2 to be exhausted and demand 3 to be satisfied; hence, we must assign one variable a value of 0. We choose x_{13} for this purpose, since c_{13} is the smallest cost in row 3 or column 1 (and we break the tie arbitrarily among $c_{13} = c_{22} = c_{24} = 10$, according to the rule: choose the variable whose first subscript is the smallest, and the one among these whose second subscript is the smallest). Now, we make no further assignments in row 2 or in columns 1 and 3. The minimum remaining cost is thus $c_{32} = c_{35} = c_{44} = 10$. Again, breaking the tie arbitrarily according to the rule mentioned above, we let $x_{32} = 10$, which satisfies

Origins \ Destinations	1	2	3	4	5	6	Supply
1	10	15	10 (0)	12	20 (5)	15 (5)	10
2	5 (5)	10	8 (10)	15	10	12	15
3	15	10 (10)	12	12	10 (2)	20	12
4	20	15	15	10 (8)	12 (1)	12	9
Demand	5	10	10	8	8	5	

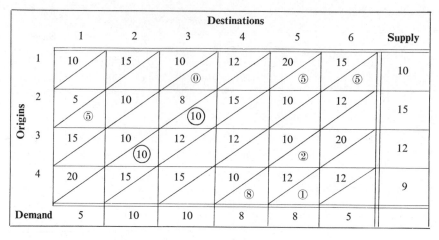

Figure 10–14 Initial solution by matrix minima for Example 2

the demand at destination 2. Now, our next minimization is over all c_{ij} excluding those in columns 1, 2, 3, and in row 2, and the resulting minimum is $c_{35} = c_{44} = 10$. Thus, we let $x_{35} = 2$, which exhausts the supply at origin 3. We then minimize over all c_{ij} except those in rows 2, 3, and in columns 1, 2, 3, which yields a minimum of $c_{44} = 10$. Hence, we let $x_{44} = 8$, satisfying the demand at destination 4. Next, we find min $\{c_{15}, c_{16}, c_{45}, c_{46}\} = c_{45} = c_{46} = 12$. Breaking the tie arbitrarily by the same rule as above, we let $x_{45} = 1$, which exhausts the supply at origin 4. Finally, we find that $x_{15} = x_{16} = 5$, completing the solution which is summarized in Figure 10–14. The value of the objective function corresponding to this solution is $z = 492$.

Method 4: Vogel's Method

A somewhat more sophisticated method for finding an initial basic feasible solution has been devised by Vogel.† This method may be described as follows:

1. For each row, find the two smallest costs, and designate their non-negative difference by R_i, $i = 1, 2, \ldots, m$.
2. For each column, find the two smallest costs, and designate their non-negative difference by C_j, $j = 1, 2, \ldots, n$.
3. Find the maximum $\{R_1, R_2, \ldots, R_m, C_1, C_2, \ldots, C_n\}$.

a) If this maximum is R_k, then allocate as much as possible to the variable in row k whose cost is the smallest. If this allocation exhausts a supply,

† N. V. Reinfeld and W. R. Vogel, *Mathematical Programming*, Prentice-Hall, Englewood Cliffs, N.J. (1958).

delete the corresponding row from further consideration; if this allocation satisfies a demand, delete the corresponding column from further consideration. If both occur simultaneously, delete only the row from further consideration.

b) If this maximum is C_p, then allocate as much as possible to the variable in column p whose cost is the smallest. If this allocation exhausts a supply, delete the corresponding row from further consideration; if this allocation satisfies a demand, delete the corresponding column from further consideration. If both occur simultaneously, delete only the column from further consideration.

4. Repeat Steps 1, 2, and 3 until all but one row (or all but one column) have been deleted. The remaining allocations in this one row (or column) will then be deterministic. Make these allocations and delete this last row (or column).

The number R_i represents a lower bound on the unit cost increase of *not* allocating as much as possible to the variable whose cost is the smallest in row i; C_j has a similar meaning for column j. In other words, each unit in row i not allocated to the variable with the smallest cost in row i will cost *at least* R_i dollars more allocated elsewhere.

EXAMPLE 7

To find an initial basic solution for the problem of Example 2, we begin by computing the R_i and C_j:

$$R_1 = c_{11} - c_{13} = 10 - 10 = 0$$

$$R_2 = c_{23} - c_{21} = 8 - 5 = 3$$

$$R_3 = c_{32} - c_{35} = 10 - 10 = 0$$

$$R_4 = c_{45} - c_{44} = c_{46} - c_{44} = 12 - 10 = 2$$

$$C_1 = c_{11} - c_{21} = 10 - 5 = 5$$

$$C_2 = c_{32} - c_{22} = 10 - 10 = 0$$

$$C_3 = c_{13} - c_{32} = 10 - 8 = 2$$

$$C_4 = c_{14} - c_{44} = c_{34} - c_{44} = 12 - 10 = 2$$

$$C_5 = c_{25} - c_{35} = 10 - 10 = 0$$

$$C_6 = c_{26} - c_{46} = 12 - 12 = 0$$

Hence, the maximum occurs at $C_1 = 5$; since the smallest cost in column 1 is $c_{21} = 5$, we allocate as much as possible to x_{21}: $x_{21} = 5$. Since this allocation satisfies the demand at destination 1, we delete column 1 from further consideration.

Recomputing the R_i and C_j yields:

i	1	2	3	4
R_i	2	2	0	2

j	1	2	3	4	5	6
C_j	-	0	2	2	0	0

We arbitrarily break the tie for the maximum $\{R_i, C_j\}$ by choosing $R_1 = 2$. The smallest cost in row 1 is $c_{13} = 2$ (since column 1 has been deleted, c_{11} is not considered here). Allocating as much as possible to x_{13} yields $x_{13} = 10$, which simultaneously exhausts the supply at origin 1 and satisfies the demand at destination j. We delete only row 1 from further consideration.

The new R_i and C_j are

i	1	2	3	4
R_i	-	2	0	2

j	1	2	3	4	5	6
C_j	-	0	4	2	0	0

Hence, $C_3 = 4$ is the maximum, and the smallest cost in column 3 is $c_{23} = 8$. We allocate as much as possible to x_{23}. which is $x_{23} = 0$, and delete

Origins \ Destinations	1	2	3	4	5	6	Supply
1	10	15	10 (10)	12	20	15	10
2	5 (5)	10 (10)	8 (0)	15	10	12	15
3	15	10 (0)	12	12	10 (8)	20 (4)	12
4	20	15	15	10 (8)	12	12 (1)	9
Demand	5	10	10	8	8	5	

Figure 10–15 Initial solution by Vogel's method for Example 2

column 3 from further consideration. The new R_i and C_j are

i	1	2	3	4
R_i	–	0	0	0

j	1	2	3	4	5	6
C_j	–	0	–	2	0	0

Thus, we find the smallest cost in column 4, which is $c_{44} = 10$, and allocate as much as possible to x_{44}, which yields $x_{44} = 8$, satisfying the demand at destination 4. Deleting column 4 from further consideration yields the following:

i	1	2	3	4
R_i	–	0	0	0

j	1	2	3	4	5	6
C_j	–	0	–	–	0	0

Again we break the tie arbitrarily by choosing row 2; the smallest cost in row 2 is $c_{22} = c_{25} = 10$. We arbitrarily allocate as much as possible to x_{22}, yielding $x_{22} = 10$. We then delete row 2 from further consideration, and obtain

i	1	2	3	4
R_i	–	–	0	0

j	1	2	3	4	5	6
C_j	–	5	–	–	2	8

Thus, we wish to allocate as much as possible to the variable in column 6 with the smallest cost, which is $c_{46} = 12$. Hence, $x_{46} = 1$ and row 4 is deleted from further consideration.

We now have only one row (3) remaining, and the allocations to be made in row 3 are automatically determined by the remaining demands in columns 2, 5 and 6; namely, we must have $x_{32} = 0$, $x_{35} = 8$, $x_{36} = 4$. This completes the initial basic feasible solution, which is shown in Figure 10–15. The corresponding value of the objective function is $z = 477$.

A summary discussion of all these methods for finding an initial basic feasible solution is probably in order. First of all, each of the methods does in fact find a basic feasible solution; that is, a solution with exactly $(m + n - 1)$ allocated (assigned) variables whose corresponding cells do not form a loop.

This is so because each method can be considered as a permutation of the Northwest Corner Rule, in which we have merely interchanged the ordering of the rows and columns. The fundamental principle in all these methods, including the Northwest Corner Rule, is the same. At each step, allocate as much as possible, and delete either a row or a column (but not both) from further consideration. This process must lead to $(m + n - 1)$ allocations, and cannot lead to a loop being formed, because the very first allocation leads to the deletion of either a row or a column, and the deleted row or column therefore cannot have a second allocation. Recall that a loop can be formed only if every row or column appears an even number of times, if at all.

Examples 2 through 6 do not provide us with conclusive evidence as to which method tends to yield better initial solutions. The matrix minima method yielded a solution almost as poor as the Northwest Corner Rule! In general, though, the methods of this section will provide better initial solutions than the Northwest Corner Rule. However, among these four methods, none has proved to be clearly superior to any of the others, and so the determination of which method to use is merely a matter of personal preference.

PROBLEMS

1. In many applications involving transportation-type problems the total of supplies available exceeds the demand, as factories may produce more than is currently needed. Hence, the amount to be shipped from any given origin may be less than the amount available. The resulting problem becomes:

$$\text{Minimize } z = \sum_{i=1}^{\hat{m}} \sum_{j=1}^{\hat{n}} c_{ij} x_{ij} \qquad \text{(10-24a)}$$

Subject to:

$$\sum_{j=1}^{\hat{n}} \leq a_i \qquad i = 1, 2, \ldots, \hat{m} \qquad \text{(10–24b)}$$

$$\sum_{i=1}^{\hat{m}} x_{ij} = b_j, \qquad j = 1, 2, \ldots, \hat{n} \qquad \text{(10–24c)}$$

$$x_{ij} \geq 0, \qquad i = 1, 2, \ldots, \hat{m} \text{ and } j = 1, 2, \ldots, \hat{n} \qquad \text{(10–24d)}$$

Show how to convert the above problem into the transportation problem defined by equations (10–1) through (10–4):

Assume that total supply exceeds total demand; i.e., $\sum_{i=1}^{\hat{m}} a_i > \sum_{j=1}^{\hat{n}} b_j$, and be certain that in your converted problem, (10–5) is satisfied.

2. Suppose that $\sum_{i=1}^{\hat{m}} a_i = \sum_{j=1}^{\hat{n}} b_j$ in the problem defined by equations (10–24). Show that the constraints (10–24b) will hold as equalities in any feasible solution.

3. Consider the problem resulting from (10–24) when the constraints (10–24c) are replaced by

$$\sum_{i=1}^{\hat{m}} x_{ij} \geq b_j, \qquad j = 1, 2, \ldots, \hat{n} \tag{10–25}$$

Show that, if all $c_{ij} \geq 0$, and $\sum a_i \geq \sum b_j$, the constraints (10–25) will hold as strict equalities in the optimal solution. This means that no more will be shipped than is necessary.

4. Consider the problem defined by (10–24a, b, d) and (10–25). Suppose not all $c_{ij} \geq 0$. Convert the constraints to a set of equations and introduce two additional constraints to ensure that $\sum a_i = \sum b_j$. Discuss the differences between the resulting problem and the transportation problem defined by (10–1) through (10–4). Describe how the former can be solved by a modification of the algorithm presented in Chapter 10. In particular, show that, although some of the variables in the above problem have coefficients of -1 in some of the constraints, the results of Theorems 10–1 and 10–2 are still valid; show also how to modify Theorem 10–3 for this problem.

5. Formulate the dual of the linear programming problem defined by (10–1) through (10–4). Show that the variables u_i, u_j of the "*uv* method" are the dual variables.

6. Suppose a constant α is added to each cost in one row of the transportation problem tableau. How are the optimal solution and optimal value of the objective function changed?

7. Find an initial basic feasible solution for each of the following transportation problems by the Northwest Corner Rule:

(a)

a_i	7	9	11		
b_j	4	6	4	8	5

$$C = \begin{bmatrix} 7 & 9 & 5 & 3 & 6 \\ 6 & 4 & 6 & 7 & 5 \\ 8 & 6 & 4 & 5 & 7 \end{bmatrix}$$

(b)

a_i	20	20	20	20		
b_j	15	15	15	15	15	15

$$C = \begin{bmatrix} 9 & 5 & 9 & 6 & 5 & 5 \\ 6 & 5 & 7 & M & 5 & 8 \\ 6 & 7 & 6 & 9 & 8 & 5 \\ 7 & 8 & 4 & 6 & 6 & 5 \end{bmatrix}$$

(Note: $c_{24} = M$, where M is a very large positive number, means that no shipments from origin 2 to destination 4 are allowed.)

(c)

a_i	120	50	80	140				
b_j	30	20	10	10	80	40	30	20

$$C = \begin{bmatrix} 10 & 5 & 8 & 20 & 18 & 10 & 5 & 10 \\ 20 & 10 & 12 & 5 & 16 & 12 & 6 & 12 \\ 30 & 15 & 20 & 15 & 14 & 20 & 7 & 12 \\ 25 & 20 & 26 & 10 & 12 & 15 & 8 & 20 \end{bmatrix}$$

8. Find an initial basic feasible solution for each of the problems of Exercise 7 by the row minima method.

9. Find an initial basic feasible solution for each of the problems of Exercise 7 by the column minima method.

10. Find an initial basic feasible solution for each of the problems of Exercise 7 by Vogel's method.

11. Solve the transportation problem of Exercise 7(a), using the initial basic feasible solution obtained in Exercise 7. For the first two iterations, compute the $z_{ij} - c_{ij}$ by both methods discussed in Section 10–3.

12. Solve the transportation problem of Exercise 7(b).

13. Solve the transportation problem of Exercise 7(c).

14. Formulate as a transportation problem:
A certain department chairman at a large midwestern university has 19 graduate assistants whom he would like to assign to various courses offered by the department. Each assistant can only be assigned to one course. There are five courses which currently are in need of assistants, as follows:

Course	A	B	C	D	E
Maximum no. of assistants needed	3	2	4	3	5

However, not all of the 19 graduate students are capable of assisting in every course. In particular, the first four students are qualified to assist in courses A, C, and D; the next 10 students are qualified to assist in courses B, C, and E; the remaining 5 students can assist in course C only.
The department chairman would like to maximize the number of graduate students who are assigned to a course.

15. An alternate approach to that developed in Chapter 10 for proving that transportation problems with integer supplies and demands have only integer basic feasible solutions is as follows:

(a) Let B be any square submatrix of order k of the coefficient matrix A of a transportation problem (formed by taking any k rows and any k columns of A, $k \leq m + n$). Show that *either* det $(B) = 0$, or at least one column of B contains exactly one 1. Use this result to prove that det (B) is either 0, $+1$, or -1.
The fact that every square submatrix of A has determinant equal to 0, $+1$, or -1 is called the *unimodularity* property of A.
(b) Prove that every basic feasible solution for a transportation problem is integral, provided that the supplies and demands are integers, using the unimodularity property of A.
(c) Use the results of part (a) to prove that all y_{ij}^k are 0, $+1$, or -1.

16. Find all alternative optima for the problem of Exercise 7(c).

11

The Assignment Problem

11-1 INTRODUCTION

In this chapter we shall consider yet another special type of linear programming problem: the assignment problem. Basically, the assignment problem may be stated as follows: Given n individuals and n jobs, we wish to assign each individual to exactly one job so that each job is covered by exactly one individual, in such a way that the total cost is minimized. Assume that c_{ij} is the cost of assigning the ith individual to the jth job.

We can describe the problem mathematically by defining

$$x_{ij} = \begin{cases} 1, & \text{if individual } i \text{ is assigned to job } j \\ 0, & \text{otherwise} \end{cases} \tag{11–1}$$

Then, the total cost of a given assignment is

$$z = \sum_{i=1}^{n} \sum_{j=1}^{n} c_{ij} x_{ij}$$

and the constraints are

$$\sum_{j=1}^{n} x_{ij} = 1, \qquad i = 1, 2, \ldots, n \tag{11–2}$$

$$\sum_{i=1}^{n} x_{ij} = 1, \qquad j = 1, 2, \ldots, n \tag{11–3}$$

The constraints (11–2) insure that each individual is assigned to exactly one job, and the constraints (11–3) insure that each job is covered by one individual.

The problem as stated above is not a linear programming problem because of the requirement that each x_{ij} can assume only the values 0 and 1.

However, suppose that we replace (11–1) with the set of non-negativity restrictions

$$x_{ij} \geq 0 \qquad i = 1, 2, \ldots, n, \quad \text{and} \quad j = 1, 2, \ldots, n$$

We now have the following *transportation problem:*

$$\text{Minimize } z = \sum_{i=1}^{n} \sum_{j=1}^{n} c_{ij} x_{ij}$$

Subject to

$$\begin{aligned} \sum_{j=1}^{n} x_{ij} &= 1, \qquad i = 1, 2, \ldots, n \\ \sum_{i=1}^{n} x_{ij} &= 1, \qquad j = 1, 2, \ldots, n \\ x_{ij} \geq 0, \quad i &= 1, 2, \ldots, n \quad \text{and} \quad j = 1, 2, \ldots, n \end{aligned} \tag{11–4}$$

In Chapter 10 we proved† that all basic feasible solutions to such a transportation problem will be integer-valued. Note that the constraints of (11–4) restrict the possible integer values of each variable to either 0 or 1. Hence, an optimal solution for (11–4) will also be an optimal solution for the assignment problem. Because of this equivalence in the two problems, we shall henceforth consider (11–4) as the mathematical formulation of the assignment problem.

As we have observed above the assignment problem (11–4) is merely a special case of the transportation problem‡ in which $m = n$ and $a_i = 1$, $i = 1, 2, \ldots, n$, $b_j = 1, j = 1, 2, \ldots, n$. Hence, we could solve assignment problems by using the transportation problem algorithm. However, just as we found in Chapter 10 that using the simplex method to solve transportation methods is inefficient, so now we shall discover that a special algorithm for the assignment problem will be better than trying to use a transportation problem algorithm directly.

To motivate the need for such a special algorithm, we observe that a basic feasible solution for (11–4) consists of $(2n - 1)$ variables; however, it should be clear from the constraints of (11–4) that every basic feasible solution will contain exactly n basic variables equal to 1, and $(n - 1)$ basic variables equal to 0. This high level of degeneracy suggests that, if we were to use the transportation problem algorithm of Chapter 10, we might have to perform a large number of iterations in which the values of the variables do not change (i.e., in which basic and nonbasic variables, both equal to 0, are being interchanged).

---

† See Theorem 10–4 and the ensuing discussion.

‡ Dantzig[2] has shown that the transportation problem can also be considered as a special case of the assignment problem; that is, any transportation problem may be formulated as an assignment problem.

11-2 EXAMPLES OF ASSIGNMENT PROBLEMS

Before turning to the development of an assignment problem algorithm, we present several examples of problems which can be formulated as assignment problems.

EXAMPLE 1

A small clerical office of a certain company hires typists on a part-time basis, as needed. The office currently has three tasks suitable for these typists:

1. A batch of ten letters, which would take an average typist a total of one hour to complete;
2. A technical report, which could be completed in three hours by an average typist;
3. An updating of the company's inventory files, which consists primarily of typing long columns of combinations of numbers and alphabetic characters (denoting item catalog numbers and quantities). This task should take one hour for an average typist.

There are three typists available for these tasks, and the office manager wishes to assign one task to each typist in order that the tasks may be performed simultaneously. The first of the three typists is the so-called average typist mentioned above in the task descriptions; the second typist is 25 per cent faster than average on letter typing, 10 per cent above average on technical typing, and average on the third type of task; the third typist is 15 per cent below average, 10 per cent above average, and 10 per cent below average, respectively, on the three types of tasks.

In other words, it will take the second typist 25 per cent less time to perform task 1 (or 45 minutes), 10 per cent less time for the second task (162 minutes), and 60 minutes for the third task. The times required for the typists to perform each task are summarized below.

Time (in minutes)

	Task	1	2	3
Typist	1	60	180	60
	2	45	162	60
	3	74	162	66

The typists charge the same rate per hour, so that minimizing the total time required to complete the three tasks will also minimize the total cost. Letting $i = 1, 2, 3$ denote the three typists, and $j = 1, 2, 3$ denote the three tasks, respectively, the table above then provides the values of the c_{ij}, and the problem is completely described.

EXAMPLE 2

Suppose in the previous example the office has the same three tasks, but has a pool of five typists from which any three can be chosen. Below is a table giving the time required by typists 4 and 5 for each of the three tasks.

Time (in minutes)

	Task	1	2	3
Typist	4	60	180	54
	5	54	171	60

In order to make the number of tasks equal to the number of typists, as required by the assignment problem formulation, we introduce two dummy "tasks," Task 4 and Task 5, and let $c_{i4} = c_{i5} = 0$, $i = 1, 2, 3, 4, 5$. This means that these tasks require no time to complete. In the optimal solution, three of the five typists will be assigned to the first three tasks, and two will be "assigned" to Tasks 4 and 5, which is interpreted to mean these two typists are actually given no task.

EXAMPLE 3

The chairman of the mathematics department at a small midwestern university has five graduate students who have been granted teaching assistantships. Each of these students must be assigned to assist with exactly one course. Fortunately, there are only five courses currently in need of a teaching assistant. The professor for each of these five courses has ranked the five graduate students in order of his preference, with a ranking of 1 denoting first choice, a ranking of 2, second choice, etc. Because of differences in their academic backgrounds, not every student is qualified to assist in every course. In such a case, the student has been given a ranking of "M," instead of a numerical ranking. The rankings are summarized below.

The department chairman has decided to assign the students by minimizing the sum of the rankings corresponding to assigned students. To do so, he merely lets the c_{ij} be the ranking numbers (or M) and lets M assume a very

Student \ Course	Elementary Calculus	Computer Science	Linear Programming	Differential Equations	Complex Variables
A	1	1	1	4	2
B	3	2	*M*	1	*M*
C	2	*M*	*M*	3	1
D	5	3	2	5	*M*
E	4	*M*	3	2	3

large, positive value, to exclude the possibility of assigning students to courses for which they are unqualified.

11-3 FUNDAMENTAL THEORY FOR AN ALGORITHM

As the examples of the preceding section suggest, once the costs c_{ij} are given, an assignment problem is completely determined. Hence, in our development of some useful properties of optimal solutions to assignment problems, we shall concentrate our attention on the cost matrix $C = [c_{ij}]$.

We begin with the following observation. Suppose in a particular assignment problem all c_{ij} are non-negative and a feasible assignment exists for which all corresponding c_{ij} are equal to zero. Then, this solution must be optimal, since its objective function value is zero, which is clearly minimal.

Although the above fact may seem to be of rather limited applicability, we shall soon see that we can develop an extremely good general algorithm for solving assignment problems, based upon it. Toward this end, the following theorem is most useful.

Theorem 11-1 If a constant is added to any row and/or column of the cost matrix of an assignment problem, the resulting assignment problem has the same optimal solution (or same set of optimal solutions) as the original problem.

Proof: Let $C = [c_{ij}]$ and suppose we add α_i to row i, $i = 1, 2, \ldots, n$, and add β_j to column j, $j = 1, 2, \ldots, n$. Thus, the new cost matrix is $\hat{C} = [\hat{c}_{ij}]$, where

$$\hat{c}_{ij} = c_{ij} + \alpha_i + \beta_j$$

If we denote by z, $\hat{z}$ the values of the objective functions corresponding to C and $\hat{C}$, respectively, then

$$\begin{aligned}
\hat{z} &= \sum_{i=1}^{n} \sum_{j=1}^{n} \hat{c}_{ij} x_{ij} \\
&= \sum_{i=1}^{n} \sum_{j=1}^{n} (c_{ij} + \alpha_i + \beta_j) x_{ij} \\
&= \sum_{i=1}^{n} \sum_{j=1}^{n} c_{ij} x_{ij} + \sum_{i=1}^{n} \sum_{j=1}^{n} \alpha_i x_{ij} + \sum_{i=1}^{n} \sum_{j=1}^{n} \beta_j x_{ij} \\
&= z + \sum_{i=1}^{n} \alpha_i \left(\sum_{j=1}^{n} x_{ij} \right) + \sum_{j=1}^{n} \beta_j \left(\sum_{i=1}^{n} x_{ij} \right) \\
&= z + \sum_{i=1}^{n} \alpha_i + \sum_{j=1}^{n} \beta_j,
\end{aligned}$$

since

$$\sum_{i=1}^{n} x_{ij} = \sum_{j=1}^{n} x_{ij} = 1.$$

Hence, we see that z and $\hat{z}$ differ by a constant amount, independent of the values of the variables; therefore, an optimal solution to one problem must be optimal for the other, and vice-versa.

We can (and will) make use of this theorem in two ways. First of all, if we are given an assignment problem some of whose costs are negative, we can instead deal with a related assignment problem whose costs are all non-negative by merely adding a large enough constant to every cost. Secondly, if we choose the α_i and β_j properly, we can introduce many zeros into the cost matrix, and thereby—hopefully—find a feasible assignment for which the corresponding c_{ij} are all zero; that is, an optimal solution.

Essentially, this is what our proposed algorithm will do. In fact, in order to completely describe the algorithm, we need only develop a systematic procedure for introducing the zeros into the cost matrix.

Let us define more precisely some of the terminology which we have been using. A cell (i, j) is said to be *assigned* if the corresponding x_{ij} is to be equal to one. A feasible solution for the assignment problem shall be called a *complete assignment*. Hence, a complete assignment consists of exactly one assigned cell in each row and in each column, with no two assigned cells in the same row or in the same column. Since our goal is to find a complete assignment with cost equal to zero, we shall only assign cells whose corresponding c_{ij} is zero. We shall refer to such assigned cells as *assigned zeros*.

If for a given cost matrix we cannot assign enough zeros to form a complete assignment, then we will make use of Theorem 11–1 to increase the number of zeros available for assignment.

EXAMPLE 4

Suppose we are given an assignment problem whose cost matrix is:

$$C = \begin{bmatrix} 5 & 2 & 3 & 4 \\ 7 & 8 & 4 & 5 \\ 6 & 3 & 5 & 6 \\ 2 & 2 & 3 & 5 \end{bmatrix}$$

We can begin by introducing zeros, by subtracting 2 from row 1, 4 from row 2, 3 from row 3, and 2 from row 4, yielding a cost matrix of

$$\begin{bmatrix} 3 & 0 & 1 & 2 \\ 3 & 4 & 0 & 1 \\ 3 & 0 & 2 & 3 \\ 0 & 0 & 1 & 3 \end{bmatrix}$$

This matrix now has at least one zero in every row, and also at least one zero in every column except column 4. We can easily remedy that by subtracting 1 from every element in column 4, yielding the cost matrix

$$\begin{bmatrix} 3 & 0 & 1 & 1 \\ 3 & 4 & 0 & 0 \\ 3 & 0 & 2 & 2 \\ 0 & 0 & 1 & 2 \end{bmatrix}$$

Now, this matrix has at least one zero in every row and column; however, it is not possible to find a complete assignment consisting of zero cells. To see this, observe that row 1 contains only one zero; if we assign this cell, then none of the remaining zeros in column 2 can be assigned. We indicate this by drawing a box around cell (1, 2) and by drawing a line through column 2.

$$\begin{bmatrix} 3 & \boxed{0} & 1 & 1 \\ 3 & 4 & 0 & 0 \\ 3 & 0 & 2 & 2 \\ 0 & 0 & 1 & 2 \end{bmatrix}$$

Observe, now, that there is no remaining zero for assignment in row 3, so that a complete assignment of zeros is not possible. However, for purposes of illustration, let us continue assigning zeros until no more can be assigned. Since row 4 has only one zero available for assignment, let us assign it, and cross out column 1 (even though there do not happen to be any remaining zeros in column 1):

$$\begin{bmatrix} 3 & \boxed{0} & 1 & 1 \\ 3 & 4 & 0 & 0 \\ 3 & 0 & 2 & 2 \\ \boxed{0} & 0 & 1 & 2 \end{bmatrix}$$

Next, we see that column 3 has only one zero; assigning it and drawing a line through row 2 yields

$$\begin{bmatrix} 3 & \boxed{0} & 1 & 1 \\ 3 & 4 & \boxed{0} & 0 \\ 3 & 0 & 2 & 2 \\ \boxed{0} & 0 & 1 & 2 \end{bmatrix}$$

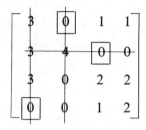

Now, all of the zeros are covered by lines, and none remains for assignment. The order in which we made assignments above was arbitrary.

Note that the lines themselves suggest a method of introducing more zeros: The minimum of the uncovered elements is one; if we subtract this number from every uncovered row, and add the same number to every covered column, then the assigned zeros will remain zero (since one will be both added and subtracted to such cells). The resulting cost matrix is

$$\begin{bmatrix} 3 & 0 & 0 & 0 \\ 4 & 5 & 0 & 0 \\ 3 & 0 & 1 & 1 \\ 0 & 0 & 0 & 1 \end{bmatrix}$$

This procedure must produce at least one new zero; in our example we have actually produced three new zeros, and an optimal assignment may now be easily found. Two such optimal assignments are given below:

$$\begin{bmatrix} 3 & 0 & \boxed{0} & 0 \\ 4 & 5 & 0 & \boxed{0} \\ 3 & \boxed{0} & 1 & 1 \\ \boxed{0} & 0 & 0 & 1 \end{bmatrix} \quad \begin{bmatrix} 3 & 0 & 0 & \boxed{0} \\ 4 & 5 & \boxed{0} & 0 \\ 3 & \boxed{0} & 1 & 1 \\ \boxed{0} & 0 & 0 & 1 \end{bmatrix}$$

Example 4 suggests a method by which zeros may be introduced into the cost matrix. We began by subtracting the smallest element in each row from every element in that row (which yields at least one zero in each row) and then we repeated this procedure for the columns, yielding at least one zero in each column. We then assigned as many zeros as we could; next, we covered *all* the zeros with lines, so that all uncovered elements of the cost matrix are strictly positive. Choosing the smallest of these uncovered elements, we subtracted this quantity from all elements in uncovered rows and added the same quantity to all elements in covered columns. The resulting cost matrix has at least one new zero, and (if the covering lines are chosen properly) all previously assigned zeros remain.

Before developing this procedure a bit further, let us consider

Theorem 11–2 Suppose the zeros of a cost matrix C are covered by k lines, and let m be the minimum of the uncovered elements of C. If m is subtracted from every row containing uncovered elements and added to every column covered by a line, resulting in a new cost matrix $\hat{C}$, then the sum of the elements of $\hat{C}$ is $mn(n - k)$ less than the sum of the elements of C.

Proof: First, we note that the subtraction and addition procedure described in the theorem is equivalent to the procedure of adding m to every element covered by a line (elements covered by both a row line and a column line being increased by a total of $2m$) and then subtracting m from every element of C. The net effect in both procedures is to decrease uncovered elements by m, leave unchanged elements covered by one line, and increase by m elements covered by both a row and a column line.

If each element in C is decreased by m, the total decrease in the sum of the elements is mn^2; if we then add m to each of the n elements in every line, then the total increase in the sum of the elements is nm per line, or nmk for the k lines. Hence, the total net decrease in the sum of the elements is $mn^2 - nmk = mn(n - k)$.

According to Theorem 11–2, if we can cover all the zeros by fewer than n lines ($k < n$), then the decrease in the sum of all the elements is a strictly positive quantity. Moreover, since the elements of the cost matrix are always non-negative, we can only decrease the sum of the elements until eventually the entire sum is equal to 0, which implies all the elements are zero and an optimal assignment can be trivially found. Of course, we would terminate the procedure as soon as enough zeros are introduced to yield a complete assignment.

If we assume that all of the original elements of C are non-negative *integers*, then the sum of the elements decreases by an integral amount, and the procedure can be repeated only a finite number of times before reaching the state in which all the elements are zero.

(Requiring the elements of C to be integers is not unduly restrictive, so far as applying an algorithm is concerned. If the original elements are rational numbers, then multiplying every element by the common denominator of all the elements will yield a matrix whose elements are all integers; and, of course, digital computers cannot store irrational numbers, but rather must approximate such a number by a rational number.)

Let us summarize the discussion of this section. Given a cost matrix for an assignment problem, our goal is to introduce zeros into the cost matrix so that we may find a complete assignment consisting entirely of zeros. If the original elements of the cost matrix are not already non-negative, we add a constant (e.g., the absolute value of the most negative element) to each row containing negative elements so that the resulting cost matrix contains only non-negative elements. We then introduce at least one zero in every row and in every column by subtracting appropriate constants from the elements in each row and each column. Next, we assign as many zeros as possible; if the resulting set of assignments constitutes a complete assignment, it is optimal. If a complete assignment is not found, we cover the zero elements with lines and introduce at least one new zero by the method described in Theorem 11–2. Again, we assign as many zeros as possible, and repeat the sequence of assigning zeros and introducing new zeros until a complete assignment is found.

In order to introduce a new zero, we need only be able to cover the existing zeros with fewer than n lines. A theorem credited to the Hungarian mathematician D. König guarantees that this is always possible:

Theorem 11–3 (König) If k is the maximum number of zeros which can be assigned, then there exists a set of k lines which will cover all the zeros.

The truth of König's theorem is intuitively obvious, since if k is the maximum number of zeros which can be assigned, each unassigned zero must lie in either the same row or the same column as an assigned zero. Hence, if the lines are chosen properly, there must be at least one set of k lines which covers all the zeros. However, König's theorem does not tell us how to find such a set of lines.

In order to complete the description of the algorithm we have been developing, we must describe procedures for

(a) finding the maximum number of zeros which can be assigned for a given cost matrix, and
(b) finding a set of k lines which covers all the zeros of the cost matrix, where k is the number of zeros assigned in (a).

11–4 A METHOD FOR ASSIGNING ZEROS

The material in this section, and of the next two sections as well, was originally developed by H. Kuhn,[3,4] who called the resulting algorithm "The Hungarian Method for the Assignment Problem," since it is based on König's theorem and on the work of another Hungarian mathematician, J. Egerváry.

Throughout the ensuing discussion we shall denote an assigned zero by 0* and an unassigned zero by 0. We shall assume that we have a cost matrix with non-negative elements and which contains some zeros; we already know how to obtain such a matrix from any given cost matrix. All searches of rows (or columns) are performed in increasing order of the row (or column) numbers.

The procedure described below shall yield, upon its completion, a *maximal assignment*; that is, an assignment such that no larger number of assignments can be made for the current cost matrix.

STEP 1. Assign in each row the first 0, if any, not in the column of a previously assigned zero.

STEP 2. Search each column for a 0*. If a 0* is found, proceed to the next column; if a 0* is found in every column, then a complete assignment has been found. If a 0* is not found in a column, this column is called *eligible* and is searched for an unassigned zero (0). If a 0 is not found, proceed to the next column (if there is no next column, there are no more assignable zeros). If a 0 is found, proceed to Step 3.

STEP 3. A 0 has been found in cell (i_1, j_0). Find a sequence of cells of the following form:

$$\begin{array}{ll} 0 & \text{in } (i_1, j_0) \\ 0^* & \text{in } (i_1, j_1) \\ 0 & \text{in } (i_2, j_1) \\ 0^* & \text{in } (i_2, j_2) \\ & \quad\vdots \\ 0 & \text{in } (i_p, j_{p-1}) \\ 0^* & \text{in } (i_p, j_p) \\ 0 & \text{in } (i_{p+1}, j_p) \\ & \quad\vdots \end{array}$$

The discussion of how this sequence is generated can be divided into two cases:

Case A: Going from Cell (i_p, j_{p-1}) to Cell (i_p, j_p)

A 0 has just been found in cell (i_p, j_{p-1}). There are currently p 0's and $(p - 1)$ 0*'s in the sequence. If no 0* exists in row i_p, then we can increase the number of assigned zeros by interchanging the 0's and 0*'s in the above sequence: The resulting assigned zeros will be in rows $i_1, i_2, \ldots, i_p$, whereas the original sequence contained no assigned zero in row i_p. Hence, we perform this interchange (called a *transfer* of assignments) and return to Step 2.

If, on the other hand, a 0* is found in row i_p, say in column j_p, then we are in Case *B*.

Case B: Going from Cell (i_p, j_p) *to Cell* (i_{p+1}, j_p)

A 0* has just been found in cell (i_p, j_p), and there are currently p assigned 0's and p unassigned 0's in the sequence. Column j_p is now searched for a 0; if such a 0 is found, say in row i_{p+1}, then cell (i_{p+1}, j_p) is added to the sequence and we are back in Case *A*.

If, on the other hand, column j_p contains no unassigned 0, then row i_p is labeled *essential*, signifying the fact that no transfer of assignments can be made involving row i_p. The cells (i_p, j_p) and (i_p, j_{p-1}) are deleted from the sequence; column j_{p-1} is searched for another 0, beginning with row (i_{p+1}).

If the unassigned 0 which is found in this Case is in a row which already appears in the sequence, it is discarded and the search continues for another unassigned 0.

Step 3 terminates in one of two ways. The first is mentioned in Case *A*, in which a transfer of assignments is performed, and the procedure returns to Step 2. The second termination condition occurs if the deletion of cells procedure of Case *B* causes all cells to be deleted from the sequence; in this case, we return to Step 2 and attempt to find another unassigned 0 in column j_0 (starting with row $i_1 + 1$). Hence, in both cases, Step 3 is followed by a return to Step 2. When Step 2 is completed, there are no more eligible columns remaining with unassigned zeros. Therefore, the current assignment is maximal.

EXAMPLE 5

After introducing zeros to the cost matrix of Example 4

$$\begin{bmatrix} 5 & 2 & 3 & 4 \\ 7 & 8 & 4 & 5 \\ 6 & 3 & 5 & 6 \\ 2 & 2 & 3 & 5 \end{bmatrix}$$

(by subtracting the smallest element in each row from every element in that row and then subtracting the smallest element in each column from every element in that column) we obtained the cost matrix

$$\begin{bmatrix} 3 & 0 & 1 & 1 \\ 3 & 4 & 0 & 0 \\ 3 & 0 & 2 & 2 \\ 0 & 0 & 1 & 2 \end{bmatrix}$$

Using Step 1 of the zero-assigning procedure of this section, we assign the zero in cell (1, 2), the zero in cell (2, 3), the zero in cell (4, 1). The zero in row 3 cannot be assigned, because of the previously assigned zero in the same column. Thus, we proceed to Step 2, and search each column for a 0*. Columns 1, 2, and 3 each have a 0*, but column 4 does not, and hence is an eligible column. We search this column for an unassigned 0; there is such a 0 in cell (2, 4). Hence, we proceed to Step 3.

In Step 3, we begin generating a sequence of cells which contain, alternatingly, a 0 and a 0*. The first cell is the one found in Step 2, the cell (2, 4). We now search row 2 for a 0*, and find one in cell (2, 3). Next, we search column 3 for a 0 (Case *B*) and find that column 3 contains no unassigned 0. Hence, row 2 is labeled an essential row. The last two cells in the sequence are now deleted, leaving us with no remaining cells in the sequence. We must return to Step 2, and look for another unassigned 0 in column 4. There being none, the procedure terminates. Hence, no more zeros can be assigned in the current cost matrix (a fact which we already knew from the discussion in Example 4).

EXAMPLE 6

Consider the cost matrix

$$\begin{bmatrix} 0 & 0 & 1 & 2 \\ 4 & 3 & 0 & 0 \\ 0 & 3 & 2 & 2 \\ 0 & 3 & 1 & 1 \end{bmatrix}$$

This matrix was obtained from the matrix of Example 5 by interchanging rows 1 and 4, and interchanging columns 1 and 2. Since such interchanges cannot alter the number of assignments possible, we know that an assignment consisting of three assigned zeros exists.

In Step 1 of our procedure, we can only assign zeros in cells (1, 1) and (2, 3):

$$\begin{bmatrix} 0^* & 0 & 1 & 2 \\ 4 & 3 & 0^* & 0 \\ 0 & 3 & 2 & 2 \\ 0 & 3 & 1 & 1 \end{bmatrix}$$

The zeros in rows 3 and 4 are in column 1, which already has an assigned 0. Moving to Step 2, we find that column 1 contains a 0*, but column 2 does not.

We search column 2 for an unassigned 0 and find one in cell (1, 2). We therefore proceed to Step 3 and generate the sequence of cells:

$$\begin{array}{l} 0 \ \text{in} \ (1, 2) \\ 0^* \ \text{in} \ (1, 1) \\ 0 \ \text{in} \ (3, 1) \end{array}$$

This sequence is generated as follows: starting with the 0 in cell (1, 2) found in Step 2, we look for a 0* in row 1, and find it in cell (1, 1). Next, we look in column 1 for an unassigned 0. The first such unassigned 0 is in cell (3, 1). We then search row 3 for a 0*, and find none. As indicated in Case A of Step 3, this means the number of assigned zeros can be increased by one, by performing a transfer of assignments.

Hence, we assign the zeros in cells (1, 2) and (3, 1) and remove the assignment from the zero in cell (1, 1). We now have three assigned zeros:

$$\begin{bmatrix} 0 & 0^* & 1 & 2 \\ 4 & 3 & 0^* & 0 \\ 0^* & 3 & 2 & 2 \\ 0 & 3 & 1 & 1 \end{bmatrix}$$

Returning to Step 2, the first column without a 0* is column 4, which has an unassigned 0 in cell (2, 4). Proceeding to Step 3, we generate the sequence:

$$\begin{array}{l} 0 \ \text{in} \ (2, 4) \\ 0^* \ \text{in} \ (2, 3) \end{array}$$

We are now in Case B, trying to find an unassigned 0 in column 3. As column 3 contains no unassigned 0, row 2 is essential, and the last two cells are deleted from the sequence. Now, we attempt to find another unassigned 0 in column 4. There being none, the procedure terminates, having found the maximum number of assigned zeros.

The procedure described in this section will always find the maximum number of assignable zeros, because the sequences of cells generated in Step 3 determine all possible transfers of assignments which involve eligible columns (columns containing at least one 0 but no assigned 0). Of course, a transfer of assignment is actually performed only if it increases the number of assigned zeros.

11-5 A METHOD FOR COVERING THE ZEROS

In the procedure of the previous section, we labeled a row *essential* if it contained an assigned 0 and no other assignable zeros. Let us define an *essential column* as being a column which contains a 0* in a nonessential row. Given a cost matrix with a maximal assignment* consisting of k zeros, as determined by the procedure of Section 11–4, then the total of the number of essential rows plus the number of essential columns is equal to k. This is so because each 0* is in either an essential row or an essential column, but not both.

Theorem 11–4 In a maximal assignment, every 0 of the cost matrix is either in an essential row or an essential column.

Proof: Obviously every 0* lies in either an essential row or an essential column. Suppose that a 0 (unassigned zero) in cell (i, j) lies in neither an essential row nor an essential column. Then, this zero could be assigned: Column j cannot contain an assigned 0, since such a 0* would have to be in an essential row; similarly, row i cannot contain a 0*, for if row i is nonessential and contains a 0* in cell (i, j), then column j is essential. Thus, an unassigned 0 which lies in neither an essential row nor an essential column can be assigned, which would increase the total number of assigned zeros by one. However, we are assuming that we had already found the maximum number of assigned zeros; hence, all zeros must lie in either an essential row or an essential column.

Combining Theorem 11–4 with the observations preceding it yields

Theorem 11–5 Given a cost matrix and a maximal assignment consisting of k assigned zeros, then:

(1) there are a total of k essential rows and essential columns;
(2) the k lines which cover the essential rows and essential columns cover all the zeros of the cost matrix; and
(3) no 0* lies in the intersection of two of these lines.

Note that Theorem 11–5 is a stronger version of König's theorem, in that it tells us not only that the k lines exist but also how to find them. Since these lines cover the essential rows and the essential columns, we shall call them essential lines.

Now, as we have seen in Section 11–3, if all the zeros of a cost matrix are covered, we can introduce new zeros into the cost matrix by subtracting the minimum uncovered element from every element of the cost matrix and adding the same quantity to the elements in every covered line (see Theorem 11–2 and its proof.) Although this process obviously introduces at least one

*A *maximal assignment* is one in which the number of assigned zeros cannot be increased.

new zero, it may eliminate some of the original zeros. Recall that elements which lie at the intersection of two of the essential lines are increased, while all other covered elements remain unchanged. Hence, if a zero lies at the intersection of two lines, it will be eliminated. However, the important fact is that the assigned zeros do not lie at the intersection of two essential lines, by part (3) of Theorem 11–5. Hence, the resulting cost matrix contains all the assigned zeros of the preceding cost matrix, plus at least one new zero. Therefore, we have

Theorem 11–6 The number of zeros in a maximal assignment for the cost matrix resulting from the procedure described in Theorem 11–2 is at least equal to the number of zeros in a maximal assignment for the preceding cost matrix, provided only that the lines used to cover the zeros are the essential lines.

EXAMPLE 7

Let us take the cost matrix and assigned zeros obtained in Example 5:

$$\begin{bmatrix} 3 & 0^* & 1 & 1 \\ 3 & 4 & 0^* & 0 \\ 3 & 0 & 2 & 2 \\ 0^* & 0 & 1 & 2 \end{bmatrix}$$

In Example 5, we determined that row 2 is an essential row. Now, we observe that columns 1 and 2 are essential columns, since each contains a 0* in a nonessential row. Covering the cost matrix with these three lines yields:

$$\begin{bmatrix} 3 & 0^* & 1 & 1 \\ 3 & 4 & 0^* & 0 \\ 3 & 0 & 2 & 2 \\ 0^* & 0 & 1 & 2 \end{bmatrix}$$

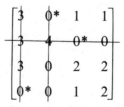

This is the same set of lines we found, quite by accident, in Example 4, and the resulting cost matrix is the same as that obtained in that example.

11–6 THE HUNGARIAN METHOD AND SOME VARIATIONS

With the detailed procedures developed in the preceding two sections on assigning zeros and determining a set of covering lines, we have completely described an algorithm for solving assignment problems. This algorithm, called the Hungarian method, may be conveniently summarized by the flow diagram given in Figure 11–1.

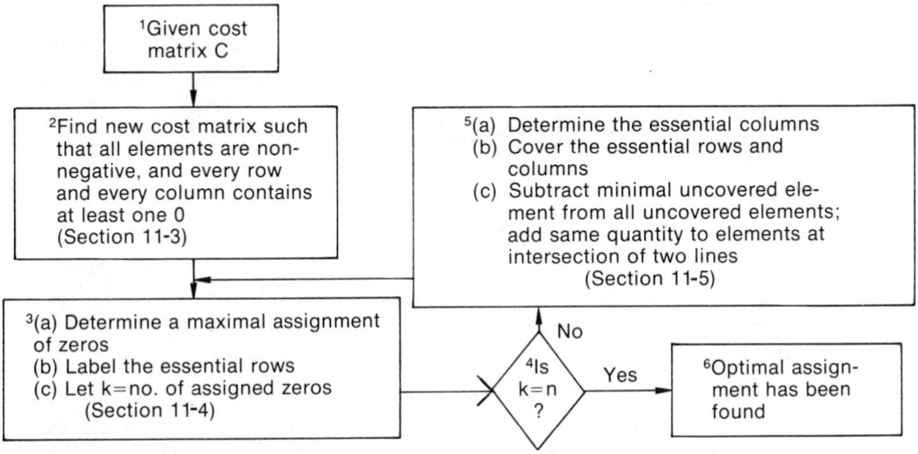

Figure 11–1 Flow diagram of the Hungarian method

As we have previously remarked, the Hungarian method is a finite algorithm, since the "k" of Step 3 is non-decreasing, and the sum of the elements of the cost matrix generated in Step 5 is strictly decreasing, at each iteration, and is bounded from below by zero. Moreover, it is obvious that the algorithm can be (and should be) implemented using integer arithmetic exclusively.

Historically, the Hungarian method led to the later development of the primal-dual algorithm discussed in Chapter 9; the latter is in fact a generalization of the Hungarian method. Box 3, in Figure 11–1, corresponds to the solution of the "restricted primal," in which only certain variables are allowed to become positive (in the assignment problem, these are the variables whose $c_{ij} = 0$). Box 5 of Figure 11–1 corresponds to finding a new dual solution in such a way that at least one variable previously restricted to a value of zero is now allowed to become positive.

Computationally, the slowest part of the Hungarian method involves the generation of all possible transfers, in the search for a maximal assignment. This procedure is described in Step 3 of Section 11–4.

There are several ways in which this procedure can be modified, along with the determination of the zero-covering lines. Several such modifications are described by Kuhn.[4] In one of these, Kuhn suggests a procedure for finding $n - 1$ lines to cover all the zeros (which can always be done unless a complete assignment exists). Kuhn's modification still leads to a finite algorithm, and although it reduces the computation involving the assignment of zeros at each iteration, the total number of iterations may increase. This is so because the amount of decrease in the sum of the elements will probably be less per iteration than in the original method.

Another modification to the Hungarian method which should tend to make the assignment of zeros procedure more efficient is to modify Box 3 of Figure 11–1 as follows:

1. Search each row (in order) for an unmarked 0.† If a row contains only one unmarked 0, assign that 0, and mark it 0*. Mark each zero in the column containing this 0* with an m (0^m). Such zeros cannot be assigned.

   If a row contains no unmarked zeros or more than one, proceed to the next row. Continue until all rows have been searched.
2. Repeat the procedure of Step 1 for the columns.
3. Alternately repeat Step 1 and Step 2, until either:
   (a) All zeros have been marked. In this case, a maximal assignment has been found. All rows containing a 0* are essential rows. Proceed to Box 4 of Figure 11–1.
   (b) Every row and every column containing unmarked zeros has at least two such unmarked zeros. All rows containing a 0* should be labeled essential. To assign the remaining zeros, delete the rows and columns which contain an assigned 0. Then use the original procedure (of Section 11–4) on this reduced matrix, before proceeding to Box 4 of Figure 11–1.

In the above procedure, it may be necessary to repeat Steps 1 and 2 several times, because each time Step 1 is implemented the number of marked zeros may change; and similarly with Step 2. In each of these steps, a zero is assigned only if it is the only unmarked zero in the particular row or column under consideration. Hence, there is no way to increase the number of assigned zeros with a different assignment.

In many cases the procedure will lead to Step 3(a), in which case a maximal assignment will have been found, and there will be no need to search for possible transfers of assignment which would increase the number of assigned zeros. Even when this procedure leads to Step 3(b), the complicated and lengthy zero assigning procedure of Section 11–4 will need to be performed on a smaller matrix. Frequently, this matrix will be considerably smaller, since most of the zeros will have been assigned or eliminated from consideration in the repetitions of Steps 1 and 2.

EXAMPLE 8

Consider the assignment problem whose cost matrix is

$$C = \begin{bmatrix} 0 & 0 & 2 & 3 & 2 & 4 \\ 3 & 4 & 0 & 0 & 0 & 5 \\ 0 & 2 & 0 & 2 & 0 & 0 \\ 0 & 4 & 2 & 0 & 3 & 1 \\ 0 & 2 & 0 & 4 & 0 & 2 \\ 4 & 3 & 0 & 2 & 0 & 4 \end{bmatrix}$$

† An unmarked zero is one that is not assigned (0*) or is unassignable (0^m).

This matrix is already in the form required by Box 3, namely, there is at least one zero in every row and column, and the elements are all non-negative.

If we assign zeros by the method of Section 11–4, we find that initially there are only 3 assigned zeros: 0* in (1, 1), 0* in (2, 3), 0* in (3, 5).

$$\begin{bmatrix} 0^* & 0 & 2 & 3 & 2 & 4 \\ 3 & 4 & 0^* & 0 & 0 & 5 \\ 0 & 2 & 0 & 2 & 0^* & 0 \\ 0 & 4 & 2 & 3 & 0 & 1 \\ 0 & 2 & 0 & 4 & 0 & 2 \\ 4 & 3 & 0 & 2 & 0 & 4 \end{bmatrix}$$

Following through with the method of Section 11–4, we next search each column until we find one containing no 0*. The first such column is column 2. The next step is to generate the sequence of cells:

0 in (1, 2)
0* in (1, 1)
0 in (3, 1)
0* in (3, 5)
0 in (4, 5)

As there is no 0* in row 4, a transfer of assignments is possible, yielding

$$\begin{bmatrix} 0 & 0^* & 2 & 3 & 2 & 4 \\ 3 & 4 & 0^* & 0 & 0 & 5 \\ 0^* & 2 & 0 & 2 & 0 & 1 \\ 0 & 4 & 2 & 3 & 0^* & 0 \\ 0 & 2 & 0 & 4 & 0 & 2 \\ 4 & 3 & 0 & 2 & 0 & 4 \end{bmatrix}$$

We again search the columns until finding one with no 0*, namely, column 4. We now generate the sequence

0 in (2, 4)
0* in (2, 3)
0 in (3, 3)
0* in (3, 1)
0 in (1, 1)
0* in (1, 2)

There is no unassigned 0 in column 2; hence, row 1 is labeled an essential row, and the last two cells are deleted from the sequence. Now, the last cell in the sequence is (3, 1), containing a 0*, and we search column 1 for an unassigned 0, beginning in row 2. We find such a 0 in (4, 1) and continue the sequence:

0 in (4, 1)
0* in (4, 5)
0 in (5, 5)

There being no 0* in row 5, a transfer of assignments is again possible, involving the entire 7 cell sequence generated above:

$$\begin{bmatrix} 0 & 0^* & 2 & 3 & 2 & 4 \\ 3 & 4 & 0 & 0^* & 0 & 5 \\ 0 & 2 & 0^* & 2 & 0 & 1 \\ 0^* & 4 & 2 & 3 & 0 & 0 \\ 0 & 2 & 0 & 4 & 0^* & 2 \\ 4 & 3 & 0 & 2 & 0 & 4 \end{bmatrix}$$

We now have 5 assigned zeros, and continue: The first (and only) column without a 0* is column 6. We generate the sequence of cells:†

0 in (4, 6)
0* in (4, 1)
0 in (3, 1)
0* in (3, 3)
0 in (2, 3)
0* in (2, 4)

There is no unassigned 0 in column 4; hence, row 2 is labeled an essential row, cells (2, 3) and (2, 4) are deleted from the sequence, and we search for another unassigned 0 in column 3, finding one in (5, 3). The sequence continues:

0 in (5, 3)
0* in (5, 5)
0 in (4, 5)
0* in (4, 1)

† Since row 1 has been labeled an essential row, we know that no transfer of assignments involving row 1 is possible. Hence, row 1 is excluded from the cell generation process.

Now, we must search column 1 for an unassigned 0. However, there are none in rows which either are not already in the sequence, or are not in essential rows. Hence, we delete cells (4, 5) and (4, 1), label row 4 as an essential row, and return to column 5, searching it for another 0. We find this 0 in cell (5, 6). The entire sequence so far is

$$\begin{array}{l} 0 \text{ in } (4, 6) \\ 0^* \text{ in } (4, 1) \\ 0 \text{ in } (3, 1) \\ 0^* \text{ in } (3, 3) \\ 0 \text{ in } (5, 3) \\ 0^* \text{ in } (5, 5) \\ 0 \text{ in } (5, 6) \end{array}$$

There is no 0* in row 6, and so a transfer of assignments is possible, yielding at last an optimal assignment:

$$\begin{bmatrix} 0 & 0^* & 2 & 3 & 2 & 4 \\ 3 & 4 & 0 & 0^* & 0 & 5 \\ 0^* & 2 & 0 & 2 & 0 & 1 \\ 0 & 4 & 2 & 3 & 0 & 0^* \\ 0 & 2 & 0^* & 4 & 0 & 2 \\ 4 & 3 & 0 & 2 & 0^* & 4 \end{bmatrix}$$

EXAMPLE 9

Let us re-solve the assignment problem of Example 8, by using the modification described in this section.

We begin by searching each row for a single unmarked 0. There being no such row, we proceed to Step 2, and search the columns in like manner. Column 2 has only 1 unmarked zero, in (1, 2). We assign it, and mark all other zeros in row 1 by 0^m. Next, we see that column 4 has only one zero, which we assign, and mark all other zeros in row 2 by 0^m. Finally, we find that column 6 contains only one zero. Assigning it and marking all other zeros in row 4 by 0^m yields:

$$\begin{bmatrix} 0^m & 0^* & 2 & 3 & 2 & 4 \\ 3 & 4 & 0^m & 0^* & 0^m & 5 \\ 0 & 2 & 0 & 2 & 0 & 1 \\ 0^m & 4 & 2 & 3 & 0^m & 0^* \\ 0 & 2 & 0 & 4 & 0 & 2 \\ 4 & 3 & 0 & 2 & 0 & 4 \end{bmatrix}$$

Now, we return to Step 1, the row-searching procedure, and find that all remaining rows contain at least two unmarked zeros. Hence, we are in Step 3(b), in which we label rows 1, 2, 4 essential and delete these rows, along with the columns containing the 0*'s (columns 2, 4, 6). The reduced matrix thus obtained is

$$\begin{array}{c} \\ (3) \\ (5) \\ (6) \end{array}\begin{array}{c} \begin{array}{ccc} (1) & (3) & (5) \end{array} \\ \begin{bmatrix} 0 & 0 & 0 \\ 0 & 0 & 0 \\ 4 & 0 & 0 \end{bmatrix} \end{array}$$

We have indicated the original row and column numbers above, and shall refer to these numbers.

The remaining three assigned zeros are easily found now, by merely implementing Step 1 of the original procedure of Section 11–4, which yields a 0* in cells (3, 1), (5, 3), and (6, 5).

Hence, we have found the same optimal assignment as in Example 8, but with considerably less effort.

EXAMPLE 10

Let us solve, by the modified method of this section, the assignment problem whose cost matrix is

$$C = \begin{bmatrix} 4 & 0 & 0 & 0 & 1 & 5 \\ 3 & 2 & 3 & 0 & 4 & 3 \\ 1 & 0 & 0 & 0 & 3 & 0 \\ 0 & 3 & 3 & 5 & 0 & 0 \\ 2 & 0 & 2 & 3 & 0 & 4 \\ 2 & 5 & 0 & 0 & 2 & 4 \end{bmatrix}$$

In the modified method, we only assign a zero in a given row (or column) if the zero is the only unmarked zero in that row (or column). Below is the sequence of marked zeros produced by applying this method to the above matrix. An arrow is used to indicate the row or column in which the new marked zeros appear. Remember that when an arrow points to a *row*, the 0* in that row was assigned because it was the only unmarked zero in its corresponding *column*, and vice-versa. Note that Step 1 (row searches) has been performed twice and Step 2 (column searches) has been performed once. The resulting assignment is optimal.

$$\begin{array}{c}\begin{matrix} & & & \downarrow & & \end{matrix}\\ \begin{bmatrix} 4 & 0 & 0 & 0^m & 1 & 5 \\ 3 & 2 & 3 & 0^* & 4 & 3 \\ 1 & 0 & 0 & 0^m & 3 & 0 \\ 0 & 3 & 3 & 5 & 0 & 0 \\ 2 & 0 & 2 & 3 & 0 & 4 \\ 2 & 5 & 0 & 0^m & 2 & 4 \end{bmatrix}\end{array}, \quad \begin{array}{c}\begin{matrix} & & \downarrow & & & \end{matrix}\\ \begin{bmatrix} 4 & 0 & 0^m & 0^m & 1 & 5 \\ 3 & 2 & 3 & 0^* & 4 & 3 \\ 1 & 0 & 0^m & 0^m & 3 & 0 \\ 0 & 3 & 3 & 5 & 0 & 0 \\ 2 & 0 & 2 & 3 & 0 & 4 \\ 2 & 5 & 0^* & 0^m & 2 & 4 \end{bmatrix}\end{array}$$

$$\begin{matrix} \\ \\ \\ \rightarrow \\ \\ \\ \end{matrix}\begin{bmatrix} 4 & 0 & 0^m & 0^m & 1 & 5 \\ 3 & 2 & 3 & 0^* & 4 & 3 \\ 1 & 0 & 0^m & 0^m & 3 & 0 \\ 0^* & 3 & 3 & 5 & 0^m & 0^m \\ 2 & 0 & 2 & 3 & 0 & 4 \\ 2 & 5 & 0^* & 0^m & 2 & 4 \end{bmatrix}, \quad \begin{array}{c}\begin{matrix} \\ \\ \\ \\ \rightarrow \\ \\ \end{matrix}\begin{bmatrix} 4 & 0 & 0^m & 0^m & 1 & 5 \\ 3 & 2 & 3 & 0^* & 4 & 3 \\ 1 & 0 & 0^m & 0^m & 3 & 0 \\ 0^* & 3 & 3 & 5 & 0^m & 0^m \\ 2 & 0^m & 2 & 3 & 0^* & 4 \\ 2 & 5 & 0^* & 0^m & 2 & 4 \end{bmatrix}\\ \begin{matrix} & \downarrow & & & & \end{matrix}\end{array},$$

$$\begin{matrix} \\ \\ \rightarrow \\ \\ \\ \\ \end{matrix}\begin{bmatrix} 4 & 0 & 0^m & 0^m & 1 & 5 \\ 3 & 2 & 3 & 0^* & 4 & 3 \\ 1 & 0^m & 0^m & 0^m & 3 & 0^* \\ 0^* & 3 & 3 & 5 & 0^m & 0^m \\ 2 & 0^m & 2 & 3 & 0^* & 4 \\ 2 & 5 & 0^* & 0^m & 2 & 4 \end{bmatrix}, \quad \begin{bmatrix} 4 & 0^* & 0^m & 0^m & 1 & 5 \\ 3 & 2 & 3 & 0^* & 4 & 3 \\ 1 & 0^m & 0^m & 0^m & 3 & 0^* \\ 0^* & 3 & 3 & 5 & 0^m & 0^m \\ 2 & 0^m & 2 & 3 & 0^* & 4 \\ 2 & 5 & 0^* & 0^m & 2 & 4 \end{bmatrix}$$

11-7 THE TRAVELING SALESMAN PROBLEM

We shall conclude this chapter with a brief discussion of the so-called "traveling salesman problem," which is structurally quite similar to the assignment problem.

The traveling salesman problem may be stated as follows: Given a set of n cities, numbered from 1 to n, a salesman wishes to determine a route (called a "tour") which begins in city 1, goes through each of the other $(n - 1)$ cities exactly once, and then returns to city 1, in such a way that the total distance traveled is a minimum. There are $(n - 1)!$ possible tours, so that total enumeration of these to find the tour of minimum distance is impractical, even for a relatively small value of n.

If we let c_{ij} be the distance between city i and city j, and let

$$x_{ij} = \begin{cases} 1, & \text{if salesman goes from city } i \text{ to city } j \\ 0, & \text{otherwise} \end{cases}$$

then the total distance of any tour is given by

$$z = \sum_{i=1}^{n} \sum_{j=1}^{n} c_{ij}x_{ij} \tag{11–5}$$

Furthermore, since each city is to be included in the tour one and only one time,

$$\sum_{j=1}^{n} x_{ij} = 1, \qquad i = 1, 2, \ldots, n \tag{11–6}$$

$$\sum_{i=1}^{n} x_{ij} = 1, \qquad j = 1, 2, \ldots, n \tag{11–7}$$

The constraints (11–6) insure that for each i, the salesman will travel from city i to exactly one city; the constraints (11–7) insure that the salesman will travel to each city j, exactly once.

The similarity between the traveling salesman problem and the assignment problem should now be apparent; namely, the constraints (11–6) and (11–7), along with the objective function (11–4) are exactly the same as those of the assignment problem.

However, the constraints (11–6) and (11–7) by themselves do not guarantee a feasible solution for the traveling salesman problem, as they do for the assignment problem.

For example, the solution $x_{11} = x_{22} = \ldots = x_{nn} = 1$, all other $x_{ij} = 0$, satisfies the constraints (11–6) and (11–7). This solution is feasible for the assignment problem, but is totally meaningless for the traveling salesman problem. If the above solution represented the only difference between the two problems, we could easily exclude it by setting all $c_{ii} = M$, where M is a very large positive number. However, this step does not resolve the difficulty. Consider a seven city traveling salesman problem. Then, the following is a feasible solution for the related assignment problem but not for the traveling salesman problem:

$$x_{12} = x_{26} = x_{67} = x_{71} = x_{34} = x_{45} = x_{53} = 1$$
$$\text{all other } x_{ij} = 0$$

This solution actually defines two disjoint "subtours" for the traveling salesman problem, as illustrated in Figure 11–2.

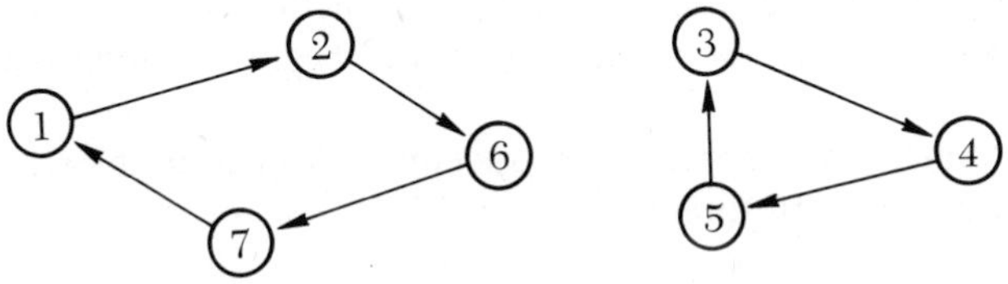

Figure 11–2 Two subtours for a seven city traveling salesman problem

Unfortunately, there is no easy method of eliminating such subtours from consideration. Hence, despite the striking similarities between the traveling salesman problem and the assignment problem, the former cannot be solved directly by the Hungarian method (or even, so far as is known, by the simplex method).

Some algorithms have been devised to solve traveling salesman problems. Many of these are described by Bellmore and Nemhauser.[1] The most promising exact methods seem to be those of Little et al.[5] and Shapiro.[6] However, to date, these methods seem to be impractical for problems involving more than 100 cities. Contrast this with the Hungarian method, which is limited in its applicability only by the available storage capacity.

Another type of model which can be formulated mathematically as a traveling salesman problem is a sequencing problem, in which a given set of n jobs must be performed one at a time, and it is desired to find the order which produces the minimum total cost. Such a problem is illustrated below.

EXAMPLE 11

A small metal parts manufacturer has orders for five types of parts. The manufacturer only has the manpower and equipment to fill one order at a time. After each order is filled, the equipment must be re-tooled to prepare for the next order to be filled. Because of varying degrees of similarity between the five types of parts, the amount of re-tooling necessary between each order depends upon the sequence in which these orders are filled. In any case, the re-tooling process consists entirely of dismantling and re-assembling various pieces of machinery, all involving equipment already on hand. Hence, the re-tooling costs are directly proportional to the time required for the re-tooling (i.e., the costs are exclusively manpower costs). The manufacturer estimates the equipment change-over times are as follows:

Change-over Times (man-hours)

From \ To	*A*	*B*	*C*	*D*	*E*
A	—	6	4	7	5
B	5	—	3	4	6
C	3	5	—	5	5
D	7	4	5	—	3
E	10	6	7	5	—

The equipment is already set up for production of part type A, from a previous order. Hence, the manufacturer will begin by filling the order for part A.

To formulate this problem as a traveling salesman problem, we need to specify which part corresponds to "City 1" (which is obviously part A) and the cost matrix. Since, the manufacturer does not wish to re-tool again after the last order has been filled (i.e., he does not wish to return to City 1) we shall let the corresponding costs of re-tooling to part A be zero.

Thus, if we denote parts A, B, C, D, and E by 1, 2, 3, 4, and 5, respectively, and let

$$c_{ij} = \text{change-over time from part } i \text{ to part } j,$$

then

$$C = \begin{bmatrix} - & 6 & 4 & 7 & 5 \\ 0 & - & 3 & 4 & 6 \\ 0 & 5 & - & 5 & 5 \\ 0 & 4 & 5 & - & 3 \\ 0 & 6 & 7 & 5 & - \end{bmatrix}$$

The formulation of this model as a traveling salesman problem is now complete. The reader should be certain he understands why this problem cannot be formulated as an assignment problem.

PROBLEMS

1. Solve the assignment problem whose cost matrix is:

$$C = \begin{bmatrix} 3 & 6 & 5 & 7 \\ 3 & 7 & 2 & 4 \\ 4 & 3 & 4 & 5 \\ 6 & 4 & 5 & 6 \end{bmatrix}$$

2. Solve the assignment problem whose cost matrix is:

$$C = \begin{bmatrix} 2 & 3 & 4 & 4 \\ 7 & 3 & 1 & 5 \\ 4 & 4 & 2 & 3 \\ 4 & 1 & 4 & 5 \end{bmatrix}$$

Prove that the optimal solution is unique.

3. Solve by the Hungarian method, using the method of Section 11–4 for assigning zeros, the assignment problem whose cost matrix is:

$$C = \begin{bmatrix} 3 & 7 & 5 & 7 & 5 & 6 \\ 8 & 6 & 6 & 8 & 6 & 6 \\ 7 & 4 & 6 & 4 & 8 & 6 \\ 7 & 5 & 5 & 8 & 5 & 5 \\ 6 & 7 & 2 & 3 & 4 & 6 \\ 5 & 8 & 5 & 5 & 5 & 9 \end{bmatrix}$$

4. Solve the assignment problem of Exercise 3, using the modification described in Section 11–6 for assigning zeros.

5. Solve by the Hungarian method, using the method of Section 11–4 for assigning zeros, the assignment problem whose cost matrix is:

$$C = \begin{bmatrix} 2 & 6 & 7 & 3 & 8 & 7 \\ 6 & 1 & 3 & 9 & 7 & 3 \\ 3 & 6 & 5 & 7 & 3 & 5 \\ 2 & 2 & 7 & 8 & 4 & 8 \\ 4 & 9 & 6 & 8 & 7 & 6 \\ 7 & 5 & 5 & 7 & 7 & 5 \end{bmatrix}$$

6. Solve the assignment problem of Exercise 5, using the modification described in Section 11–6 for assigning zeros.

7. Show how the following problem can be solved by the Hungarian method:

$$\text{Maximize } z = \sum_{i=1}^{n} \sum_{j=1}^{n} c_{ij} x_{ij}$$

$$\sum_{j=1}^{n} x_{ij} = 1, \qquad i = 1, 2, \ldots, n$$

$$\sum_{i=1}^{n} x_{ij} = 1, \qquad j = 1, 2, \ldots, n$$

$$x_{ij} \geq 0 \qquad i = 1, 2, \ldots, n \quad \text{and} \quad j = 1, 2, \ldots, n$$

Assume that not all $c_{ij} \geq 0$.

8. Describe how the following problem can be solved by the Hungarian method:

A company has m men and n tasks, $m < n$. Each man may be assigned to at most one task; hence, not all tasks will be accomplished. The cost of assigning man i to task j is c_{ij} dollars. If task j is not accomplished, the cost to the company is d_j dollars. The company wishes to assign the men to the tasks in such a way that the total cost is minimized.

9. Formulate the model of Exercise 14, Chapter 10, as an assignment problem.

10. Show how to formulate the transportation problem below as an assignment problem:

$$\text{Minimize } z = \sum_{i=1}^{m} \sum_{j=1}^{n} c_{ij}x_{ij}$$

$$\sum_{j=1}^{n} x_{ij} = a_i, \qquad i = 1, 2, \ldots, m$$

$$\sum_{j=1}^{n} x_{ij} = b_j, \qquad j = 1, 2, \ldots, n$$

$$x_{ij} \geq 0, \qquad i = 1, 2, \ldots, m \text{ and } j = 1, 2, \ldots, n$$

REFERENCES

1. Bellmore M., and G. L. Nemhauser: The traveling salesman problem: A survey. *Operations Research*, **16**:538–558, 1968.
2. Dantzig, G. B.: *Linear Programming and its Extensions.* Princeton University Press, Princeton, N.J., 1963.
3. Kuhn, H. W.: The Hungarian method for the assignment problem. *Naval Research Logistics Quarterly*, **2**:83–97, 1955.
4. Kuhn, H. W.: Variants of the Hungarian method for the assignment problem. *Naval Research Logistics Quarterly*, **3**:253–258, 1956.
5. Little, J. D. C., G. Marty, D. W. Sweeney, and C. Karel: An algorithm for the traveling salesman problem. *Operations Research*, **11**:972–989, 1963.
6. Shapiro, D. M.: Algorithms for the solution of the optimal cost and bottleneck traveling salesman problems, D.Sc. Dissertation, Washington University, Department of Applied Mathematics and Computer Science, St. Louis, Mo. (1967).

12

Parametric and Post-optimal Analysis

12–1 INTRODUCTION

When a practical problem has been formulated and solved as a linear programming problem, it is frequently the case that not all of the input parameters (i.e., the elements of the coefficient matrix A, the requirements vector $\bar{b}$, and the cost vector $\bar{c}$) are known exactly. Typically, some of these parameters have been estimated or calculated only approximately. Thus, it is important to know how sensitive the optimal solution is to small changes in such parameters. For example, if it is known that a particular c_j is accurate to within ± 5 per cent, then we must also determine whether the computed optimal solution remains optimal for all values of c_j in this range; without this additional information it is not at all certain that the computed optimal solution is indeed the true optimal solution to the actual problem.

There are also many practical situations which arise in which a linear programming model is used repeatedly on a regular basis to find, for example, the optimal weekly (or monthly) production quantities. In such cases, a few changes in the c_j and/or the b_i are not uncommon.

Another type of modification in the linear programming model which sometimes occurs is the addition of a new constraint or variable to the original formulation, either because the original formulation was erroneous, or because the model situation has changed.

The study of how sensitive a given optimal solution is to various changes in the input parameters is usually called *sensitivity analysis* or *parametric analysis*. Sensitivity analysis, along with the investigation of how specific changes in the input parameters (e.g., a change in one of the c_j or b_i, the addition of a constraint) affect the optimal solution, is called *post-optimal analysis*. It is the purpose of this chapter to develop a methodology for performing such a post-optimal analysis. As we shall see, it is frequently not necessary to solve the modified problem from the beginning to obtain the desired information; instead this information can be found by performing relatively few computations using the data in the optimal tableau to the originally solved problem.

We shall begin by investigating how specific changes in the parameters affect the optimal solution, before turning to the problem of studying the effects of *varying* the parameters from their current values (parametric analysis).

12–2 CHANGES IN THE OBJECTIVE FUNCTION

Suppose we have solved the linear programming problem, maximize $z = \bar{c}'\bar{x}$, subject to $A\bar{x} = \bar{b}$, $\bar{x} \geq \bar{0}$, and have obtained an optimal basic feasible solution, $\bar{x}_B^*$. If we denote by $\bar{c}_B$ the cost vector corresponding to $\bar{x}_B^*$, then the current values of $z_j - c_j$ are

$$z_j^* - c_j = \bar{c}_B'\bar{y}_j - c_j \qquad j = 1, 2, \ldots, n \tag{12–1}$$

where the $\bar{y}_j$ also correspond to the current optimal solution.

Now, suppose we wish to change one of the c_j, say c_k, to a new value, $\hat{c}_k$, where

$$\hat{c}_k = c_k + \delta_k \tag{12–2}$$

Since a change in the objective function in no way alters the set of feasible solutions, the solution $\bar{x}_B^*$ will remain optimal provided that the new values of $z_j - c_j$ (resulting from the change in c_k) are still non-negative.

If c_k corresponds to a variable which is currently nonbasic, then it is obvious from equation (12–1) that only $z_k - c_k$ will be changed; all other $z_j - c_j$ will remain unchanged (since $\bar{c}_B$ has not been changed) and hence non-negative. Moreover,

$$\begin{aligned}\hat{z}_k - \hat{c}_k &= \bar{c}_B'\bar{y}_k - \hat{c}_k \\ &= \bar{c}_B'\bar{y}_k - (c_k + \delta_k) \\ &= (\bar{c}_B'\bar{y}_k - c_k) - \delta_k \\ &= (z_k^* - c_k) - \delta_k\end{aligned}$$

Thus, $\hat{z}_k - \hat{c}_k \geq 0$ if

$$\delta_k \leq (z_k^* - c_k) \tag{12–3}$$

According to (12–3), if the cost c_k of any nonbasic variable x_k is increased by an amount up to $(z_k^* - c_k)$, the current optimal solution will remain optimal. If such a cost c_k is increased by more than the quantity $(z_k^* - c_k)$ then the resulting $z_k - c_k$ will become negative, and a few more simplex iterations may be needed to determine the new optimal solution. Note that the cost of any nonbasic variable can be *decreased* without bound, without affecting the optimality of $\bar{x}_B^*$.

Consider now the case in which we wish to change the cost c_k corresponding to a basic variable x_k. Suppose that x_k is the pth basic variable x^*_{Bp}. Let

$$\hat{c}_k = c_k + \delta_k = \hat{c}_{Bp} = c_{Bp} + \delta_k.$$

Then, for $j \neq k$,

$$\begin{aligned}\hat{z}_j - \hat{c}_j &= \hat{c}'_B \bar{y}_j - c_j \\ &= \sum_{i=1}^{m} \hat{c}_{Bi} y_{ij} - c_j \\ &= \sum_{i=1}^{m} c_{Bi} y_{ij} + \delta_k y_{pj} - c_j \\ &= z^*_j + \delta_k y_{pj} - c_j\end{aligned}$$

Hence,

$$\hat{z}_j - \hat{c}_j = \begin{Bmatrix} (z^*_j - c_j) + \delta_k y_{pj}, & j \neq k \\ 0, & j = k \end{Bmatrix} \tag{12–4}$$

($\hat{z}_k - \hat{c}_k = 0$ because x_k is basic). In order that all $\hat{z}_j - \hat{c}_j \geq 0$, it is necessary that

$$-\delta_k y_{pj} \leq (z^*_j - c_j), \qquad j = 1, 2, \ldots, n, \qquad j \neq k \tag{12–5}$$

Note that δ_k must satisfy *all* of the inequalities (12–5) simultaneously. For each $y_{pj} > 0$, we have that

$$\delta_k \geq (z^*_j - c_j)/y_{pj}$$

and hence,

$$\delta_k \geq \max \left\{ \frac{z^*_j - c_j}{y_{pj}} \,\middle|\, y_{pj} > 0 \right\} \tag{12–6}$$

Similarly, we obtain

$$\delta_k \leq \min \left\{ \frac{z^*_j - c_j}{y_{pj}} \,\middle|\, y_{pj} < 0 \right\} \tag{12–7}$$

Thus, if δ_k lies in the range determined by (12–6) and (12–7), then $\bar{x}^*_B$ remains optimal. If δ_k falls outside this range, at least one $\hat{z}_j - c_j$ will be negative.

EXAMPLE 1

Suppose we have the following optimal tableau for a linear programming problem.

		c_j	2	1	1	2	0
$\bar{c}_B$	Basis vectors	$\bar{x}_B$	$\bar{y}_1$	$\bar{y}_2$	$\bar{y}_3$	$\bar{y}_4$	$\bar{y}_5$
2	$\bar{a}_1$	3	1	0	-1	3	2
1	$\bar{a}_2$	4	0	1	4	-1	-2
	z	10	0	0	1	3	2

According to the inequality (12–3), we can increase each c_j corresponding to a nonbasic variable by as much as its corresponding $z_j^* - c_j$, and the current optimal solution will remain optimal for the resulting problem. Hence, since x_3, x_4, x_5 are currently nonbasic in the above optimal tableau, we have that the current solution remains optimal so long as

$$-\infty \leq c_3 \leq 1 + 1 = 2$$
$$-\infty \leq c_4 \leq 2 + 3 = 5$$
$$-\infty \leq c_5 \leq 0 + 2 = 2$$

However, if we wished to increase c_3 to $+3$, for example, then $z_3 - c_3 = -1$, and the current solution is no longer optimal. Since $z_3 - c_3$ is the only negative $z_j - c_j$ (the others remaining unchanged) we bring $\bar{a}_3$ into the basis, replacing $\bar{a}_2$. The resulting tableau is

		c_j	2	1	3	2	0
$\bar{c}_B$	Basis vectors	$\bar{x}_B$	$\bar{y}_1$	$\bar{y}_2$	$\bar{y}_3$	$\bar{y}_4$	$\bar{y}_5$
2	$\bar{a}_1$	$\frac{13}{4}$	1	$\frac{1}{4}$	0	$\frac{11}{4}$	$\frac{3}{2}$
3	$\bar{a}_3$	1	0	$\frac{1}{4}$	1	$-\frac{1}{4}$	$-\frac{1}{2}$
	z	11	0	$\frac{1}{4}$	0	$\frac{11}{4}$	$\frac{3}{2}$

Hence, we have found the new optimal solution in just one iteration.

EXAMPLE 2

Starting with the optimal tableau given at the beginning of Example 1, let us investigate the effect of changing c_1. Since x_1 is the basic variable x_{B1}, we must recompute all the $z_j - c_j$; if $\hat{c}_1 = \hat{c}_{B1} = c_1 + \delta_1 = c_{B1} + \delta_1$, then

$$\begin{aligned}\hat{z}_j - c_j &= \hat{c}_{B1}y_{1j} + c_{B2}y_{2j} - c_j \\ &= (c_{B1}y_{1j} + c_{B2}y_{2j} - c_j) + \delta_1 y_{1j} \\ &= (z_j^* - c_j) + \delta_1 y_{1j}, \qquad j = 2, 3, 4, 5\end{aligned}$$

Hence,

$$\hat{z}_2 - c_2 = 0 + \delta_1 0 = 0$$
$$\hat{z}_3 - c_3 = 1 + \delta_1(-1) = 1 - \delta_1$$
$$\hat{z}_4 - c_4 = 3 + \delta_1(3) = 3 + 3\delta_1$$
$$\hat{z}_5 - c_5 = 2 + \delta_1(2) = 2 + 2\delta_1$$

If all $\hat{z}_j - c_j \geq 0$, then

$$1 - \delta_1 \geq 0,\ 3 + 3\delta_1 \geq 0,\ 2 + 2\delta_1 \geq 0$$

from which we see that $\delta_1 \leq 1$, $\delta_1 \geq -1$, $\delta_1 \geq -1$, respectively. Hence if $-1 \leq \delta_1 \leq 1$, and therefore, $1 \leq \hat{c}_1 \leq 3$, then the current solution remains optimal. If $\hat{c}_1 > 3$, then $\hat{z}_3 - c_3 < 0$; if $c_1 < 1$, then both $\hat{z}_4 - c_4 < 0$ and $\hat{z}_5 - c_5 < 0$. In either case, more simplex iterations will be necessary to find the new optimal solution. For example, if we increase c_1 to $\hat{c}_1 = 5$ ($\delta_1 = 3$), then $\hat{z}_3 - c_3 = -2$. Bringing $\bar{a}_3$ into the basis and removing $\bar{a}_2$ yields the tableau

		c_j	5	1	1	2	0
$\bar{c}_B$	Basis vectors	$\bar{x}_B$	$\bar{y}_1$	$\bar{y}_2$	$\bar{y}_3$	$\bar{y}_4$	$\bar{y}_5$
5	$\bar{a}_1$	$\frac{13}{4}$	1	$\frac{1}{4}$	0	$\frac{11}{4}$	$\frac{3}{2}$
1	$\bar{a}_3$	1	0	$\frac{1}{4}$	1	$-\frac{1}{4}$	$-\frac{1}{2}$
	z	$\frac{69}{4}$	0	$\frac{1}{2}$	0	$\frac{23}{2}$	7

Hence, we have again found the new optimal solution in just one additional iteration.

EXAMPLE 3

Suppose we have found an optimal solution to a linear programming problem and then decide to replace the objective function with an entirely different one. The only quantities affected by such a change are the $z_j - c_j$, and the simplest way to handle such a case (i.e., many simultaneous changes in the c_j) is to recompute the new $z_j - c_j$, from $z_j - c_j = \bar{c}_B'\bar{y}_j - c_j$.

For example, consider again the original problem of Example 1, whose objective function is $z = 2x_1 + x_2 + x_3 + 2x_4 + 0x_5$. Suppose we wish to replace this objective function with $z = 3x_1 + 4x_2 + 2x_3 + x_4 + 2x_5$.

Then, using the original optimal tableau of Example 1, we find that

$$\bar{c}_B = (c_1, c_2) = (3, 4)$$
$$\bar{y}_3 = \begin{pmatrix} -1 \\ 4 \end{pmatrix}, \quad \bar{y}_4 = \begin{pmatrix} 3 \\ -1 \end{pmatrix}, \quad \bar{y}_5 = \begin{pmatrix} 2 \\ -2 \end{pmatrix}$$

Hence,

$$z_3 - c_3 = \bar{c}_B'\bar{y}_3 - c_3 = (3, 4)\begin{pmatrix} -1 \\ 4 \end{pmatrix} - 2 = 11$$

$$z_4 - c_4 = \bar{c}_B'\bar{y}_4 - c_4 = (3, 4)\begin{pmatrix} 3 \\ -1 \end{pmatrix} - 1 = 4$$

$$z_5 - c_5 = \bar{c}_B'\bar{y}_5 - c_5 = (3, 4)\begin{pmatrix} 2 \\ -2 \end{pmatrix} - 2 = -4$$

The current solution is therefore no longer optimal; we bring $\bar{a}_5$ into the basis, and remove $\bar{a}_1$ from the basis, yielding the new optimal tableau:

		c_j	3	4	2	1	2
$\bar{c}_B$	Basis vectors	$\bar{x}_B$	$\bar{y}_1$	$\bar{y}_2$	$\bar{y}_3$	$\bar{y}_4$	$\bar{y}_5$
2	$\bar{a}_5$	$\frac{3}{2}$	$\frac{1}{2}$	0	$-\frac{1}{2}$	$\frac{3}{2}$	1
4	$\bar{a}_2$	7	1	1	3	2	0
	z	31	2	0	9	10	0

Although in each of these three examples we have only required at most one additional iteration to find the new optimal solution, such is not always the case. Typically, though, if the changes are not too great, only a relatively few iterations are required to find the new optimal solution (if the current solution is in fact no longer optimal).

12–3 CHANGES IN THE REQUIREMENTS VECTOR

Suppose we have found an optimal solution to the linear programming problem

$$\text{Maximize } z = \bar{c}'\bar{x}$$
$$A\bar{x} = \bar{b}$$
$$\bar{x} \geq \bar{0}$$

and then decide that we wish to modify the requirements vector $\bar{b}$.

Recall from Chapter 8 that the requirements vector of a primal linear programming problem is the cost vector of the corresponding dual problem. Hence, we could treat the analysis of changes in the requirements vector by applying the results of the previous section to the dual problem.

Instead, let us attack the problem directly.

If we wish to change the ith component of $\bar{b}$ by an amount f_i (positive or negative), then $\hat{b}_i = b_i + f_i$, $i = 1, 2, \ldots, m$; or, in vector notation, we have $\hat{\bar{b}} = \bar{b} + \bar{f}$.

Now, we must recompute the values of the basic variables corresponding to the vectors in the current basic feasible solution. If we denote the current basis matrix by B, then

$$\begin{aligned}\hat{\bar{x}}_B &= B^{-1}\hat{\bar{b}} \\ &= B^{-1}[\bar{b} + \bar{f}] \\ &= B^{-1}\bar{b} + B^{-1}\bar{f} \\ &= \bar{x}_B^* + B^{-1}\bar{f}\end{aligned}$$

Hence, depending on $B^{-1}\bar{f}$, the new basic solution may or may not be feasible. However, the $z_j^* - c_j$ are unaffected by a change in the requirements vector, and therefore if $\hat{\bar{x}}_B$ is feasible, it is also optimal. If $\hat{\bar{x}}_B$ is not feasible (one or more $\hat{x}_{Bi} < 0$) we can use the dual simplex method to find the new optimal solution.

Observe that in order to compute $\hat{\bar{x}}_B$ it is necessary to have B^{-1}. But, B^{-1} will be available in the final tableau either if we have used the revised simplex method (see Chapter 6) or if we have used the simplex method and have performed all computations on the columns which contained the identity matrix used to obtain the initial basic feasible solution. In the latter case, these columns will currently contain B^{-1} (the reader is asked to prove this fact in the exercises). In other words, if the columns of A are ordered so that the first m columns constitute an identity matrix, then

$$B^{-1} = [\bar{y}_1, \bar{y}_2, \ldots, \bar{y}_m]$$

Thus,

$$\begin{aligned}\hat{\bar{x}}_B &= \bar{x}_B^* + B^{-1}\bar{f} \\ &= \bar{x}_B^* + \sum_{j=1}^{m} f_j \bar{y}_j\end{aligned}$$

Or,

$$\hat{x}_{Bi} = x_{Bi} + \sum_{j=1}^{m} f_j y_{ij}$$

EXAMPLE 4

Consider the following linear programming problem.

$$\text{Maximize } z = 4x_1 + 3x_2$$

$$\begin{aligned} x_1 + x_2 &\le 5 \\ 3x_1 + x_2 &\le 7 \\ x_1 + 2x_2 &\le 10 \\ x_1, x_2 &\ge 0 \end{aligned}$$

After adding slack variables x_3, x_4, x_5 and using the corresponding columns $\bar{a}_3$, $\bar{a}_4$, $\bar{a}_5$ as our initial basis matrix, we obtain the optimal solution in two iterations. Below is the tableau for this solution:

		c_j	4	3	0	0	0
$\bar{c}_B$	Basis vectors	$\bar{x}_B$	$\bar{y}_1$	$\bar{y}_2$	$\bar{y}_3$	$\bar{y}_4$	$\bar{y}_5$
3	$\bar{a}_2$	4	0	1	$\frac{3}{2}$	$-\frac{1}{2}$	0
4	$\bar{a}_1$	1	1	0	$-\frac{1}{2}$	$\frac{1}{2}$	0
0	$\bar{a}_5$	1	0	0	$-\frac{5}{2}$	$\frac{1}{2}$	1
	z	16	0	0	$\frac{5}{2}$	$\frac{1}{2}$	0

According to our remarks earlier in this section, $B^{-1} = [\bar{y}_3, \bar{y}_4, \bar{y}_5]$, since these columns contained the identity matrix used to form the initial basic solution. Since we know $B = [\bar{a}_2, \bar{a}_1, \bar{a}_5]$ it is an easy matter to verify (and the skeptical reader should do so!) that indeed,

$$B^{-1} = [y_3, y_4, y_5] = \begin{bmatrix} \frac{3}{2} & -\frac{1}{2} & 0 \\ -\frac{1}{2} & \frac{1}{2} & 0 \\ -\frac{5}{2} & \frac{1}{2} & 1 \end{bmatrix}$$

Suppose now we wish to increase the original b_1 by one unit, and decrease the original b_3 by one unit. Hence, $\bar{f} = (1, 0, -1)$ and

$$\hat{\bar{x}}_B = \bar{x}_B^* + B^{-1}\bar{f}$$

$$= \begin{bmatrix} 4 \\ 1 \\ 1 \end{bmatrix} + \begin{bmatrix} \frac{3}{2} & -\frac{1}{2} & 0 \\ -\frac{1}{2} & \frac{1}{2} & 0 \\ -\frac{5}{2} & \frac{1}{2} & 1 \end{bmatrix} \begin{bmatrix} 1 \\ 0 \\ -1 \end{bmatrix}$$

$$= \begin{bmatrix} 4 \\ 1 \\ 1 \end{bmatrix} + \begin{bmatrix} \frac{3}{2} \\ -\frac{1}{2} \\ -\frac{7}{2} \end{bmatrix} = \begin{bmatrix} \frac{11}{2} \\ \frac{1}{2} \\ -\frac{5}{2} \end{bmatrix}$$

Hence, the current basic solution is no longer feasible, and we must use the dual simplex method to find the new optimal solution. Accordingly, we choose $\bar{a}_5$ to leave the basis and $\bar{a}_3$ to enter, yielding the feasible (and hence optimal tableau):

	c_j		4	3	0	0	0
$\bar{c}_B$	Basis vectors	$\bar{x}_B$	$\bar{y}_1$	$\bar{y}_2$	$\bar{y}_3$	$\bar{y}_4$	$\bar{y}_5$
3	$\bar{a}_2$	4	0	1	0	$-\frac{1}{5}$	$\frac{3}{5}$
4	$\bar{a}_1$	1	1	0	0	$\frac{2}{5}$	$-\frac{1}{5}$
0	$\bar{a}_3$	1	0	0	1	$-\frac{1}{5}$	$-\frac{2}{5}$
	z	16	0	0	0	1	1

Again, we remark that one will not always be able to find the new optimal solution in just one iteration.

The reader should observe also that a change in only one of the b_i may result in a change in all of the $\hat{x}_{Bi}$. In the above example we changed two of the b_i, but all three of the basic variables changed in value.

As another example, suppose we return to our original optimal tableau, and decrease b_2 by two units. Thus, now $\bar{f} = (0, -2, 0)$, and

$$\hat{\bar{x}}_B = x_B^* + B^{-1}\bar{f}$$

$$= \begin{bmatrix} 4 \\ 1 \\ 1 \end{bmatrix} + \begin{bmatrix} \frac{3}{2} & -\frac{1}{2} & 0 \\ -\frac{1}{2} & \frac{1}{2} & 0 \\ -\frac{5}{2} & \frac{1}{2} & 1 \end{bmatrix} \begin{bmatrix} 0 \\ -2 \\ 0 \end{bmatrix}$$

$$= \begin{bmatrix} 4 \\ 1 \\ 1 \end{bmatrix} + \begin{bmatrix} 1 \\ -1 \\ -1 \end{bmatrix} = \begin{bmatrix} 5 \\ 0 \\ 0 \end{bmatrix}$$

Here, $\hat{\bar{x}}_B$ is still feasible, and therefore also still optimal.

12–4 ADDITION OF A VARIABLE

Consider the situation in which an optimal solution to the linear programming problem

$$\begin{aligned} \text{Maximize } z &= \bar{c}'\bar{x} \\ A\bar{x} &= \bar{b} \\ \bar{x} &\geq \bar{0} \end{aligned} \tag{12–8}$$

has been found, and it is then somehow determined that a variable had been left out of the formulation of the problem. Hence, it is now desired to add a new variable, call it x_{n+1}, and its associated activity vector $\bar{a}_{n+1}$ and cost c_{n+1} to the problem yielding

$$\text{Maximize } z = \bar{c}'\bar{x} + c_{n+1}x_{n+1}$$

$$[A, \bar{a}_{n+1}]\begin{bmatrix} \bar{x} \\ x_{n+1} \end{bmatrix} = \bar{b} \tag{12–9}$$

$$\bar{x} \geq \bar{0}, \qquad x_{n+1} \geq 0$$

Now, we already have a basic feasible solution to (12–9); namely, the optimal solution to the original problem (12–8), $\bar{x}_B^*$, is a feasible solution of (12–9), with the new variable x_{n+1} currently nonbasic (and hence zero). Moreover, this solution will be optimal for (12–9) if $z_{n+1} - c_{n+1} \geq 0$, since the remaining $z_j - c_j, j = 1, 2, \ldots, n$, are already non-negative and are not changed by the addition of a new nonbasic variable.

Thus, we must compute

$$z_{n+1} - c_{n+1} = \bar{c}_B' B^{-1}\bar{a}_{n+1} - c_{n+1}$$

If this quantity is negative, then the current solution is no longer optimal, and we must calculate $\bar{y}_{n+1} = B^{-1}\bar{a}_{n+1}$, and proceed with the simplex method (or revised simplex, of course), by bringing $\bar{a}_{n+1}$ into the basis.

We remind the reader to refer back to the remarks made in the preceding section regarding the availability of B^{-1}.

EXAMPLE 5

Consider the modified problem of Example 4, whose optimal tableau is

		c_j	4	3	0	0	0
$\bar{c}_B$	Basis vectors	$\bar{x}_B$	$\bar{y}_1$	$\bar{y}_2$	$\bar{y}_3$	$\bar{y}_4$	$\bar{y}_5$
3	a_2	4	0	1	0	$-\frac{1}{5}$	$\frac{3}{5}$
4	a_1	1	1	0	0	$\frac{2}{5}$	$-\frac{1}{5}$
0	a_3	1	0	0	1	$-\frac{1}{5}$	$-\frac{2}{5}$
	z	16	0	0	0	1	1

We also know from Example 4 that

$$B^{-1} = \begin{bmatrix} 0 & -\frac{1}{5} & \frac{3}{5} \\ 0 & \frac{2}{5} & -\frac{1}{5} \\ 1 & -\frac{1}{5} & -\frac{2}{5} \end{bmatrix}$$

Suppose now we wish to add a new variable x_6, where $c_6 = 4$, and

$$\bar{a}_6 = \begin{bmatrix} -1 \\ 1 \\ 1 \end{bmatrix}$$

Thus, $z_6 - c_6 = \bar{c}'_B B^{-1} \bar{a}_6 - c_6$

$$= (3, 4, 0) \begin{bmatrix} 0 & \frac{1}{5} & \frac{3}{5} \\ 0 & \frac{2}{5} & -\frac{1}{5} \\ 1 & -\frac{1}{5} & -\frac{2}{5} \end{bmatrix} \begin{bmatrix} -1 \\ 1 \\ 1 \end{bmatrix} - 4$$

$$= (3, 4, 0) \begin{bmatrix} \frac{2}{5} \\ \frac{1}{5} \\ -\frac{8}{5} \end{bmatrix} - 4$$

$$= \tfrac{6}{5} + \tfrac{4}{5} + 0 - 4$$

$$= -2$$

Therefore, since the current solution is no longer optimal, we compute

$$\bar{y}_6 = B^{-1}\bar{a}_6 = \begin{bmatrix} 0 & -\frac{1}{5} & \frac{3}{5} \\ 0 & \frac{2}{5} & -\frac{1}{5} \\ 1 & -\frac{1}{5} & -\frac{2}{5} \end{bmatrix} \begin{bmatrix} -1 \\ 1 \\ 1 \end{bmatrix} = \begin{bmatrix} \frac{2}{5} \\ \frac{1}{5} \\ -\frac{8}{5} \end{bmatrix}$$

The addition of the $\bar{a}_6$ column to the simplex tableau yields:

		c_j	4	3	0	0	0	4
$\bar{c}_B$	Basis vectors	$\bar{x}_B$	$\bar{y}_1$	$\bar{y}_2$	$\bar{y}_3$	$\bar{y}_4$	$\bar{y}_5$	$\bar{y}_6$
3	$\bar{a}_2$	4	0	1	0	$-\frac{1}{5}$	$\frac{3}{5}$	$\frac{2}{5}$
4	$\bar{a}_1$	1	1	0	0	$\frac{2}{5}$	$-\frac{1}{5}$	$\frac{1}{5}$
0	$\bar{a}_3$	1	0	0	1	$-\frac{1}{5}$	$-\frac{2}{5}$	$-\frac{8}{5}$
	z	16	0	0	0	1	1	-2

Now, $\bar{a}_6$ enters the basis, and $\bar{a}_1$ leaves, yielding the solution:

		c_j	4	3	0	0	0	4
$\bar{c}_B$	Basis vectors	$\bar{x}_B$	$\bar{y}_1$	$\bar{y}_2$	$\bar{y}_3$	$\bar{y}_4$	$\bar{y}_5$	$\bar{y}_6$
3	$\bar{a}_2$	2	−2	1	0	−1	1	0
4	$\bar{a}_6$	5	5	0	0	2	−1	1
0	$\bar{a}_3$	9	8	0	1	5	−2	0
	z	26	10	0	0	5	−1	0

Hence, our solution is still not optimal. Upon bringing $\bar{a}_5$ into the basis and removing $\bar{a}_2$ we do obtain the optimal solution:

		c_j	4	3	0	0	0	4
$\bar{c}_B$	Basis vectors	$\bar{x}_B$	$\bar{y}_1$	$\bar{y}_2$	$\bar{y}_3$	$\bar{y}_4$	$\bar{y}_5$	$\bar{y}_6$
0	a_5	2	−2	1	0	−1	1	0
4	a_6	7	3	1	0	1	0	1
0	a_3	13	4	2	1	3	0	0
	z	28	8	1	0	4	0	0

12–5 ADDITION OF A CONSTRAINT

The addition of a new constraint to the constraint set $A\bar{x} = \bar{b}$, $\bar{x} \geq 0$, typically also means the addition of a new variable; namely, the slack or surplus variable associated with this new constraint. Hence, we shall consider this case first. Let us assume then, the constraint we wish to add is

$$\sum_{j=1}^{n} a_{m+1,j}x_j + x_{n+1} = b_{m+1} \tag{12–10}$$

We do not assume that $b_{m+1} > 0$, so that a constraint which was originally either of the "less than or equal to" or of the "greater than or equal to" variety, can be expressed in the form (12–10). Since x_{n+1} is assumed to be a slack or surplus variable, we also assume that $c_{n+1} = 0$.

The first step is to determine whether the current optimal solution satisfies (12–10). If it does, then it is still optimal, since the objective function is unchanged, and the addition of a constraint cannot increase the size of the feasible region.

If the current optimal solution does not satisfy (12–8), we must find the new optimal solution. Let us proceed to do so.

Now, our new constraint set contains $m + 1$ equations, and therefore requires a basis consisting of $m + 1$ vectors. The optimal basis we have already found for the original problem contains m vectors, and a convenient choice for the $(m + 1)^{\text{st}}$ vector is $\bar{e}_{m+1}$, the vector corresponding to x_{n+1}.

If we denote by $\tilde{B}$ the resulting $(m + 1)^{\text{st}}$ order basis matrix, then

$$\tilde{B} = \begin{bmatrix} B & \bar{0} \\ \bar{\alpha}' & 1 \end{bmatrix}$$

where B is the basis matrix for the current optimal basic solution to the original problem, and $\bar{\alpha}'$ is an m-component row vector containing the elements of $[a_{m+1,1}, a_{m+1,2}, \ldots, a_{m+1,n}]$ corresponding to the columns of B.

It is easy to verify that

$$\tilde{B}^{-1} = \begin{bmatrix} B^{-1} & \bar{0} \\ -\bar{\alpha}'B^{-1} & 1 \end{bmatrix}$$

Hence, if B^{-1} is available, we can easily compute $\tilde{B}^{-1}$. Moreover, if we denote by $\tilde{\bar{a}}_j$ the $(m + 1)$ component activity vector of x_j, then

$$\tilde{\bar{a}}_j = \begin{bmatrix} \bar{a}_j \\ a_{m+1,j} \end{bmatrix}$$

Thus,

$$\begin{aligned} \tilde{\bar{y}}_j &\equiv \tilde{B}^{-1}\tilde{\bar{a}}_j \\ &= \begin{bmatrix} B^{-1} & \bar{0} \\ -\bar{\alpha}'B^{-1} & 1 \end{bmatrix} \begin{bmatrix} \bar{a}_j \\ a_{m+1,j} \end{bmatrix} \\ &= \begin{bmatrix} B^{-1}\bar{a}_j \\ (-\bar{\alpha}'B^{-1}\bar{a}_j + a_{m+1,j}) \end{bmatrix} \qquad (12\text{–}11) \\ &= \begin{bmatrix} \bar{y}_j \\ (-\bar{\alpha}'\bar{y}_j + a_{m+1,j}) \end{bmatrix} \end{aligned}$$

Hence, in order to compute the $\tilde{\bar{y}}_j$ for our new, enlarged basis, we need only compute the $(m + 1)^{\text{st}}$ component of each $\tilde{\bar{y}}_j$, since the first m components are unchanged. Thus, from the last component of the vector equation (12–11), we compute

$$y_{m+1,j} = -\bar{\alpha}'\bar{y}_j + a_{m+1,j} \qquad (12\text{–}12)$$

for each nonbasic variable x_j.

We must also compute the value of the $(m + 1)^{st}$ basic variable, $x_{B,m+1} = x_{n+1}$. Let $\tilde{\bar{x}}_B = \begin{bmatrix} \bar{x}_B \\ x_{B,m+1} \end{bmatrix}$ and $\tilde{\bar{b}} = \begin{bmatrix} \bar{b} \\ b_{m+1} \end{bmatrix}$. Then,

$$\begin{aligned}
\tilde{\bar{x}}_B &= \tilde{B}^{-1}\tilde{\bar{b}} \\
&= \begin{bmatrix} B^{-1} & \bar{0} \\ -\bar{\alpha}'B^{-1} & 1 \end{bmatrix} \begin{bmatrix} \bar{b} \\ b_{m+1} \end{bmatrix} \\
&= \begin{bmatrix} B^{-1}\bar{b} \\ (-\bar{\alpha}'B^{-1}\bar{b} + b_{m+1}) \end{bmatrix} \\
&= \begin{bmatrix} \bar{x}_B \\ -\bar{\alpha}'\bar{x}_B + b_{m+1} \end{bmatrix}
\end{aligned}$$

Thus,

$$x_{n+1} = x_{B,m+1} = b_{m+1} - \bar{\alpha}'\bar{x}_B \tag{12–13}$$

Observe that the only computations necessary to obtain the $(m + 1)^{st}$ row are those indicated in equations (12–12) and (12–13). Observe also that these computations do not require a knowledge of B^{-1}.

Since the first m basic variables are unchanged, and if the solution does not satisfy the new constraint, it must be true that $x_{n+1} < 0$. However, none of the $z_j - c_j$ has been changed (since $c_{B,m+1} = 0$), so that the solution satisfies the optimality criterion that all $z_j - c_j \geq 0$. Hence, we can use the dual simplex method to find the new optimal solution.

EXAMPLE 6

Let us again consider the modified problem of Example 4, whose optimal tableau is given by

		c_j	4	3	0	0	0
$\bar{c}_B$	Basis vectors	$\bar{x}_B$	$\bar{y}_1$	$\bar{y}_2$	$\bar{y}_3$	$\bar{y}_4$	$\bar{y}_5$
3	$\bar{a}_2$	4	0	1	0	$-\frac{1}{5}$	$\frac{3}{5}$
4	$\bar{a}_1$	1	1	0	0	$\frac{2}{5}$	$-\frac{1}{5}$
0	$\bar{a}_3$	1	0	0	1	$-\frac{1}{5}$	$-\frac{2}{5}$
	z	16	0	0	0	1	1

Suppose we wish to add the constraint $2x_1 + x_2 \leq 5$.

First, we test to see if the current solution satisfies this constraint. Upon substituting $x_{B1} = x_2 = 4$, $x_{B2} = x_1 = 1$ into the constraint, we see that it is not satisfied. Hence, we add the slack variable, x_6, and proceed to calculate y_{4_j} and x_{B_4} from equations (12–12) and (12–13).

Since $[a_{41}, a_{42}, a_{43}, a_{44}, a_{45}, a_{46}] = [2, 1, 0, 0, 0, 1]$, and the current basis columns are 2, 1, and 3, respectively, we see that $\bar{\alpha}' = [a_{42}, a_{41}, a_{43}] = [1, 2, 0]$.

Thus,

$$
\begin{aligned}
y_{44} &= -\bar{\alpha}'\bar{y}_4 + a_{44} \\
&= -[1, 2, 0]\begin{bmatrix} -\frac{1}{5} \\ \frac{2}{5} \\ -\frac{1}{5} \end{bmatrix} + 0 \\
&= -(-\tfrac{1}{5} + \tfrac{4}{5} + 0) \\
&= -\tfrac{3}{5} \\
y_{45} &= -\bar{\alpha}'\bar{y}_5 + a_{45} \\
&= -[1, 2, 0]\begin{bmatrix} \frac{3}{5} \\ -\frac{1}{5} \\ -\frac{2}{5} \end{bmatrix} + 0 \\
&= -(\tfrac{3}{5} - \tfrac{2}{5} + 0) \\
&= -\tfrac{1}{5} \\
x_{B4} &= b_{m+1} - \bar{\alpha}'\bar{x}_B \\
&= 5 - [1, 2, 0]\begin{bmatrix} 4 \\ 1 \\ 1 \end{bmatrix} \\
&= 5 - (4 + 2) \\
&= -1
\end{aligned}
$$

The extended tableau is therefore

		c_j	4	3	0	0	0	0
$\bar{c}_B$	Basis vectors	$\bar{x}_B$	$\bar{y}_1$	$\bar{y}_2$	$\bar{y}_3$	$\bar{y}_4$	$\bar{y}_5$	$\bar{y}_6$
3	$\bar{a}_2$	4	0	1	0	$-\frac{1}{5}$	$\frac{3}{5}$	0
4	$\bar{a}_1$	1	1	0	0	$\frac{2}{5}$	$-\frac{1}{5}$	0
0	$\bar{a}_3$	1	0	0	1	$-\frac{1}{5}$	$-\frac{2}{5}$	0
0	$\bar{a}_6$	−1	0	0	0	$-\frac{3}{5}$	$-\frac{1}{5}$	1
	z	16	0	0	0	1	1	0

Using the dual simplex method, we choose $\bar{a}_6$ to leave the basis and $\bar{a}_4$ to enter, yielding the new optimal tableau given below:

	c_j		4	3	0	0	0	0
$\bar{c}_B$	Basis vectors	$\bar{x}_B$	$\bar{y}_1$	$\bar{y}_2$	$\bar{y}_3$	$\bar{y}_4$	$\bar{y}_5$	$\bar{y}_6$
3	$\bar{a}_2$	$\frac{13}{3}$	0	1	0	0	$\frac{2}{3}$	$-\frac{1}{3}$
4	$\bar{a}_1$	$\frac{1}{3}$	1	0	0	0	$-\frac{1}{3}$	$\frac{2}{3}$
0	$\bar{a}_3$	$\frac{4}{3}$	0	0	1	0	$-\frac{1}{3}$	$-\frac{1}{3}$
0	$\bar{a}_4$	$\frac{5}{3}$	0	0	0	1	$\frac{1}{3}$	$-\frac{5}{3}$
0	z	$\frac{43}{3}$	0	0	0	0	$\frac{2}{3}$	$\frac{5}{3}$

The above example illustrates another point; we can obtain the new $(m + 1)^{\text{st}}$ row of the simplex tableau by merely adding appropriate multiples of the other rows to the $(m+ 1)^{\text{st}}$ constraint, so that the coefficients of the original m basic variables vanish from the $(m + 1)^{\text{st}}$ row. The reader is asked to discuss this point in the exercises, but let us verify it for the problem of Example 6.

The constraint we wish to add, $2x_1 + x_2 + x_6 = 5$, has coefficients $[5, 2, 1, 0, 0, 0, 1]$, beginning with $b_{m+1} = 5$. The basic variables are currently $x_{B1} = x_2$, $x_{B2} = x_1$, $x_{B3} = x_3$. Hence, we wish to eliminate the coefficients of x_2, x_1 by adding (-1) times row 1 and (-2) times row 2, respectively, of the simplex tableau. Schematically, these calculations may be summarized thusly:

$$\begin{aligned} (-1) \times \text{row } 1 &= [-4, 0, -1, 0, \tfrac{1}{5}, -\tfrac{3}{5}, 0] \\ (-2) \times \text{row } 2 &= [-2, -2, 0, 0, -\tfrac{4}{5}, \tfrac{2}{5}, 0] \\ \text{new constraint} &= [5, 2, 1, 0, 0, 0, 1] \\ \hline \text{sum} &= [-1, 0, 0, 0, -\tfrac{3}{5}, -\tfrac{1}{5}, 1] \end{aligned}$$

This last is precisely the set of coefficients computed previously and inserted into the fourth row of the simplex tableau.

Let us now turn to the problem of adding an equality constraint to the optimal tableau. Suppose we wish to add the constraint

$$\sum_{j=1}^{n} a_{m+1,j} x_j = b_{m+1} \tag{12–14}$$

The first step is to see whether the current solution satisfies equation (12–14); if so, the current solution is still optimal. If not, we must find the new optimal solution.

As there is no convenient variable to select as the $(m + 1)^{st}$ basic variable, let us introduce an artificial variable into the constraint (12–14). Before doing so, however, we observe that if this artificial variable is negative and if we assign a price of $c_{m+1} = 0$ to it, then the resulting situation is identical to that discussed previously, of adding a slack or surplus variable. After one iteration of the dual simplex method, the artificial variable will be removed from the basis, and we can then delete its column from the tableau and continue with the dual simplex method (if we have not already found the optimal solution after this first iteration).

Having made the above observation, let us now determine how to introduce the artificial variable at a negative level. By equation (12–13), we see that the value of the $(m + 1)^{st}$ basic variable is

$$x_{B,m+1} = b_{m+1} - \bar{\alpha}'\bar{x}_B$$

where the components of $\bar{\alpha}$ are the $a_{m+1,j}$ which correspond to the basic variables. Hence, if $(b_{m+1} - \bar{\alpha}'\bar{x}_B) < 0$, we add the artificial variable x_{n+1} to the new constraint (12–14), yielding

$$\sum_{j=1}^{n} a_{m+1,j}x_j + x_{n+1} = b_{m+1}$$

and proceed as before, treating x_{n+1} as a slack variable.

If, on the other hand, $(b_{m+1} - \bar{\alpha}'\bar{x}_B) > 0$, then we subtract the artificial variable from the new constraint (12–14), yielding

$$-\sum_{j=1}^{n} a_{m+1,j}x_j + x_{n+1} = -b_{m+1}$$

and again proceed as before.

12–6 CHANGES IN THE COEFFICIENT MATRIX

When it is necessary to change one or more of the elements of the coefficient matrix A, the situation becomes much more complicated than we have thus far encountered when making changes in $\bar{c}$ or $\bar{b}$, or in adding a variable or a constraint. This is particularly true if we wish to change an element from a column of A which is basic in the optimal solution, since in this case the optimal basis matrix must be recomputed. Moreover, the resulting m columns corresponding to the current basic variables may no longer be linearly independent. Even if these columns are still linearly independent, B^{-1} must be recomputed, and so must all $\bar{y}_j$ and all $z_j - c_j$. Hence, we must either devise a computationally reasonable scheme for performing these computations (or a clever way of avoiding them) or re-solve the problem from the beginning.

Before considering all of these complex questions, let us turn to the somewhat easier problem of making changes in one or more of the elements of a nonbasic column $\bar{a}_p$. Suppose we wish to replace $\bar{a}_p$ with

$$\tilde{\bar{a}}_p = \bar{a}_p + \bar{\alpha}$$

where $\bar{\alpha}$ is any m-component vector. Then, since $\bar{a}_p$ is nonbasic, the optimal basis matrix B remains unchanged, and so do $\bar{x}_B^* = B^{-1}\bar{b}$, $\bar{c}_B$, and $z_j - c_j = \bar{c}_B' B^{-1}\bar{a}_j - c_j$, for $j \neq p$. Hence, the only changes are:

$$\begin{aligned}\tilde{\bar{y}}_p &= B^{-1}\tilde{\bar{a}}_p \\ &= B^{-1}[\bar{a}_p + \bar{\alpha}] \\ &= B^{-1}\bar{a}_p + B^{-1}\bar{\alpha} \\ &= \bar{y}_p + B^{-1}\bar{\alpha}\end{aligned} \tag{12–15}$$

$$\begin{aligned}\tilde{z}_p - c_p &= \bar{c}_B' B^{-1}\tilde{\bar{a}}_p - c_p \\ &= \bar{c}_B' B^{-1}[\bar{a}_p + \bar{\alpha}] - c_p \\ &= \bar{c}_B' B^{-1}\bar{a}_p - c_p + \bar{c}_B B^{-1}\bar{\alpha} \\ &= (z_p - c_p) + \bar{c}_B' B^{-1}\bar{\alpha}\end{aligned} \tag{12–16}$$

If $\tilde{z}_p - c_p \geq 0$, the current solution is still optimal, and $\tilde{\bar{y}}_p$ need not be computed. If, however, $\tilde{z}_p - c_p < 0$, then we must compute $\tilde{\bar{y}}_p$, and perform one or more simplex operations to find the new optimal solution.

Now, let us return to the problem of changing the elements of a basic vector $\bar{a}_q$. Rather than performing the massive amount of computation involved in either re-solving the problem from the beginning, or recomputing B^{-1}, all $\bar{y}_j$ and $z_j - c_j$, we shall take the following approach instead. Before changing the elements of $\bar{a}_q$, we shall first remove this vector from the basis, and make the changes in the then nonbasic vector $\bar{a}_q$.

The only question remaining is: can we remove $\bar{a}_q$ from the basis without a great deal of computational effort? In answer to this question, consider the following. Suppose we were to solve the problem

$$\begin{aligned}\text{maximize } z &= -x_q \\ \text{subject to } A\bar{x} &= \bar{b} \\ \bar{x} &\geq \bar{0}\end{aligned} \tag{12–17}$$

We already have an initial solution for this problem, namely the current optimal solution for the original problem. Moreover, the optimal solution to

(12–17) will have $x_q = 0$, which either means $\bar{a}_q$ is nonbasic or an alternative optimal solution exists for which $\bar{a}_q$ is nonbasic.†

It is a relatively easy matter to find such a solution, beginning with the current optimal tableau. If $\bar{a}_q$ is currently the tth basic variable, then the $z_j - c_j$ for problem (12–17) are

$$\begin{aligned} z_j - c_j &= \bar{c}'_B \bar{y}_j - c_j \\ &= -y_{tj} \end{aligned} \tag{12–18}$$

since $c_j = 0, j = 1, 2, \ldots, n, j \neq q$, and $c_q = c_{Bt} = -1$. Hence, in order to drive $\bar{a}_q = \bar{b}_t$ out of the basis, we bring in any vector $\bar{a}_k$ for which $-y_{tk} < 0$; i.e., $y_{tk} > 0$. The vector to leave is determined, as usual, by

$$\frac{x_{Bk}}{y_{rk}} = \min_i \left\{ \frac{x_{Bi}}{y_{ik}} \middle| y_{ik} > 0 \right\}$$

If $r = t$, then we have removed $\bar{a}_q$ from the basis; if not, we must continue. Since our goal here is actually to remove $\bar{a}_q$ from the basis, it might be reasonable to choose the vector $\bar{a}_k$ which enters the basis, according to

$$\theta_k(z_k - c_k) = \min_j \{\theta_j(z_j - c_j)\} \tag{12–19}$$

where

$$\theta_j = \min_i \left\{ \frac{x_{Bi}}{y_{ij}} \middle| y_{ij} > 0 \right\} \tag{12–20}$$

If this method of choosing $\bar{a}_k$ is employed, then not only will we obtain the maximum decrease in x_q at each iteration, but if at any given iteration $\bar{a}_q$ can be removed from the basis, the use of equation (12–19) will tell us which entering vector $\bar{a}_k$ will accomplish this goal. That is, if

$$\theta_k = \frac{x_{Bt}}{y_{kt}}$$

then $\bar{b}_t = \bar{a}_q$ is removed from the basis.

The above procedure for removing $\bar{a}_q$ from the basis does not actually require the recomputation of $z_j - c_j$, and the $z_j - c_j$ for the original problem may be computed at each iteration as usual, although they are not used during this procedure. Moreover, although we cannot guarantee that only a few iterations will be required to drive $\bar{a}_q$ out of the basis, frequently this will be the case. In fact, in almost all cases this procedure is probably at least as good as re-solving the problem from the beginning.

† The only exceptions to these statements are when $x_q > 0$ in every basic feasible solution, or when $\bar{a}_q$ must be in every basis because all other sets of m vectors are linearly dependent. In such cases, this approach will not work.

EXAMPLE 7

Suppose we have again solved the modified linear programming problem of Example 4

$$\text{maximize } z = 4x_1 + 3x_2$$

subject to:

$$\begin{aligned} x_1 + x_2 + x_3 &= 6 \\ 3x_1 + x_2 + x_4 &= 7 \\ x_1 + 2x_2 + x_5 &= 9 \\ x_1, x_2, x_3, x_4, x_5 &\geq 0 \end{aligned}$$

The optimal tableau is:

		c_j	4	3	0	0	0
$\bar{c}_B$	Basis vectors	$\bar{x}_B$	$\bar{y}_1$	$\bar{y}_2$	$\bar{y}_3$	$\bar{y}_4$	$\bar{y}_5$
3	$\bar{a}_2$	4	0	1	0	$-\frac{1}{5}$	$\frac{3}{5}$
4	$\bar{a}_1$	1	1	0	0	$\frac{2}{5}$	$-\frac{1}{5}$
0	$\bar{a}_3$	1	0	0	1	$-\frac{1}{5}$	$-\frac{2}{5}$
	z	16	0	0	0	1	1

Suppose we wish to replace $\bar{a}_1$ with $\tilde{\bar{a}}_1 = \bar{a}_1 + \bar{\alpha}$, where $\bar{\alpha} = (1, 0, 1)'$. Then, since $\bar{a}_1$ is currently a basis vector, we will first try to remove it from the basis before making the above change in $\bar{a}_1$. Thus, according to equation (12–18), the variable $x_1 = x_{B2}$ will decrease if we choose a vector $\bar{a}_k$ to enter the basis such that $y_{2k} > 0$. Since $y_{24} = \frac{2}{5}$ is the only positive (nonbasic) y_{2k}, we therefore choose $\bar{a}_4$ to enter the basis. Then, $\bar{a}_1$ must leave the basis, and in only one iteration we have accomplished our goal of removing $\bar{a}_1$ from the basis. The resulting tableau is:

		c_j	4	3	0	0	0
c_B	Basis vectors	$\bar{x}_B$	$\bar{y}_1$	$\bar{y}_2$	$\bar{y}_3$	$\bar{y}_4$	$\bar{y}_5$
3	$\bar{a}_2$	$\frac{7}{2}$	$\frac{1}{2}$	1	0	0	$\frac{1}{2}$
0	$\bar{a}_4$	$\frac{5}{2}$	$\frac{5}{2}$	0	0	1	$-\frac{1}{2}$
0	$\bar{a}_3$	$\frac{3}{2}$	$\frac{1}{2}$	0	1	0	$-\frac{1}{2}$
	z	$\frac{31}{2}$	$-\frac{5}{2}$	0	0	0	$\frac{3}{2}$

Now, since the identity matrix used for the initial basis was $[\bar{a}_3, \bar{a}_4, \bar{a}_5]$, these corresponding columns presently contain the current B^{-1}. Hence,

$$B^{-1} = \begin{bmatrix} 0 & 0 & \frac{1}{2} \\ 0 & 1 & -\frac{1}{2} \\ 1 & 0 & -\frac{1}{2} \end{bmatrix}$$

and

$$\begin{aligned} \tilde{\bar{y}}_1 &= \bar{y}_1 + B^{-1}\bar{\alpha} \\ &= \begin{bmatrix} \frac{1}{2} \\ \frac{5}{2} \\ \frac{1}{2} \end{bmatrix} + \begin{bmatrix} 0 & 0 & \frac{1}{2} \\ 0 & 1 & -\frac{1}{2} \\ 1 & 0 & -\frac{1}{2} \end{bmatrix} \begin{bmatrix} 1 \\ 0 \\ 1 \end{bmatrix} \\ &= \begin{bmatrix} \frac{1}{2} \\ \frac{5}{2} \\ \frac{1}{2} \end{bmatrix} + \begin{bmatrix} \frac{1}{2} \\ -\frac{1}{2} \\ \frac{1}{2} \end{bmatrix} = \begin{bmatrix} 1 \\ 2 \\ 1 \end{bmatrix} \end{aligned}$$

Also, from equation (12–16),

$$\begin{aligned} \tilde{z}_1 - c_1 &= (z_1 - c_1) + \bar{c}_B' B^{-1}\bar{\alpha} \\ &= -\tfrac{5}{2} + (3, 0, 0) \begin{bmatrix} 0 & 0 & \frac{1}{2} \\ 0 & 1 & \frac{1}{2} \\ 1 & 0 & -\frac{1}{2} \end{bmatrix} \begin{bmatrix} 1 \\ 0 \\ 1 \end{bmatrix} \\ &= -\tfrac{5}{2} + \tfrac{3}{2} \\ &= -1 \end{aligned}$$

Inserting $\tilde{\bar{y}}_1$ and $\tilde{z}_1 - c_1$ into the tableau yields

		c_j	4	3	0	0	0
$\bar{c}_B$	Basis vectors	$\bar{y}_B$	$\tilde{\bar{y}}_1$	$\bar{y}_2$	$\bar{y}_3$	$\bar{y}_4$	$\bar{y}_5$
3	$\bar{a}_2$	$\frac{7}{2}$	1	1	0	0	$\frac{1}{2}$
0	$\bar{a}_4$	$\frac{5}{2}$	2	0	0	1	$-\frac{1}{2}$
0	$\bar{a}_3$	$\frac{3}{2}$	1	0	1	0	$-\frac{1}{2}$
	z	$\frac{31}{2}$	-1	0	0	0	$\frac{3}{2}$

As this solution is clearly not optimal, we proceed with the simplex method, by bringing $\tilde{a}_1$ into the basis, removing $\bar{a}_4$ and thus obtain the new optimal solution:

		c_j	4	3	0	0	0
$\bar{c}_B$	Basis vectors	$\bar{x}_B$	$\tilde{y}_1$	$\bar{y}_2$	$\bar{y}_3$	$\bar{y}_4$	$\bar{y}_5$
3	$\bar{a}_2$	$\frac{9}{4}$	0	1	0	$-\frac{1}{2}$	$\frac{3}{4}$
4	$\bar{a}_1$	$\frac{5}{4}$	1	0	0	$\frac{1}{2}$	$-\frac{1}{4}$
4	$\bar{a}_3$	$\frac{1}{4}$	0	0	1	$-\frac{1}{2}$	$-\frac{1}{4}$
	z	$\frac{67}{4}$	0	0	0	$\frac{1}{2}$	$\frac{5}{4}$

12-7 PARAMETRIC ANALYSIS

It is frequently useful to study the behavior of the optimal solution to a linear programming problem, as the entire objective function is systematically varied, or as the entire requirements vector is systematically varied. Such a study is often called a parametric analysis.

Let us investigate first the case in which the objective function is varied parametrically. As before, we assume we have already solved a given linear programming problem. We shall denote the elements of the optimal tableau by $\bar{x}_B^*$, $\bar{y}_j^*$, and $z_j^* - c_j$.

Suppose now we let

$$\tilde{\bar{c}} = \bar{c} + \lambda \bar{d}$$

where $\bar{d}$ is any fixed n-component vector,† and λ is a scalar parameter; we shall assume $\lambda > 0$. Then, as we noted in Section 12–2, the only quantities affected by a change in $\bar{c}$ are the $z_j - c_j$. If we let $\bar{d}_B$ denote the vector of components of $\bar{d}$ corresponding to the basic variables $\bar{x}_B^*$, then the new $z_j - c_j$ are:

$$\begin{aligned} \tilde{z}_j - \tilde{c}_j &= \tilde{\bar{c}}_B' \bar{y}_j^* - \tilde{c}_j \\ &= (\bar{c}_B' + \lambda \bar{d}_B')\bar{y}_j^* - (c_j + \lambda d_j) \qquad (12\text{–}21) \\ &= (\bar{c}_B' \bar{y}_j^* - c_j) + \lambda(\bar{d}_B' \bar{y}_j^* - d_j) \\ &= (z_j^* - c_j) + \lambda(\bar{d}_B' \bar{y}_j^* - d_j) \end{aligned}$$

If we let

$$f_j = \bar{d}_B' \bar{y}_j^* - d_j \qquad (12\text{–}22)$$

† The components of vector $\bar{d}$ indicate the rates with which we wish to vary the c_j. In many cases, $\bar{d}$ will probably be $\bar{d} = (1, 1, \ldots, 1)$. However, in some instances we may wish to vary the prices c_j at different rates; e.g., c_2 may increase twice as fast as c_1. Also, we may wish to increase some prices while decreasing others.

then equation (12–21) becomes

$$\tilde{z}_j - \tilde{c}_j = (z_j^* - c_j) + \lambda f_j \tag{12–23}$$

Then, the current solution will remain optimal so long as

$$\tilde{z}_j - \tilde{c}_j \geq 0$$
$$z_j^* - c_j + \lambda f_j \geq 0, \qquad j = 1, 2, \ldots, n \tag{12–24}$$

Since $\lambda > 0$, the inequalities (12–24) will be satisfied for any j for which $f_j \geq 0$. Hence, we must require

$$f_j\lambda \geq -(z_j^* - c_j)$$
$$\lambda \leq -(z_j^* - c_j)/f_j$$

for each $f_j < 0$. In other words,

$$\lambda \leq \min_j \left\{ \frac{-(z_j^* - c_j)}{f_j} \middle| f_j < 0 \right\} \tag{12–25}$$

Moreover, if

$$\frac{-(z_p^* - c_p)}{f_p} = \min_j \left\{ \frac{-(z_j^* - c_j)}{f_j} \middle| f_j < 0 \right\}$$

then it is the change in c_p which limits the maximum magnitude of λ, and hence c_p which is most sensitive to change, with respect to $\bar{d}$ (since a different $\bar{d}$ may very well produce a different result).

If λ exceeds $-(z_p^* - c_p)/f_p$, the current solution is no longer optimal. However, we may wish to study how the optimal solution changes as a function of λ. In this event, we would find the new optimal solution resulting from letting λ be slightly larger than $-(z_p^* - c_p)/f_p$, and then finding the new upper limit on λ for which this solution remains optimal. We could repeat this process for whatever range of values for λ were of interest.

A similar parametric analysis can be developed to study changes in the requirements vector of the form $\tilde{\bar{b}} = \bar{b} + \lambda\bar{g}$, where $\bar{g}$ is any vector of fixed components, and $\lambda > 0$ is a scalar parameter. Such an analysis can either be developed directly or by considering the dual linear programming problem, in which these changes become changes in the objective function. The details of these topics are left as exercises for the reader.

PROBLEMS

1. Consider the following linear programming problem:

$$\text{Maximize } z = x_1 + 3x_2$$
$$x_1 + x_2 \leq 8$$
$$4x_1 + x_2 \leq 26$$
$$-x_1 + x_2 \leq 4$$
$$x_1, x_2 \geq 0$$

After adding slack variables x_3, x_4, x_5 and solving, using the corresponding columns as our initial basis matrix, we obtain the following optimal tableau:

		c_j	1	3	0	0	0
$\bar{c}_B$	Basis vectors	$\bar{x}_B$	$\bar{a}_1$	$\bar{a}_2$	$\bar{a}_3$	$\bar{a}_4$	$\bar{a}_5$
1	$\bar{a}_1$	2	1	0	$\frac{1}{2}$	0	$-\frac{1}{2}$
0	$\bar{a}_4$	12	0	0	$-\frac{5}{2}$	1	$\frac{3}{2}$
3	$\bar{a}_2$	6	0	1	$\frac{1}{2}$	0	$\frac{1}{2}$
	z	20	0	0	2	0	1

(a) Determine the range of values for c_1 for which the above solution remains optimal.
(b) Determine the range of values for c_2 for which the above solution remains optimal.
(c) Determine the range of values for c_3, c_4, c_5, respectively, for which the above solution remains optimal.

2. Suppose the objective function in the linear programming problem of Exercise 1 is changed to $\hat{z} = 3x_1 + 4x_2$.
Is the current optimal solution still optimal? If not, find the new optimal solution.

3. Suppose the objective function in the linear programming problem of Exercise 1 is changed to $\hat{z} = 3x_1 + 2x_2$.
Is the current optimal solution still optimal? If not, find the new optimal solution.

4. Suppose the requirements vector $\bar{b}' = [8, 26, 4]$ in the linear programming problem of Exercise 1 is changed to $\hat{\bar{b}}' = [8, 26, 9]$. Is the current solution still optimal? If not, find the new optimal solution.
Illustrate the situation graphically.

5. Suppose the requirements vector in the linear programming problem of Exercise 1 is changed to $\hat{\bar{b}}' = (1, 1, 1)$. Is the current solution still optimal? If not, find the new optimal solution.
Illustrate the situation graphically.

6. If the constraint $x_1 + 2x_2 \leq 12$ is added to the linear programming problem of exercise 1, determine the optimal basic feasible solution for the resulting problem. Illustrate graphically.

7. If the constraint $2x_1 + x_2 \geq 12$ is added to the linear programming problem of exercise 1, determine the optimal basic feasible solution for the resulting problem. Illustrate graphically.

8. If the constraint $3x_1 + 4x_2 = 27$ is added to the linear programming problem of exercise 1, determine the optimal basic feasible solution for the resulting problem. Illustrate graphically.

9. Suppose the first constraint in the linear programming problem of exercise 1 is changed to $\frac{1}{2}x_1 + x_2 \leq 8$.

   Determine the optimal solution for the resulting problem. (Note: This change can be treated as a change in the column $\bar{a}_1' = [1, 4, -1]$) to $\tilde{\bar{a}}_1' = [\frac{1}{2}, 4, -1]$.

10. Suppose the first constraint in the linear programming problem of Exercise 1 is changed to $2x_1 + x_2 \leq 8$. Determine the optimal solution for the resulting problem. Illustrate graphically (see the "Note" in exercise 9).

11. Suppose we have obtained an optimal solution for a given linear programming problem, and it is then determined that one of the constraints should be removed. Describe a procedure for obtaining the optimal solution for the resulting problem. Illustrate your procedure by obtaining the optimal solution for the linear programming problem which results when the first constraint is removed from the problem of exercise 1.

12. Investigate how the optimal solution to the linear programming problem of exercise 1 changes as the parameter λ increases from 0 to $+\infty$, where

$$\bar{c} = \begin{bmatrix} 1 \\ 3 \end{bmatrix} + \lambda \begin{bmatrix} 1 \\ -1 \end{bmatrix}$$

13. Develop a procedure for performing a parametric analysis to study how the optimal solution for a linear programming problem changes, when the requirements vector $\bar{b}$ is varied according to $\tilde{\bar{b}} = \bar{b} + \lambda\bar{g}$, where $\bar{g}$ is any vector of fixed components, and λ is a non-negative scalar parameter. Illustrate your procedure on the problem of exercise 1, if $\bar{g}' = [1, -1, 2]$.

14. To investigate how changes in the requirements vector affect the optimal solution of a linear programming problem, we must know B^{-1}, the inverse of the optimal basis matrix. If the simplex method was used to solve the problem, show that B^{-1} is contained in the columns of the optimal tableau corresponding to the columns which contained the identity matrix used to obtain the initial basic solution.

15. Prove that, when a constraint is to be added to an already solved linear programming problem, the new $(m + 1)^{st}$ row of the simplex tableau can be obtained by adding appropriate multiples of the other rows of the tableau, so that the coefficients of the original in optimal basic variables vanish from the $(m + 1)^{st}$ row. (See Example 6 and discussion on page 278.)

13

Large Scale Linear Programming: Special Computational Techniques

13–1 INTRODUCTION

Perhaps the primary reason that linear programming has become one of the most important and successful tools of operations research is the ability of the simplex method to solve very large problems with a relatively small amount of computer time. In recent years, this ability has increased substantially, partly due to the increased storage capacities and speed of the more modern digital computers, and partly due to the development of some special computational techniques, to be used in conjunction with the simplex method (or the revised simplex method). The use of these techniques has greatly increased the size of problems which can be handled by the simplex method. In fact, several case studies have been reported[8] of the successful solution—in less than 60 minutes of computer time—of linear programming problems with an excess of 50,000 constraints and 250,000 variables.

The key to success in solving such large problems lies in the fact that most very large practical problems contain many constraints with a particular special structure and/or many constraints which contain a large number of zero coefficients. Several of the techniques to be presented in this chapter are designed to take advantage of these special structures. These techniques are presented in Sections 13–3 through 13–5.

The first special technique we shall discuss is the composite simplex algorithm, a new method for finding an initial basic feasible solution which is more efficient than those methods described in Chapter 5.

13–2 THE COMPOSITE SIMPLEX ALGORITHM

The first step in solving a linear programming problem is to obtain an initial basic feasible solution. Two methods for obtaining such are discussed in Chapter 5. Both of these methods require the addition of an artificial variable to each constraint except those of the form $\sum a_{ij}x_j \leq b_i$, where $b_i \geq 0$. For many large problems, this means that a large number of artificial variables will be required.

In an effort to reduce as much as possible the number of artificial variables needed, many researchers have devised algorithms which allow some of the basic variables to be negative, initially. These variables are then somehow ultimately forced to assume non-negative values, while at the same time any artificial variables in the initial basis are driven to zero.

The composite simplex algorithm is one such algorithm which has evolved from these efforts, and experimental studies seem to indicate that it is computationally superior to some of the other variations (see Wolfe[11]).

Essentially, the composite simplex algorithm is designed to find an initial basic feasible solution for a linear programming problem, and thus it can be considered as a replacement for "Phase I" of the regular two phase method.

As mentioned above, we shall allow some of the variables in the initial basis to assume negative values. In order to construct an initial basis, we first convert all inequalities to the form $\sum a_{ij}x_j \leq b_i$, where b_i may be positive or negative. We then add a slack variable to each such constraint. If all the original constraints were inequalities, we now immediately have an initial basic solution (not necessarily feasible) composed of the slack variables. If the original problem contains some equality constraints, we add an artificial variable to each such constraint,† and again immediately obtain an initial basic solution. Unless the original problem has a large number of equality constraints (which is not typical) the number of artificial variables needed will be comparatively small.

Let us assume that we now have an initial basic solution, which is not feasible. There are two possible sources of infeasibility: 1. Positive artificial variables; or 2. negative variables (artificial or legitimate).

Accordingly, we let

$$P = \{i \mid x_{Bi} \geq 0 \text{ and } x_{Bi} \text{ is artificial‡}\}$$
$$N = \{i \mid x_{Bi} < 0\}$$

† In some cases, we may be able to find an initial basic variable corresponding to an equality constraint without adding an artificial variable, which further reduces the number of artificial variables. For example, if one of the variables in an equality constraint only appears in that one constraint, this is easily accomplished.

‡ We shall also include artificial variables in the basis at a zero level in the set P, since it is desired to drive such variables out of the basis.

Now, if we let b_i be the current value of the ith basic variable, then the ith row of the simplex tableau can be expressed as the equation

$$x_{Bi} + \sum_j y_{ij}x_j = b_i,$$

or

$$x_{Bi} = b_i - \sum_j y_{ij}x_j \tag{13–1}$$

We now define the total infeasibility w of the solution as

$$\begin{aligned} w &= \sum_{i \in P \cup N} |x_{Bi}| \\ &= \sum_{i \in P} x_{Bi} - \sum_{i \in N} x_{Bi} \end{aligned} \tag{13–2}$$

It is clear that w is always non-negative, and that a solution is a feasible solution to the original problem if and only if $w = 0$. In the composite simplex method, vectors are chosen to enter and leave the basis so that w is decreased at each iteration, until an initial feasible solution is found ($w = 0$).

In order to determine the effect on w of bringing a nonbasic vector $\bar{a}_j$ into the basis, we must eliminate the basic variables from equation (13–2), by substituting equation (13–1) into (13–2):

$$\begin{aligned} w &= \sum_{i \in P} x_{Bi} - \sum_{i \in N} x_{Bi} \\ &= \sum_{i \in P} \{b_i - \sum y_{ij}x_j\} = \sum_{i \in N} \{b_i - \sum y_{ij}x_j\} \\ &\sum_{i \in P} b_i - \sum_{i \in N} b_i + \sum_j \{ \sum_{i \in N} y_{ij} - \sum_{i \in P} y_{ij}\}x_j \end{aligned}$$

The only variables x_j appearing in the third sum above are the currently nonbasic variables. If we let

$$d_j = \sum_{i \in N} y_{ij} - \sum_{i \in P} y_{ij} \tag{13–3}$$

then w can be written

$$w = \text{constant} + \sum_j d_j x_j \tag{13–4}$$

Hence, w will decrease if we choose a variable x_j to enter the basis such that $d_j < 0$. Note that the quantities d_j are analogous to the usual $z_j - c_j$ of the simplex method.

We shall choose the entering vector† $\bar{a}_k$ such that

$$d_k = \min_j \{d_j\} \tag{13–5}$$

† We assume there are no nonbasic artificial vectors; once an artificial vector has been driven from the basis, it is no longer needed, and its column may be deleted from the tableau.

This rule yields the greatest rate of decrease in the total infeasibility w, just as choosing the most negative $z_j - c_j$ in the simplex method yields the greatest rate of increase in the objective function.

If $d_k \geq 0$, but the constant term of equation (13–4) is nonzero (it is always non-negative) then the problem has no feasible solution, since $w > 0$ and cannot be decreased further.

Now, having determined the vector $\bar{a}_k$ which is to enter the basis, we must decide which vector is to leave the basis. We have (possibly) three types of basic variables; namely, two corresponding to the two types of infeasibilities described previously, and variables currently feasible. In choosing the variable to leave the basis, we wish to keep feasible all variables which are currently feasible, and we also wish to prevent any negative artificial variables from becoming positive. Furthermore, we would like as many negative legitimate variables as possible to become non-negative.

To see how these goals may be accomplished, let us examine the simplex transformation formulas; if the rth basic variable leaves the basis, then

$$\left.\begin{aligned} \hat{x}_{Bi} &= x_{Bi} - \theta y_{ik}, \qquad i \neq r \\ \hat{x}_{Br} &= x_{Br}/y_{rk} = \theta \end{aligned}\right\} \tag{13–6}$$

We consider three separate cases:

Case 1: $x_{Br} \geq 0$

This corresponds to the situation in the regular simplex method, in which no non-negative variable will become negative, so long as r is chosen such that

$$\frac{x_{Br}}{y_{rk}} = \min_i \left\{\frac{x_{Bi}}{y_{ik}} \,\middle|\, y_{ik} > 0\right\}$$

However, since not all x_{Bi} may be non-negative in the composite simplex situation, we modify the above rule slightly, and let

$$\theta_1 = \frac{x_{Br_1}}{y_{r_1k}} = \min_i \left\{\frac{x_{Bi}}{y_{ik}} \,\middle|\, x_{Bi} \geq 0 \text{ and } y_{ik} > 0\right\} \tag{13–7}$$

Furthermore, if there is a tie for the above minimum, we choose r_1 corresponding to an artificial variable if possible.

If we were to use the decision rule (13–7) exclusively, to choose the vector to leave the basis, we would ensure that no currently non-negative variable becomes negative. However, this rule does nothing to improve the status of the variables which are currently negative. Hence, let us investigate Case 2.

Case 2: $x_{Bi} \leq 0$ *and artificial*

According to the transformation formulas, we can drive a negative artificial variable x_{Br_2} to zero (out of the basis) and replace it with the legitimate variable x_k at a positive level if $y_{r_2k} < 0$. However, to ensure that no other negative artificial x_{Bi} becomes positive, we must choose r_2 such that

$$\theta_2 = \frac{x_{Br_2}}{y_{r_2k}} = \min_i \left\{ \frac{x_{Bi}}{y_{ik}} \;\middle|\; x_{Bi} \leq 0 \text{ and artificial, } y_{ik} < 0 \right\} \tag{13–8}$$

Observe that if $\theta_2 \leq \theta_1$, then we can remove the $r_2{}^{\text{th}}$ basic variable and none of the remaining non-negative basic variables will become negative. However, if $\theta_2 > \theta_1$, then we cannot remove the $r_2{}^{\text{th}}$ basic variable without introducing new infeasibilities.

Case 3: $x_{Bi} < 0$ *and legitimate*

In this situation we now consider the effect of removing a negative legitimate variable from the basis. Since it is desirable to increase these variables as much as possible, (i.e., make as many non-negative as possible), let

$$\theta_3 = \frac{x_{Br_3}}{y_{r_3k}} = \min_i \left\{ \frac{x_{Bi}}{y_{ik}} \;\middle|\; x_{Bi} < 0 \text{ and legitimate, } y_{ik} < 0 \right\} \tag{13–9}$$

Once again, if $\theta_3 \leq \theta_1$ we can remove the $r_3{}^{\text{th}}$ basic variable without introducing any new infeasibilities, whereas if $\theta_3 > \theta_1$ we cannot do so.

Thus, combining these three cases, a logical rule for choosing the variable to leave the basis is to choose r_2 if $\theta_2 \leq \theta_1 < \infty$; otherwise, choose r_1 if $\theta_1 < \theta_3$ or r_3 if $\theta_1 \geq \theta_3$. This procedure ensures that no new infeasibilities will be introduced and also that, whenever possible, either a negative variable will be driven to zero or an artificial variable will be removed from the basis (or both). Of course, there are other rules for choosing the variable to leave the basis which are also reasonable.

We have now completely described the composite simplex algorithm. To summarize, we use the ordinary simplex tableau and the regular simplex transformation formulas (with one exception, discussed below), and choose the vector to enter according to (13–5) and the vector to leave by computing θ_1, θ_2, θ_3 and employing the rule described above.

We note, however, that the objective function cannot be transformed according to the usual transformation formulas. The quantities d_j can of course be computed directly from their definition, given by equation (13–3), but a computationally more efficient procedure can also be devised. The reader is asked to do so in the exercises.

The composite simplex algorithm can also be adopted for use with a revised simplex tableau. The details of this procedure are also deferred to the exercises.

One final comment, before we illustrate the method with an example: The composite simplex algorithm, like the simplex method, can theoretically cycle, although this has rarely been known to occur in real problems. In any case, Wolfe[11] has devised a scheme for eliminating the possibilty of cycling.

EXAMPLE 1

Consider the linear programming problem

$$\text{Maximize } z = x_1 + 3x_2$$

Subject to:

$$\begin{aligned} x_1 + x_2 + x_3 &\leq 10 \\ 2x_1 + x_2 - 2x_3 &\geq 2 \\ x_1 + 2x_2 &= 4 \\ x_1 - 2x_3 &= -2 \\ x_1, x_2, x_3 &\geq 0 \end{aligned}$$

We add a slack variable x_4 to the first constraint, multiply the second constraint by -1 and then add a slack variable x_5 to it, and add artificial variables x_{a1}, x_{a2} to the third and fourth constraints, respectively. Thus, the constraints are now:

$$\begin{aligned} x_1 + x_2 + x_3 + x_4 &= 10 \\ -2x_1 - x_2 + 2x_3 + x_5 &= -2 \\ x_1 + 2x_2 + x_{a1} &= 4 \\ x_1 - 2x_3 + x_{a2} &= -2 \\ x_1, x_2, x_3, x_4, x_5, x_{a1}, x_{a2} &\geq 0 \end{aligned}$$

Our initial basic solution is $x_4 = 10$, $x_5 = -2$, $x_{a1} = 4$, $x_{a2} = -2$, and the first tableau is:

Basic vectors	$\bar{x}_B$	$\bar{y}_1$	$\bar{y}_2$	$\bar{y}_3$	$\bar{y}_4$	$\bar{y}_5$	$\bar{y}_{a1}$	$\bar{y}_{a2}$
$\bar{a}_4$	10	1	1	1	1	0	0	0
$\bar{a}_5$	-2	-2	-1	2	0	1	0	0
$\bar{a}_{a1}$	4	1	2	0	0	0	1	0
$\bar{a}_{a2}$	-2	1	0	-2	0	0	0	1

The current level of infeasibility is therefore

$$w = -x_{B2} + x_{B3} - x_{B4} = -x_5 + x_{a1} - x_{a2} = 8$$

Also,

$$P = \{i \mid x_{Bi} \geq 0 \text{ and } x_{Bi} \text{ artificial}\} = \{3\}$$
$$N = \{i \mid x_{Bi} < 0\} = \{2, 4\}$$

Hence,

$$d_1 = y_{21} + y_{41} - y_{31} = 2 + 1 - 1 = -2$$
$$d_2 = y_{22} + y_{42} - y_{32} = -1 + 0 - 2 = -3$$
$$d_3 = y_{23} + y_{43} - y_{33} = 2 + -2 - 0 = 0$$

We therefore choose x_2 to enter the basis. Now, we must determine which basic variable is to leave the basis:

$$\theta_1 = \min_i \left\{ \frac{x_{Bi}}{y_{i2}} \middle| \, x_{Bi} \geq 0 \text{ and } y_{i2} > 0 \right\}$$
$$= \min \left\{ \frac{x_{B1}}{y_{12}}, \frac{x_{B3}}{y_{32}} \right\}$$
$$= \min \{\tfrac{10}{1}, \tfrac{4}{2}\} = \tfrac{4}{2} = \boxed{\frac{x_{B3}}{y_{32}} = 2 = \theta_1}$$

$$\theta_2 = \min_i \left\{ \frac{x_{Bi}}{y_{i2}} \middle| \, x_{Bi} \leq 0 \text{ and artificial, and } y_{i2} < 0 \right\}$$
$$= \boxed{+\infty = \theta_2}, \text{ since there is no such } x_{Bi}$$

$$\theta_3 = \max_i \left\{ \frac{x_{Bi}}{y_{i2}} \middle| \, x_{Bi} < 0 \text{ and } y_{i2} < 0 \right\}$$
$$= \frac{x_{B2}}{y_{22}} = \frac{-2}{-1} = \boxed{2 = \theta_3}$$

Since $\theta_1 = \theta_3 < \theta_2$, we choose $x_{B2} = x_5$ to leave the basis. The next tableau is:

Basis vectors	$\bar{x}_B$	$\bar{y}_1$	$\bar{y}_2$	$\bar{y}_3$	$\bar{y}_4$	$\bar{y}_5$	$\bar{y}_{a1}$	$\bar{y}_{a2}$
$\bar{a}_4$	8	−1	0	3	1	1	0	0
$\bar{a}_2$	2	2	1	−2	0	−1	0	0
$\bar{a}_{a1}$	0	−3	0	4	0	2	1	0
$\bar{a}_{a2}$	−2	1	0	−2	0	0	0	1

Now, $P = \{3\}$ and $N = \{4\}$

Therefore, $w = 2$, and

$$\begin{aligned} d_1 &= y_{41} - y_{31} = 1 - -3 = 4 \\ d_3 &= y_{43} - y_{33} = -2 - 4 = -6 \\ d_5 &= y_{45} - y_{35} = 0 - 2 = -2 \end{aligned}$$

Hence, we choose x_3 to enter the basis and compute

$$\theta_1 = \min\left\{\frac{x_{B1}}{y_{13}}, \frac{x_{B3}}{y_{33}}\right\} = 0 = \frac{x_{B3}}{y_{33}}$$

$$\theta_2 = \frac{x_{B4}}{y_{34}} = \frac{-2}{-2} = 1$$

$$\theta_3 = +\infty$$

We find that $x_{B3} = x_{a1}$ leaves the basis, and determine the next tableau:

Basis vectors	$\bar{x}_B$	$\bar{y}_1$	$\bar{y}_2$	$\bar{y}_3$	$\bar{y}_4$	$\bar{y}_5$	$\bar{y}_{a2}$
$\bar{a}_4$	8	$\frac{5}{4}$	0	0	1	$-\frac{1}{2}$	0
$\bar{a}_2$	2	$\frac{1}{2}$	1	0	0	0	0
$\bar{a}_3$	0	$-\frac{3}{4}$	0	1	0	$\frac{1}{2}$	0
$\bar{a}_{a2}$	-2	$-\frac{1}{2}$	0	0	0	1	1

Note that we have deleted the column corresponding to the artificial variable x_{a1}, since it is no longer needed.

Since $P = \varnothing$, the null set, and $N = \{4\}$, we find that

$$d_1 = y_{41} = -\tfrac{1}{2},\ d_5 = y_{45} = +1$$

and therefore x_1 enters the basis. We now compute

$$\theta_1 = \min\left\{\frac{x_{B1}}{y_{11}}, \frac{x_{B2}}{y_{21}}\right\} = \min\left\{\frac{8}{\frac{5}{4}}, \frac{2}{\frac{1}{2}}\right\}$$

$$= 4 = \frac{x_{B2}}{y_{21}}$$

$$\theta_2 = \frac{x_{B4}}{y_{41}} = \frac{-2}{-\frac{1}{2}} = 4$$

$$\theta_3 = +\infty$$

According to our rules for choosing a variable to leave the basis, whenever $\theta_1 = \theta_2 < \theta_3$, we choose the variable corresponding to θ_2 to leave, since this variable will be artificial. Hence, $x_{B4} = x_{a2}$ leaves the basis. Our new tableau is thus:

Basis Vectors	$\bar{x}_B$	$\bar{y}_1$	$\bar{y}_2$	$\bar{y}_3$	$\bar{y}_4$	$\bar{y}_5$
$\bar{a}_4$	3	0	0	0	1	2
$\bar{a}_2$	0	0	1	0	0	1
$\bar{a}_3$	3	0	0	1	0	−1
$\bar{a}_1$	4	1	0	0	0	−2

We now have an initial basic feasible solution for our original problem and proceed with Phase II of the simplex method, by computing the $z_j - c_j$ for the objective function $z = x_1 + 3x_2$. As there is only one nonbasic variable, we compute

$$z_5 - c_5 = \bar{c}_B'\bar{y}_5 - c_5 = (0, 3, 0, 1)\begin{pmatrix} 2 \\ 1 \\ -1 \\ -2 \end{pmatrix} - 0 = 1$$

Hence, our current solution is optimal.

13-3 THE DECOMPOSITION PRINCIPLE OF DANTZIG AND WOLFE

Frequently, large linear programming models encountered in practice can be formulated with the following structure:

$$\text{Maximize } z = \sum_{j=1}^{r} \bar{c}_j'\bar{x}_j$$

$$\begin{bmatrix} A_1 & 0 & 0 & \cdots & 0 \\ 0 & A_2 & 0 & \cdots & 0 \\ 0 & 0 & A_3 & \cdots & 0 \\ \cdot & \cdot & \cdot & \cdot & \cdot \\ \cdot & \cdot & \cdot & \cdot & \cdot \\ \cdot & \cdot & \cdot & \cdot & \cdot \\ 0 & 0 & 0 & \cdots & A_r \\ A_{r+1} & A_{r+2} & A_{r+3} & \cdots & A_{2r} \end{bmatrix} \begin{bmatrix} \bar{x}_1 \\ \bar{x}_2 \\ \cdot \\ \cdot \\ \cdot \\ \bar{x}_r \end{bmatrix} = \begin{bmatrix} \bar{b}_1 \\ \bar{b}_2 \\ \cdot \\ \cdot \\ \cdot \\ \bar{b}_r \\ \bar{b}_{r+1} \end{bmatrix} \tag{13–10}$$

$$\bar{x}_j \geq \bar{0}, \; j = 1, 2, \ldots, r$$

Let A_i be $m_i \times n_i$, $i = 1, 2, \ldots, r$. Then, A_{r+i} is $m_{r+1} \times n_i$, $i = 1, 2, \ldots, r$, and $\bar{b}_i$ is an m_i-component vector, $i = 1, 2, \ldots, r + 1$. Also, $\bar{c}_j$ and $\bar{x}_j$ are each n_j-component vectors, $j = 1, 2, \ldots, r$.

Hence, the linear programming problem defined by (13–10) has a total of $\sum_{j=1}^{r} n_j$ variables and $\sum_{i=1}^{r+1} m_i$ constraints. However, only the first n_1 variables appear in the first m_1 constraints, only the next n_2 variables appear in the next m_2 constraints, and so forth. The entire set of variables appears simultaneously in only the last m_{r+1} constraints.

Problems with this form occur when the model is actually composed of r subsystems which are "linked" only by a few common constraints. For example, suppose a company has r plants, each of which has independent supplies of most of its resources (e.g., labor, machine availabilities, locally supplied goods). These resource constraints would be described by $A_i\bar{x}_i = \bar{b}_i$, $i = 1, 2, \ldots, r$. However, for those resources shared by all the plants (e.g., raw materials purchased and distributed by the company's headquarters) an additional set of constraints is needed: $\sum_{j=1}^{r} A_{r+j}\bar{x}_j = \bar{b}_{m+1}$.

If we were to solve the linear programming problem (13–10) by the standard simplex method (or by revised simplex) we would require a basis of size $\sum_{i=1}^{r+1} m_i$. Fortunately, however, it is possible to take advantage of the structure of (13–10) and to solve this problem without using such a large basis. The material presented below, known as the decomposition principle, was developed by Dantzig and Wolfe.[3,6,7]

We begin by assuming that, for each j, $j = 1, 2, \ldots, r$, the set of feasible solutions to $A_j\bar{x}_j = \bar{b}_j$, $\bar{x}_j \geq \bar{0}$, is strictly bounded. If this is so, then any feasible solution can be represented as a convex combination of the extreme points. Hence, if we let $\bar{p}_{kj}$ denote the extreme points of the set of solutions of $A_j\bar{x}_j = \bar{b}_j$, $\bar{x}_j \geq \bar{0}$, and let h_j denote the number of such extreme points, then any solution $\bar{x}_j$ can be written

$$\begin{aligned} \bar{x}_j &= \sum_{k=1}^{h_j} \mu_{kj}\bar{p}_{kj} \\ \sum_{k=1}^{h_j} \mu_{kj} &= 1, \ \mu_{kj} \geq 0, \ k = 1, 2, \ldots, h_j \end{aligned} \tag{13–11}$$

Moreover, it is easy to see that any point $\bar{x}_j$ satisfying (13–11) is also a solution to $A_j\bar{x}_j = \bar{b}_j$, $\bar{x}_j \geq \bar{0}$.

Suppose now that we substitute (13–11) into $\sum_{j=1}^{r} A_{j+r}\bar{x}_j = \bar{b}_{r+1}$. This yields

$$\begin{aligned} &\sum_{j=1}^{r} A_{j+r}\left\{\sum_{k=1}^{h_j} \mu_{kj}\bar{p}_{kj}\right\} = \bar{b}_{r+1} \\ &\sum_{k=1}^{h_j} \mu_{kj} = 1, j = 1, 2, \ldots, r \\ &\mu_{kj} \geq 0 \ j = 1, 2, \ldots, r \text{ and } k = 1, 2, \ldots, h_j \end{aligned} \tag{13–12a}$$

Observe that any feasible solution of (13–12a), expressed in terms of the μ_{kj}, is also a feasible solution of (13–10), where $\bar{x}_j = \sum_k \mu_{kj}\bar{p}_{kj}$, $j = 1, 2, \ldots, r$; conversely, any feasible solution of (13–10), $\bar{x} = [\bar{x}_1, \bar{x}_2, \ldots, \bar{x}_r]$, also yields a feasible solution of (13–12a), since a feasible set of μ_{kj} can always be found (not necessarily uniquely) which satisfies (13–11).

Thus, the constraints of (13–10) and (13–12a) are completely equivalent; however, (13–12a) contains only $(m_{r+1} + r)$ equations, whereas (13–10) has $\left(\sum_{j=1}^{r} m_j + m_{r+1}\right)$ constraint equations. In many cases, this will represent a considerable reduction in the number of constraints and, therefore, in the resulting basis size.

The corresponding objective function for (13–12a) is obtained by substituting $\bar{x}_j = \sum_k \mu_{kj}\bar{p}_{kj}$ into $z = \sum_j \bar{c}_j'\bar{x}_j$, yielding

$$z = \sum_{j=1}^{r} \sum_{k=1}^{h_j} \mu_{kj}(\bar{c}_j'\bar{p}_{kj}) \tag{13–12b}$$

Hence, if we solve the linear programming problem represented by (13–12), we can easily compute the corresponding optimal solution to (13–10); if μ_{kj}^* denotes an optimal solution of (13–12), then $\bar{x}_j^*$, $j = 1, 2, \ldots, r$ is an optimal solution for the original problem (13–10), where

$$\bar{x}_j^* = \sum_{k=1}^{h_j} \mu_{kj}^*\bar{p}_{kj},\ j = 1, 2, \ldots, r$$

There is one catch to all this, which the reader may have detected by now: In converting the problem from (13–10) to (13–12), we have increased tremendously the number of variables. For each j, the number of μ_{kj} is equal to the number of extreme points of $A_j\bar{x}_j = \bar{b}_j$, $\bar{x}_j \geq \bar{0}$, and we should by now be well aware that this number can be astronomically large. Hence, if we had to compute all the extreme points $\bar{p}_{kj}$ the amount of computation necessary would be prohibitive for all but the smallest of linear programming problems.

However, as we shall see, it is not necessary to know all the $\bar{p}_{kj}$, if we choose to solve (13–12) by the revised simplex method. In the latter, the reader will recall, the tableau contains only the current basic solution and basis inverse, plus a row containing the product of the basis price vector and basis inverse ($\bar{c}_B'B^{-1}$, in the notation of Chapter 6).

To obtain the corresponding quantities for the linear programming problem (13–12), we let

$$\bar{a}_{kj} = \begin{bmatrix} A_{r+j}\bar{p}_{kj} \\ \bar{e}_j \end{bmatrix} \tag{13–13a}$$

$$f_{kj} = \bar{c}_j'\bar{p}_{kj} \tag{13–13b}$$

$$\bar{b} = \begin{bmatrix} \bar{b}_{m+1} \\ \bar{1} \end{bmatrix} \tag{13–13c}$$

In (13–13c), $\bar{1}$ denotes an r-component vector of ones. Thus, the linear programming problem (13–12) may be written

$$\begin{aligned} \text{Maximize } z &= \sum_{j=1}^{r}\sum_{k=1}^{h_j} f_{kj}\mu_{kj} \\ \sum_{j=1}^{r}\sum_{k=1}^{h_j} \bar{a}_{kj}\mu_{kj} &= \bar{b} \\ \text{all } \mu_{kj} &\geq 0 \end{aligned} \tag{13–14}$$

Now, we let B denote any basis matrix (of order $m_{r+1} + r$), and let $\bar{\mu}_B$ and $\bar{f}_B$ denote the corresponding vectors of basic variables and prices for the basic variables, respectively.

Suppose we have a basic feasible solution $\bar{\mu}_B = B^{-1}\bar{b}$ along with the *row* vector $\bar{f}_B' B^{-1} = \overline{w}$. If we partition $\overline{w}$ into $\overline{w} = [\overline{w}_1, \overline{w}_2]$, where $\overline{w}_2$ contains the last r components of $\overline{w}$, then

$$\begin{aligned} z_{kj} - f_{kj} &= \bar{f}_B' B^{-1}\bar{a}_{kj} - f_{kj} \\ &= [\overline{w}_1, \overline{w}_2]\begin{bmatrix} A_{r+j}\bar{p}_{kj} \\ \bar{e}_j \end{bmatrix} - f_{kj} \\ &= \overline{w}_1 A_{r+j}\bar{p}_{kj} + w_{2j} - f_{kj} \\ &= \overline{w}_1 A_{r+j}\bar{p}_{kj} + w_{2j} - \bar{c}_j'\bar{p}_{kj} \\ &= [\overline{w}_1 A_{r+j} - \bar{c}_j']\bar{p}_{kj} + w_{2j} \end{aligned} \tag{13–15}$$

where w_{2j} is the jth component of $\overline{w}_2$.

As usual, if all $z_{kj} - f_{kj} \geq 0$, the current solution is optimal; if one or more $z_{kj} - f_{kj} < 0$, the objective function can be increased. However, we do not actually need to compute all $z_{kj} - f_{kj}$ (which would require knowing all $\bar{p}_{kj}$). Instead, we only need to find the smallest $z_{kj} - f_{kj}$:

$$\min_{\text{all } k,j} \{z_{kj} - f_{kj}\} = \min \Big[\min_k (z_{k1} - f_{k1}), \min_k (z_{k2} - f_{k2}), \ldots, \min_k (z_{kr} - f_{kr}) \Big]$$

Observe that, for each j, the $\min_k (z_{kj} - f_{kj})$ occurs at an extreme point of $A_j\bar{x}_j = \bar{b}_j$, $\bar{x}_j \geq \bar{0}$. In particular, we see from equation (13–15) that

$$\begin{aligned} \min_k (z_{kj} - f_{kj}) &= \min_k \Big[(\overline{w}_1 A_{r+j} - \bar{c}_j')\bar{p}_{kj} + w_{2j} \Big] \\ &= w_{2j} + \min_k \Big[(\overline{w}_1 A_{r+j} - \bar{c}_j')\bar{p}_{kj} \Big] \end{aligned}$$

In order to find out at which extreme point $\bar{p}_{kj}$ the quantity $(z_{kj} - f_{kj})$ assumes a minimum (for fixed j), we can solve the linear programming problem

$$\begin{aligned} \text{Minimize } g_j &= (\bar{w}_1 A_{r+j} - \bar{c}_j')\bar{x}_j \\ A_j \bar{x}_j &= \bar{b}_j \\ \bar{x}_j &\geq \bar{0} \end{aligned} \qquad (13\text{–}16)$$

The optimal solution to (13–16) will be the extreme point $\bar{p}_{kj}$ at which $\min_k (z_{kj} - f_{kj})$ occurs.

We need to solve such a linear programming problem for $j = 1, 2, \ldots, r$. If we denote the corresponding optimal solutions by g_j^*, $j = 1, 2, \ldots, r$, then the vector to enter the basis (for the problem (13–14)) is chosen from

$$\begin{aligned} z_{KJ} - f_{KJ} &= \min_j \,[g_j^* + w_{2j}] \\ &= (\bar{w}_1 A_{r+J} - c_J')\bar{p}_{KJ} + w_{2J} \end{aligned}$$

In order to determine the variable which leaves the basis, we must first compute

$$\bar{y}_{KJ} = B^{-1}\bar{a}_{KJ}$$

where

$$\bar{a}_{KJ} = \begin{bmatrix} A_{r+J}\bar{p}_{KJ} \\ \bar{e}_J \end{bmatrix}$$

The above discussion shows that, at each iteration in solving the linear programming problem (13–14), we only need to know one extreme point, $\bar{p}_{KJ}$, where $\bar{a}_{KJ}$ has been chosen to enter the basis. Of course, in order to determine $\bar{p}_{KJ}$ we have to solve r smaller linear programming problems, given by (13–16), but this involves far less computation than enumerating all the $\bar{p}_{kj}$ and enables us to use the smaller basis size for solving the original problem. It is the resultant often considerable savings in the storage required for B^{-1} which has made the decomposition algorithm so potentially effective in dealing with very large problems.

Computational experience[1,2] with the decomposition principle indicates that it is not necessarily desirable to find the optimal solutions to the subproblems (13–16) at each iteration. All that is actually needed is to find one or more negative $z_{kj} - f_{kj}$, in order to determine a vector to enter the basis which will increase the objective function. It has been found that the use of a "crashing" technique (discussed in Chapter 7), in which several such vectors are brought into the basis before the subproblems (13–16) are resolved, has proven to be more effective than optimizing each subproblem at each iteration. This is due to the fact that the signs of the $z_{kj} - f_{kj}$ oscillate considerably from one iteration to the next, particularly in the early stages of the computation. This results in many cases of a variable being brought into the basis at one iteration, only to be removed several iterations later.

Earlier in this section we imposed the restriction that each of the subproblems (13–16) be strictly bounded. However, the decomposition principle can be modified so that this restriction can be removed. A discussion of how to do so is left for the exercises (see also Dantzig and Wolfe[7]). It is important to know that this can be done, because one might reasonably expect to find situations in which one or more of the subproblems is unbounded, even though the entire problem is not.

The reader may also wish to consult Lasdon[9] and Wismer[10] for a further discussion of some of the computational aspects of the decomposition principle, as well as other methods of handling certain specially structured large linear programming problems.

13–4 COMPUTATIONAL ASPECTS OF THE DECOMPOSITION PRINCIPLE

Perhaps the best way to illustrate the computational and mechanical details of the decomposition principle is by means of a numerical example. Hence, let us consider the following linear programming problem

$$\text{Maximize } z = 4x_1 + 2x_2 + x_5 + 2x_6$$

subject to:

$$\begin{aligned} x_1 + 2x_2 &\leq 4 \\ 2x_1 + x_2 &\leq 6 \\ x_5 + x_6 &\leq 4 \\ 2x_5 + 4x_6 &\leq 10 \\ x_1 + 4x_2 + 4x_5 + 2x_6 &= 18 \\ x_1, x_2, x_5, x_6 &\geq 0 \end{aligned}$$

First, we add slack variables x_3, x_4, x_7, x_8 to the first four constraints, respectively, yielding the coefficient matrix

$$\begin{bmatrix} A_1 & 0 \\ 0 & A_2 \\ A_3 & A_4 \end{bmatrix}$$

where

$$A_1 = \begin{bmatrix} 1 & 2 & 1 & 0 \\ 2 & 1 & 0 & 1 \end{bmatrix}, \qquad A_2 = \begin{bmatrix} 1 & 1 & 1 & 0 \\ 2 & 4 & 0 & 1 \end{bmatrix}$$

$$A_3 = [1 \;\; 4 \;\; 0 \;\; 0], \qquad A_4 = [4 \;\; 2 \;\; 0 \;\; 0]$$

Also, we have

$$\bar{b}_1 = \begin{bmatrix} 4 \\ 6 \end{bmatrix}, \quad \bar{b}_2 = \begin{bmatrix} 4 \\ 10 \end{bmatrix}, \quad \bar{b}_3 = [18]$$

$$\bar{c}_1 = [4 \;\; 2 \;\; 0 \;\; 0], \quad \bar{c}_2 = [1 \;\; 2 \;\; 0 \;\; 0]$$

$$\bar{x}_1 = [x_1 \;\; x_2 \;\; x_3 \;\; x_4], \quad \bar{x}_2 = [x_5 \;\; x_6 \;\; x_7 \;\; x_8]$$

If we were to solve this problem without using the decomposition principle, we would need to add only one artificial variable and perform one Phase I iteration to find an initial basic feasible solution. However, using the decomposition principle requires finding an initial basic feasible solution involving the variables μ_{kj}, and there is no such readily available basic feasible solution. Hence, we must add artificial variables and employ a Phase I. Typically, the decomposition principle requires the use of a Phase I even if an initial basic feasible solution is available for the original problem.

As we observed in the preceding section, we must also use the revised simplex method, which requires only a knowledge of the current basis inverse, since we do not know all the columns of the coefficient matrix for the problem (13–12).

The basis size of (13–12), for our example, is 3, so we must add three artificial variables μ_{a1}, μ_{a2}, μ_{a3}. In addition, in the revised simplex method with artificial variables we have two additional "basic variables" corresponding to the Phase I and Phase II objective functions. Upon referring back to Chapter 6, we therefore find that our initial basis matrix S is

$$S = \begin{bmatrix} 1 & 0 & 0 & 0 & 0 \\ 0 & 1 & 1 & 1 & 1 \\ 0 & 0 & 1 & 0 & 0 \\ 0 & 0 & 0 & 1 & 0 \\ 0 & 0 & 0 & 0 & 1 \end{bmatrix}$$

where the columns of S correspond, respectively, to z, z_a, μ_{a1}, μ_{a2}, μ_{a3}. Hence,

$$S^{-1} = \begin{bmatrix} 1 & 0 & 0 & 0 & 0 \\ 0 & 1 & -1 & -1 & -1 \\ 0 & 0 & 1 & 0 & 0 \\ 0 & 0 & 0 & 1 & 0 \\ 0 & 0 & 0 & 0 & 1 \end{bmatrix}$$

The initial basic feasible solution is

$$\bar{\mu}_S = \begin{bmatrix} z \\ z_a \\ \mu_{a1} \\ \mu_{a2} \\ \mu_{a3} \end{bmatrix} = S^{-1} \begin{bmatrix} 0 \\ 0 \\ \\ \bar{b} \\ \\ \end{bmatrix} = S^{-1} \begin{bmatrix} 0 \\ 0 \\ 18 \\ 1 \\ 1 \end{bmatrix} = \begin{bmatrix} 0 \\ -20 \\ 18 \\ 1 \\ 1 \end{bmatrix}$$

since

$$\bar{b} = \begin{bmatrix} \bar{b}_{m+1} \\ \bar{1} \end{bmatrix} = \begin{bmatrix} 18 \\ 1 \\ 1 \end{bmatrix}$$

by equation (13–13c).

In general, we will have (see equations (6–59) through (6–61))

$$S = \begin{bmatrix} 1 & 0 & -\bar{f}_B \\ 0 & 1 & -\bar{c}_I \\ \bar{0} & \bar{0} & B \end{bmatrix}, \qquad S^{-1} = \begin{bmatrix} 1 & 0 & \bar{f}_B B^{-1} \\ 0 & 1 & \bar{c}_I B^{-1} \\ \bar{0} & \bar{0} & B^{-1} \end{bmatrix}$$

where $\bar{c}_I$ and $\bar{f}_B$ are the Phase I and Phase II objective function coefficients, respectively, of the basic variables.

If we let $\bar{\sigma}_j$ denote the $(j+1)^{st}$ column of S^{-1}, then the initial revised simplex tableau is as given in Figure 13–1 (recall that the first column is deleted for convenience):

Basic variables	$\bar{\sigma}_1$	$\bar{\sigma}_2$	$\bar{\sigma}_3$	$\bar{\sigma}_4$	$\bar{\mu}_B$
z	0	0	0	0	0
z_a	1	-1	-1	-1	-20
$\bar{\mu}_{a1}$	0	1	0	0	18
$\bar{\mu}_{a2}$	0	0	1	0	1
$\bar{\mu}_{a3}$	0	0	0	1	1

Figure 13–1 First tableau.

Observe that we have set up the initial tableau without knowing the quantities $\bar{a}_{kj}$, f_{kj} of equations (13–13), (13–14). We shall determine these quantities as we need them.

We now must compute the objective functions for the two subproblems. From equation (13–15)

$$z_{kj} - f_{kj} = \bar{f}'_B B^{-1}\bar{a}_{kj} - f_{kj} = [\bar{w}_1 A_{r+j} - \bar{c}'_j]\bar{p}_{kj} + w_{rj} \tag{13–17}$$

where $\bar{w}_1$ contains the first m_{r+1} components of $\bar{f}_B B^{-1}$ and w_{2j} is the $(m_{r+1} + j)^{\text{th}}$ component of $\bar{f}'_B B^{-1}$. In Phase I $\bar{f}'_B B^{-1} = \bar{c}'_{B_I} B^{-1}$ is the last $(m_{r+1} + r)$ components of the second row of S^{-1}. (In Phase II, $\bar{f}'_B B^{-1}$ will be found as the last $(m_{r+1} + r)$ components of the first row of S^{-1}). In our example, $r = 2$, $m_{r+1} = 1$. Hence, $\bar{w}_1 = [-1]$, $\bar{w}_2 = [-1, -1]$, $\bar{c}_1 = \bar{c}_2 = \bar{0}$ in Phase I, and

$$\begin{aligned}(z_{k1} - f_{k1})_a &= [\bar{w}_1 A_3 - \bar{c}'_1]\bar{p}_{k1} + w_{21} \\ &= \{(-1)[1 \quad 4 \quad 0 \quad 0] - [0 \quad 0 \quad 0 \quad 0]\}\bar{p}_{k1} + (-1) \\ &= [-1 \quad -4 \quad 0 \quad 0]\bar{p}_{k1} - 1 \\ (z_{k2} - f_{k2})_a &= [\bar{w}_1 A_4 - \bar{c}'_2]\bar{p}_{k2} + w_{22} \\ &= \{(-1)[4 \quad 2 \quad 0 \quad 0] - [0 \quad 0 \quad 0 \quad 0]\}\bar{p}_{k2} + (-1) \\ &= [-4 \quad -2 \quad 0 \quad 0 \quad -1]\bar{p}_{k2} - 1\end{aligned}$$

(The subscript a indicates that these are Phase I $z_{kj} - f_{kj}$).

Thus, the objective functions for the two subproblems are, by equation (13–16),

$$g_1 = -1x_1 - 4x_2$$
$$g_2 = -4x_5 - 2x_6$$

We are now ready to solve these two subproblems:

Subproblem 1

Vectors in basis	$\bar{x}_B$	$\bar{y}_1$	$\bar{y}_2$	$\bar{y}_3$	$\bar{y}_4$
$\bar{a}_3$	4	1	2	1	0
$\bar{a}_4$	6	2	1	0	1
$z_j - c_j$	0	−1	−4	0	0

Vectors in basis	$\bar{x}_B$	$\bar{y}_1$	$\bar{y}_2$	$\bar{y}_3$	$\bar{y}_4$
$\bar{a}_2$	2	$\frac{1}{2}$	1	$\frac{1}{2}$	0
$\bar{a}_4$	4	$\frac{3}{2}$	0	$-\frac{1}{2}$	1
$z_j - c_j$	8	1	0	2	0

Subproblem 2

Vectors in basis	$\bar{x}_B$	$\bar{y}_5$	$\bar{y}_6$	$\bar{y}_7$	$\bar{y}_8$
$\bar{a}_7$	4	1	1	1	0
$\bar{a}_8$	10	2	4	0	1
$z_j - c_j$	0	−4	−2	0	0

Vectors in basis	$\bar{x}_B$	$\bar{y}_5$	$\bar{y}_6$	$\bar{y}_7$	$\bar{y}_8$
$\bar{a}_5$	4	1	1	1	0
$\bar{a}_8$	2	0	2	−2	1
$z_j - c_j$	16	0	2	4	0

In solving the above problems, we have first converted each to a maximization problem. Hence, $g_1^* = -8$, $g_2^* = -16$, and

$$\begin{aligned}(z_{KJ} - f_{KJ})_a &= \text{Min } \{g_1^* + w_{21}, g_2^* + w_{22}\} \\ &= \text{Min } \{-8 + -1, -16 + -1\} \\ &= -17\end{aligned}$$

Therefore, $\bar{a}_{K2}$ enters the basis. The corresponding optimal extreme point, of Subproblem 2, is

$$\bar{p}_{K2} = \begin{bmatrix} 4 \\ 0 \\ 0 \\ 2 \end{bmatrix}$$

and, by equation (13–13a):

$$\begin{aligned}\bar{a}_{K2} &= \begin{bmatrix} A_4\bar{p}_{K2} \\ \bar{e}_2 \end{bmatrix} \\ &= \begin{bmatrix} [4 \;\; 2 \;\; 0 \;\; 0]\,[4 \;\; 0 \;\; 0 \;\; 2]' \\ \begin{bmatrix} 0 \\ 1 \end{bmatrix} \end{bmatrix} \\ &= \begin{bmatrix} 16 \\ 0 \\ 1 \end{bmatrix}\end{aligned}$$

Since $\bar{p}_{K2}$ is the first extreme point of Subproblem 2 we have used, we shall set $K = 1$.

In order to determine the vector which leaves the basis, we must first compute

$$\bar{y}_{12} = B^{-1}\bar{a}_{12} = I_3\bar{a}_{12} = \begin{bmatrix} 16 \\ 0 \\ 1 \end{bmatrix}$$

The vector which leaves the basis is then found from

$$\text{Min } \{\tfrac{18}{16}, \tfrac{1}{1}\} = 1$$

Therefore, μ_{a3} leaves the basis. In the notation of Chapter 6, the vector

$$\bar{\eta}_{12} = \begin{bmatrix} z_{12} - f_{12} \\ (z_{12} - f_{12})_a \\ \bar{y}_{12} \end{bmatrix}$$

is used to transform the tableau. The quantity $(z_{12} - f_{12})_a = -17$, as found above; the quantity $(z_{12} - f_{12})$ is obtained from equation (13–17):

$$z_{12} - f_{12} = [\bar{w}_1 A_4 - \bar{c}_2']\bar{p}_{kj} + w_{22},$$

where for Phase II, $\bar{w}_1 = [0]$, $w_{22} = 0$ (these are found in the first row of S^{-1}). Hence,

$$\begin{aligned} z_{21} - f_{21} &= -\bar{c}_2'\bar{p}_{12} \\ &= -[1 \quad 2 \quad 0 \quad 0]\,[4 \quad 0 \quad 0 \quad 2]' \\ &= -4 \end{aligned}$$

and

$$\bar{\eta}_{12} = \begin{bmatrix} -4 \\ -17 \\ 16 \\ 0 \\ 1 \end{bmatrix}$$

Transforming the tableau of Figure 13–1 yields the tableau given in Figure 13–2.

Basic variables	$\bar{\sigma}_1$	$\bar{\sigma}_2$	$\bar{\sigma}_3$	$\bar{\sigma}_4$	$\bar{\mu}_B$
z	0	0	0	4	4
z_a	1	−1	−1	16	−3
μ_{a1}	0	1	0	−16	2
μ_{a2}	0	0	1	0	1
μ_{12}	0	0	0	1	1

Figure 13–2 Tableau 2.

Now, from the last three components of the second row of S^{-1} (found in Tableau 2), $\overline{w}_1 = [-1]$, $\overline{w}_2 = [-1, 16]$. Since $\overline{w}_1$ is unchanged from the previous iteration, the objective functions for the two subproblems are also unchanged, and therefore the previously obtained optimal solutions are still optimal, with $g_1^* = -8$, $g_2^* = -16$. Thus,

$$
\begin{aligned}
(z_{KJ} - f_{KJ})_a &= \text{Min}\,\{g_1^* + w_{21}, g_2^* + w_{22}\} \\
&= \text{Min}\,\{-8 + -1, \ -16 + 16\} \\
&= -9
\end{aligned}
$$

Hence, $\bar{a}_{K1}$ enters the basis corresponding to the optimal solution to Subproblem 1, which is

$$
\bar{p}_{K1} = \begin{bmatrix} 0 \\ 2 \\ 0 \\ 4 \end{bmatrix}
$$

Since this is the first extreme point from Subproblem 1 used, we set $K = 1$, and compute, from equation (13–13a),

$$
\begin{aligned}
\bar{a}_{11} &= \begin{bmatrix} A_3\bar{p}_{11} \\ \bar{e}_1 \end{bmatrix} \\
&= \begin{bmatrix} [1 \quad 4 \quad 0 \quad 0]\,[0 \quad 2 \quad 0 \quad 4]' \\ \begin{bmatrix} 1 \\ 0 \end{bmatrix} \end{bmatrix} \\
&= \begin{bmatrix} 8 \\ 1 \\ 0 \end{bmatrix}
\end{aligned}
$$

As before, we next compute

$$
\begin{aligned}
\bar{y}_{11} &= B^{-1}\bar{a}_{11} \\
&= \begin{bmatrix} 1 & 0 & -16 \\ 0 & 1 & 0 \\ 0 & 0 & 1 \end{bmatrix} \begin{bmatrix} 8 \\ 1 \\ 0 \end{bmatrix} \\
&= \begin{bmatrix} 8 \\ 1 \\ 0 \end{bmatrix}
\end{aligned}
$$

The vector which leaves the basis is determined by

$$\text{Min}\ \{\tfrac{2}{8}, \tfrac{1}{1}\} = \tfrac{1}{4}$$

Hence, μ_{a1} leaves the basis, and

$$\bar{\eta}_{11} = \begin{bmatrix} z_{11} - f_{11} \\ (z_{11} - f_{11})_a \\ \bar{y}_{11} \end{bmatrix}$$

$$= \begin{bmatrix} [\bar{w}_1 A_3 - \bar{c}_1']\bar{p}_{11} + w_{21} \\ -9 \\ \begin{bmatrix} 8 \\ 1 \\ 0 \end{bmatrix} \end{bmatrix}$$

$$= \begin{bmatrix} (\bar{0}'A_3 - [4 \quad 2 \quad 0 \quad 0])\ [0 \quad 2 \quad 0 \quad 4]' + 0 \\ -9 \\ \begin{bmatrix} 8 \\ 1 \\ 0 \end{bmatrix} \end{bmatrix}$$

$$= \begin{bmatrix} 4 \\ -9 \\ 8 \\ 1 \\ 0 \end{bmatrix}$$

The resulting transformed tableau is given in Figure 13–3.

Basic variables	$\bar{\sigma}_1$	$\bar{\sigma}_2$	$\bar{\sigma}_3$	$\bar{\sigma}_4$	μ_B
z	0	$\frac{1}{2}$	0	-4	5
z_a	1	$\frac{1}{8}$	-1	-2	$-\frac{3}{4}$
$\bar{\mu}_{11}$	0	$\frac{1}{8}$	0	-2	$\frac{1}{4}$
$\bar{\mu}_{a2}$	0	$-\frac{1}{8}$	1	2	$\frac{3}{4}$
$\bar{\mu}_{12}$	0	0	0	1	1

Figure 13–3 Tableau 3.

Now, from the last three components of the second row of Tableau 3, we find that $\overline{w}_1 = [\frac{1}{8}]$, $\overline{w}_2 = [-1, -2]$. Hence,

$$\begin{aligned}(z_{k1} - f_{k1})_a &= \overline{w}_1 A_3 \bar{p}_{k1} + w_{21}\\ &= \tfrac{1}{8}[1 \quad 4 \quad 0 \quad 0]\bar{p}_{k1} - 1\\ (z_{k2} - f_{k2})_a &= \overline{w}_1 A_4 \bar{p}_{k2} + w_{22}\\ &= \tfrac{1}{8}[4 \quad 2 \quad 0 \quad 0]\bar{p}_{k2} - 2\end{aligned}$$

Thus, the objective functions for our two subproblems have now become, by equation (13–16),

$$\begin{aligned}g_1 &= \tfrac{1}{8}x_1 + \tfrac{1}{2}x_2\\ g_2 &= \tfrac{1}{2}x_5 + \tfrac{1}{4}x_6\end{aligned}$$

The optimal solutions for the two subproblems are easily found to be

$$g_1^* = 0, \quad \text{with corresponding extreme point } \bar{p}_{K1} = \begin{bmatrix}0\\0\\4\\6\end{bmatrix}$$

$$g_2^* = 0, \quad \text{with corresponding extreme point } \bar{p}_{K2} = \begin{bmatrix}0\\0\\4\\10\end{bmatrix}$$

Hence,

$$\begin{aligned}(z_{KJ} - f_{KJ})_a &= \text{Min}\ \{g_1^* + w_{21}, g_2^* + w_{22}\}\\ &= \text{Min}\ \{0 + -1, 0 + -2\}\\ &= -2\end{aligned}$$

Therefore, $\bar{a}_{K2}$ enters the basis. Since this is the second vector corresponding to Subproblem 2 we have used, we let $K = 2$, and, by equation

(13–13a):

$$\bar{a}_{22} = \begin{bmatrix} A_4\bar{p}_{22} \\ \bar{e}_2 \end{bmatrix}$$

$$= \begin{bmatrix} [4 \;\; 2 \;\; 0 \;\; 0]\,[0 \;\; 0 \;\; 4 \;\; 10]' \\ \begin{bmatrix} 0 \\ 1 \end{bmatrix} \end{bmatrix}$$

$$= \begin{bmatrix} 0 \\ 0 \\ 1 \end{bmatrix}$$

Next, we compute

$$\bar{y}_{22} = B^{-1}\bar{a}_{22}$$

$$= \begin{bmatrix} \frac{1}{8} & 0 & -2 \\ -\frac{1}{8} & 1 & 2 \\ 0 & 0 & 1 \end{bmatrix} \begin{bmatrix} 0 \\ 0 \\ 1 \end{bmatrix}$$

$$= \begin{bmatrix} -2 \\ 2 \\ 1 \end{bmatrix}$$

Then, the vector which leaves the basis is determined from

$$\text{Min}\left\{\frac{\frac{3}{4}}{2}, \frac{1}{1}\right\} = \tfrac{3}{8}$$

Hence, μ_{a2} leaves the basis. Then,

$$\bar{\eta}_{22} = \begin{bmatrix} z_{22} - f_{22} \\ (z_{22} - f_{22})_a \\ \bar{y}_{22} \end{bmatrix}$$

$$= \begin{bmatrix} (\bar{w}_1 A_4 - \bar{c}_2')\bar{p}_{22} + w_{22} \\ -2 \\ \begin{bmatrix} -2 \\ 2 \\ 1 \end{bmatrix} \end{bmatrix}$$

$$= \begin{bmatrix} \tfrac{1}{2}[4\ \ 2\ \ 0\ \ 0] - [1\ \ 2\ \ 0\ \ 0])\,[0\ \ 0\ \ 4\ \ 10]' - 4 \\ -2 \\ \begin{bmatrix} -2 \\ 2 \\ 1 \end{bmatrix} \end{bmatrix}$$

$$= \begin{bmatrix} -4 \\ -2 \\ -2 \\ 2 \\ 1 \end{bmatrix}$$

The resulting tableau is given in Figure 13–4.

Basic variables	$\bar{\sigma}_1$	$\bar{\sigma}_2$	$\bar{\sigma}_3$	$\bar{\sigma}_4$	μ_B
z	0	$\frac{1}{4}$	2	0	$\frac{13}{2}$
z_a	1	0	0	0	0
μ_{11}	0	0	1	0	1
μ_{22}	0	$-\frac{1}{16}$	$\frac{1}{2}$	1	$\frac{3}{8}$
μ_{12}	0	$\frac{1}{16}$	$-\frac{1}{2}$	0	$\frac{5}{8}$

Figure 13–4 Tableau 4.

We now have a feasible solution, having driven all the artificial variables from the basis. Hence, Phase I is completed, and we now use the last three components of row 1 of S^{-1} to find $\overline{w}$ and to compute the objective functions for the two subproblems. Therefore, $\overline{w}_1 = [\frac{1}{4}]$, $\overline{w}_2 = [2, 0]$, and, from equation (13–17),

$$\begin{aligned} z_{k1} - f_{k1} &= [\overline{w}_1 A_3 - \bar{c}_1']\bar{p}_{k1} + w_{21} \\ &= (\tfrac{1}{4}[1\ \ 4\ \ 0\ \ 0] - [4\ \ 2\ \ 0\ \ 0])\bar{p}_{k1} + 2 \\ &= [-\tfrac{15}{4}, -1, 0, 0]\bar{p}_{k1} + 2 \\ z_{k2} - f_{k2} &= [\overline{w}_1 A_1 - \bar{c}_2']\bar{p}_{k2} + w_{22} \\ &= (\tfrac{1}{4}[4\ \ 2\ \ 0\ \ 0] - [1\ \ 2\ \ 0\ \ 0])\bar{p}_{k2} + 0 \\ &= [0, -\tfrac{3}{2}, 0, 0]\bar{p}_{k2} \end{aligned}$$

Hence, the objective functions for the two subproblems are

$$g_1 = \tfrac{15}{4} x_1 - x_2$$
$$g_2 = \tfrac{3}{2} x_6$$

We now solve the two subproblems:

Subproblem 1

Vectors in basis	$\bar{x}_B$	$\bar{y}_1$	$\bar{y}_2$	$\bar{y}_3$	$\bar{y}_4$
$\bar{a}_3$	4	1	2	1	0
$\bar{a}_4$	6	2	1	0	1
$z_j - c_j$	0	$-\frac{15}{4}$	-1	0	0

Vectors in basis	$\bar{x}_B$	$\bar{y}_1$	$\bar{y}_2$	$\bar{y}_3$	$\bar{y}_4$
$\bar{a}_3$	1	0	$\frac{3}{2}$	1	$-\frac{1}{2}$
$\bar{a}_1$	3	1	$\frac{1}{2}$	0	$\frac{1}{2}$
$z_j - c_j$	$\frac{45}{4}$	0	$\frac{15}{8}$	0	$\frac{15}{8}$

Subproblem 2

Vectors in basis	$\bar{x}_B$	$\bar{y}_5$	$\bar{y}_6$	$\bar{y}_7$	$\bar{y}_8$
$\bar{a}_7$	4	1	1	1	0
$\bar{a}_8$	10	2	4	0	1
$z_j - c_j$	0	0	$-\frac{3}{2}$	0	0

Vectors in basis	$\bar{x}_B$	$\bar{y}_5$	$\bar{y}_6$	$\bar{y}_7$	$\bar{y}_8$
$\bar{a}_7$	$\frac{3}{2}$	$\frac{1}{2}$	0	1	$-\frac{1}{4}$
$\bar{a}_6$	$\frac{5}{2}$	$\frac{1}{2}$	1	0	$\frac{1}{4}$
$z_j - c_j$	$\frac{15}{4}$	$\frac{8}{4}$	0	0	$\frac{3}{8}$

Thus, $g_1^* = -\frac{45}{4}$, $g_2^* = -\frac{15}{4}$ (since in solving the above problems we have first converted each to a maximization problem). We now compute

$$\begin{aligned} z_{KJ} - f_{KJ} &= \text{Min}\,\{g_1^* + w_{21}, g_2^* + w_{22}\} \\ &= \text{Min}\,\{-\tfrac{45}{4} + 2, -\tfrac{15}{4} + 0\} \\ &= -\tfrac{37}{4} \end{aligned}$$

Therefore, $\bar{a}_{K1}$ enters the basis; since this is the second extreme point from subproblem 1, we let $K = 2$, and

$$\bar{p}_{21} = \begin{bmatrix} 3 \\ 0 \\ 1 \\ 0 \end{bmatrix}$$

We next compute

$$\bar{a}_{21} = \begin{bmatrix} A_3\bar{p}_{21} \\ \bar{e}_1 \end{bmatrix} = \begin{bmatrix} [1 \quad 4 \quad 0 \quad 0]\,[3 \quad 0 \quad 1 \quad 0]' \\ \begin{bmatrix} 1 \\ 0 \end{bmatrix} \end{bmatrix} = \begin{bmatrix} 3 \\ 1 \\ 0 \end{bmatrix}$$

Then,

$$\bar{y}_{21} = B^{-1}\bar{a}_{21} = \begin{bmatrix} 0 & 1 & 0 \\ -\frac{1}{16} & \frac{1}{2} & 1 \\ \frac{1}{16} & -\frac{1}{2} & 0 \end{bmatrix} \begin{bmatrix} 3 \\ 1 \\ 0 \end{bmatrix} = \begin{bmatrix} 1 \\ \frac{5}{16} \\ -\frac{5}{16} \end{bmatrix}$$

The vector which leaves the basis is determined from

$$\text{Min} \left\{ \frac{1}{1}, \frac{\frac{3}{8}}{\frac{5}{16}} \right\} = 1$$

Hence, μ_{11} leaves the basis, and

$$\bar{\eta}_{21} = \begin{bmatrix} z_{21} - f_{21} \\ (z_{21} - f_{21})_a \\ \bar{y}_{21} \end{bmatrix} = \begin{bmatrix} -\frac{37}{4} \\ 0 \\ 1 \\ \frac{5}{16} \\ -\frac{5}{16} \end{bmatrix}$$

(In Phase II, $(z_{kj} - f_{kj})_a$ will always be zero).
The next tableau is given in Figure 13–5.

Basic variables	$\bar{\sigma}_1$	$\bar{\sigma}_2$	$\bar{\sigma}_3$	$\bar{\sigma}_4$	μ_B
z	0	$\frac{1}{4}$	$\frac{45}{4}$	0	$\frac{63}{4}$
z_a	1	0	0	0	0
μ_{21}	0	0	1	0	1
μ_{22}	0	$-\frac{1}{16}$	$\frac{3}{16}$	1	$\frac{1}{16}$
μ_{12}	0	$\frac{1}{16}$	$-\frac{3}{16}$	0	$\frac{15}{16}$

Figure 13–5 Tableau 5.

The next iteration is summarized below:

$$\overline{w}_1 = [\tfrac{1}{4}],\ \overline{w}_2 = [\tfrac{45}{4}, 0]$$

$$z_{k1} - f_{k1} = [\overline{w}_1 A_3 - \bar{c}_1']\bar{p}_{k1} + w_{21}$$

$$= [-\tfrac{15}{4}, -1, 0, 0]\bar{p}_{k1} + \tfrac{45}{4}$$

$$z_{k2} - f_{k2} = [\overline{w}_1 A_4 - \bar{c}_2']\bar{p}_{k2} + w_{22}$$

$$= [0, -\tfrac{3}{2}, 0, 0]\bar{p}_{k2} + 0$$

Thus,

$$g_1 = -\tfrac{15}{4}x_1 - x_2$$

$$g_2 = \tfrac{3}{2}x_6$$

Since g_1 and g_2 are unchanged from the previous iteration, $g_1^* = -\frac{45}{4}$, $g_2^* = -\frac{15}{4}$, and

$$z_{KJ} - f_{KJ} = \text{Min}\,\{g_1^* + w_{21}, g_2^* + w_{22}\}$$

$$= \text{Min}\,\{-\tfrac{45}{4} + \tfrac{45}{4}, -\tfrac{15}{4} + 0\}$$

$$= -\tfrac{15}{4}$$

Therefore, $\bar{a}_{K2}$ enters the basis, where $K = 3$, and

$$\bar{p}_{32} = \begin{bmatrix} 0 \\ \frac{5}{2} \\ \frac{3}{2} \\ 0 \end{bmatrix}, \quad \bar{a}_{32} = \begin{bmatrix} A_4\bar{p}_{32} \\ \bar{e}_2 \end{bmatrix} = \begin{bmatrix} 5 \\ 0 \\ 1 \end{bmatrix}$$

$$\bar{y}_{32} = \begin{bmatrix} 0 & 1 & 0 \\ -\frac{1}{16} & \frac{3}{16} & 1 \\ \frac{1}{16} & -\frac{3}{16} & 0 \end{bmatrix} \begin{bmatrix} 5 \\ 0 \\ 1 \end{bmatrix} = \begin{bmatrix} 0 \\ \frac{11}{16} \\ \frac{5}{16} \end{bmatrix}$$

$$\text{Min}\left\{\frac{\frac{1}{16}}{\frac{11}{16}}, \frac{\frac{15}{16}}{\frac{5}{16}}\right\} = \frac{1}{11}$$

Hence, μ_{22} leaves the basis, and

$$\bar{\eta}_{32} = \begin{bmatrix} -\frac{15}{4} \\ 0 \\ 0 \\ \frac{11}{16} \\ \frac{5}{16} \end{bmatrix}$$

The resulting tableau is given in Figure 13–6.

Basic variables	$\bar{\sigma}_1$	$\bar{\sigma}_2$	$\bar{\sigma}_3$	$\bar{\sigma}_4$	μ_B
z	0	$-\frac{1}{11}$	$\frac{135}{11}$	$\frac{60}{11}$	$\frac{177}{11}$
z_a	1	0	0	0	0
μ_{21}	0	0	1	0	1
μ_{32}	0	$\frac{1}{11}$	$\frac{3}{11}$	$\frac{16}{11}$	$\frac{1}{11}$
μ_{12}	0	$\frac{1}{11}$	$-\frac{3}{11}$	$-\frac{5}{11}$	$\frac{10}{11}$

Figure 13–6 Tableau 6.

The next iteration yields:

$$\begin{aligned} \overline{w}_1 &= [-\tfrac{1}{11}],\ \overline{w}_2 = [\tfrac{135}{11}, \tfrac{60}{11}] \\ z_{k1} - f_{k1} &= [\overline{w}_1 A_3 - \bar{c}_1']\bar{p}_{k1} + w_{21} \\ &= [-\tfrac{45}{11}, -\tfrac{25}{11}, 0, 0]\bar{p}_{k1} + \tfrac{135}{11} \\ z_{k2} - f_{k2} &= [\overline{w}_1 A_4 - \bar{c}_2']\bar{p}_{k2} + w_{22} \\ &= [-\tfrac{15}{11}, -\tfrac{24}{11}, 0, 0]\bar{p}_{k2} + \tfrac{60}{11} \\ g_1 &= -\tfrac{45}{11}x_1 - \tfrac{26}{11}x_2 \\ g_2 &= -\tfrac{15}{11}x_5 - \tfrac{24}{11}x_6 \end{aligned}$$

The optimal solutions for the subproblems are:

$$\begin{aligned} g_1^* &= -\tfrac{135}{11}, \quad \text{at } \bar{p}_{k1} = [3, 0, 1, 0] \\ g_2^* &= -\tfrac{69}{11}, \quad \text{at } \bar{p}_{k2} = [3, 1, 0, 0] \\ z_{KJ} - f_{KJ} &= \text{Min}\,\{g_1^* + w_{21}, g_2^* + w_{22}\} \\ &= \text{Min}\,\{-\tfrac{135}{11} + \tfrac{135}{11}, -\tfrac{69}{11} + \tfrac{60}{11}\} \\ &= -\tfrac{9}{11}, \end{aligned}$$

and $\bar{a}_{K2}$ enters the basis, with $K = 4$.

$$\bar{a}_{42} = \begin{bmatrix} A_4\bar{p}_{42} \\ \bar{e}_2 \end{bmatrix} = \begin{bmatrix} 14 \\ 0 \\ 1 \end{bmatrix}$$

$$\bar{y}_{42} = B^{-1}\bar{a}_{42} = \begin{bmatrix} 0 & 1 & 0 \\ -\frac{1}{11} & \frac{3}{11} & \frac{16}{11} \\ \frac{1}{11} & -\frac{3}{11} & -\frac{5}{11} \end{bmatrix} \begin{bmatrix} 14 \\ 0 \\ 1 \end{bmatrix} = \begin{bmatrix} 0 \\ \frac{2}{11} \\ \frac{9}{11} \end{bmatrix}$$

$$\text{Min}\left\{\frac{\frac{2}{11}}{\frac{18}{11}}, \frac{\frac{9}{11}}{\frac{4}{11}}\right\} = \frac{1}{9}$$

Hence, μ_{32} leaves the basis, and

$$\eta_{42} = \begin{bmatrix} -\frac{9}{11} \\ 0 \\ 0 \\ \frac{2}{11} \\ \frac{9}{11} \end{bmatrix}$$

The resulting tableau is given in Figure 13–7.

Basic variables	$\bar{\sigma}_1$	$\bar{\sigma}_2$	$\bar{\sigma}_3$	$\bar{\sigma}_4$	μ_B
z	0	$-\frac{1}{2}$	$\frac{27}{2}$	12	$\frac{33}{2}$
z_a	1	0	0	0	0
μ_{21}	0	0	1	0	1
μ_{42}	0	$-\frac{1}{2}$	$\frac{3}{2}$	8	$\frac{1}{2}$
μ_{12}	0	$\frac{1}{2}$	$-\frac{3}{2}$	−7	$\frac{1}{2}$

Figure 13–7 Tableau 7.

We next calculate:

$$\begin{aligned}
\overline{w}_1 &= [-\tfrac{1}{2}],\ \overline{w}_2 = [\tfrac{27}{2}, 12] \\
z_{k1} - f_{k1} &= [\overline{w}_1 A_3 - \bar{c}_1']\bar{p}_{k1} + w_{21} \\
&= [\tfrac{7}{2}, 0, 0, 0]\bar{p}_{k1} + \tfrac{27}{2} \\
z_{k2} - f_{k2} &= [\overline{w}_1 A_4 - \bar{c}_2']\bar{p}_{k2} + w_{22} \\
&= [-1, 1, 0, 0]\bar{p}_{k2} + 12 \\
g_1 &= \tfrac{7}{2}x_1 \\
g_2 &= -x_5 + x_6
\end{aligned}$$

The optimal solutions for the subproblems are:

$$\begin{aligned}
g_1^* &= -\tfrac{21}{2},\ \text{at } \bar{p}_{k1} = [3, 0, 1, 0] \\
g_2^* &= -\tfrac{5}{2},\ \text{at } \bar{p}_{k2} = [0, \tfrac{5}{2}, \tfrac{3}{2}, 0]
\end{aligned}$$

Hence,

$$\begin{aligned}
z_{KJ} - f_{KJ} &= \text{Min}\ \{g_1^* + w_{21}, g_2^* + w_{22}\} \\
&= \text{Min}\ \{-\tfrac{21}{2} + \tfrac{27}{2}, -\tfrac{5}{2} + 12\} \\
&= 3
\end{aligned}$$

Therefore, we see that all $z_{kj} - f_{kj} \geq 0$, and our current solution is optimal, with $z^* = \frac{33}{2}$, $\mu_{21}^* = 1$, $\mu_{12}^* = \mu_{42}^* = \frac{1}{2}$. Thus,

$$\bar{x}_1^* = \begin{bmatrix} x_1^* \\ x_2^* \\ x_3^* \\ x_4^* \end{bmatrix} = \mu_{21}^*\bar{p}_{21} = \begin{bmatrix} 3 \\ 0 \\ 1 \\ 0 \end{bmatrix}$$

$$\bar{x}_2^* = \begin{bmatrix} x_5^* \\ x_6^* \\ x_7^* \\ x_8^* \end{bmatrix} = \mu_{12}^*\bar{p}_{12} + \mu_{42}^*\bar{p}_{42} = \frac{1}{2}\begin{bmatrix} 4 \\ 0 \\ 0 \\ 2 \end{bmatrix} + \frac{1}{2}\begin{bmatrix} 3 \\ 1 \\ 0 \\ 0 \end{bmatrix} = \begin{bmatrix} \frac{7}{2} \\ \frac{1}{2} \\ 0 \\ 1 \end{bmatrix}$$

Observe that $\bar{x}_2^*$ is not an extreme point of the feasible region for subproblem 2 ($A_4\bar{x}_2 \leq \bar{b}_2, \bar{x}_2 \geq \bar{0}$) :

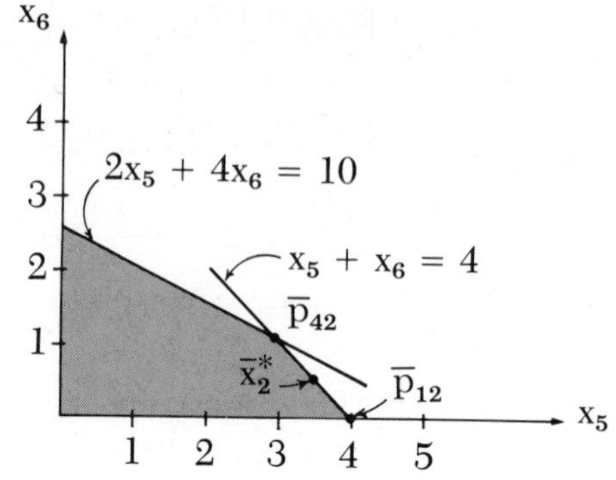

Figure 13–8.

13–5 UPPER BOUND CONSTRAINTS

Consider the following linear programming problem:

$$\text{Maximize } z = \bar{c}'\bar{x}$$

subject to:

$$A\bar{x} = \bar{b}$$
$$0 \leq x_j \leq u_j, j = 1, 2, \ldots, n$$

The coefficient matrix A is $m \times n$, and the u_j are positive constants. The constraints $x_j \leq u_j$ are called upper bound constraints. Linear programming problems in which all or many of the variables have upper bounds occur frequently in practice. If not all the variables have an upper bound, we can let $u_j = \infty$ for each such unrestricted variable. Thus, the above formulation is completely general.

If we were to solve this problem by a standard simplex procedure, we would have to add a slack variable s_j to each of the upper bound constraints, and the resulting problem would have $m + n$ constraints and $2n$ variables. However, by taking advantage of the special structure of the upper bound constraints, we can instead solve the problem by modifying the simplex method so that only the first m constraints $A\bar{x} = \bar{b}$, need be handled explicitly, while the upper bound constraints are treated in a special way. In this manner, only a basis matrix of order m is needed. If, as is usually the case, n is much larger than m, this will result in a considerable savings both in storage requirements and in computational effort expended.

Before developing such a procedure, let us investigate in more detail the properties of the constraint set for a linear programming problem with upper bound constraints. We begin by rewriting the constraint set as follows:

$$\left(\begin{array}{llllll} \bar{a}_1x_1 + \bar{a}_2x_2 + \ldots + \bar{a}_nx_n & & & & = \bar{b} \\ x_1 & + s_1 & & & = u_1 \\ \quad x_2 & & + s_2 & & = u_2 \\ & & & & \cdot \\ & & & & \cdot \\ & & & & \cdot \\ \qquad\qquad x_n & & & + s_n & = u_n \\ x_j \geq 0, \quad s_j \geq 0, \quad j = 1, 2, \ldots, n & & & & \end{array}\right) \qquad (13\text{–}20)$$

The vector $\bar{a}_j$ is the jth column of A, and a slack variable s_j has been added to each constraint $x_j \leq u_j$. We shall denote by F the set of feasible solutions of (13–20). A basic feasible solution of F contains $(m + n)$ basic variables, whose corresponding columns are linearly independent, and $2n - (m + n) = n - m$ nonbasic variables. Since each u_j is strictly positive, at least one of x_j, s_j must be positive—and hence basic—for each of the last n constraints. Therefore, only m of these last n constraints can contain two basic variables; otherwise, there would be more than $m + n$ basic variables. Suppose now that the first m upper bound constraints contain two basic variables (and the remaining $(n - m)$ upper bound constraints contain one basic variable). Thus, the variables $x_1, x_2, \ldots, x_m$ and $s_1, s_2, \ldots, s_m$ are among the basic variables. The columns corresponding to these variables are:

$$\bar{p}_j = \begin{bmatrix} \bar{a}_j \\ \bar{e}_j \end{bmatrix}, j = 1, 2, \ldots, m$$

$$\bar{q}_j = \begin{bmatrix} \bar{0} \\ \bar{e}_j \end{bmatrix}, j = 1, 2, \ldots, m$$

where $\bar{p}_j$ corresponds to x_j and $\bar{q}_j$ corresponds to s_j. It is easy to show that these columns are linearly independent if and only if $\bar{a}_1, \bar{a}_2, \ldots, \bar{a}_m$ are linearly independent (and hence a basis for $A\bar{x} = \bar{b}$). The reader is asked to do so in the exercises.

The above discussion shows that a basic feasible solution for F consists of m x_j's not at their upper bound (i.e., both x_j and s_j are basic) plus $n - m$ x_j's which are either 0 or at their upper bound (i.e., either $x_j = 0$ and $s_j = u_j$, or $x_j = u_j$ and $s_j = 0$). Furthermore, the columns of A corresponding to those basic x_j not at their upper bounds are linearly independent, and therefore constitute a basis B for $A\bar{x} = \bar{b}$. This is an extremely useful result, for it will enable us to solve the linear programming problem using only the basis B plus a knowledge of which of the remaining x_j are 0 and which are equal to u_j. Thus, the effective basis size is m. Moreover, as we shall see, the slack variables s_j need not be considered explicitly, and so the total number of variables is actually only n, rather than $2n$. Instead, we shall deal only with the original variables, the x_j. In any given basic feasible solution, we shall divide these variables into two sets. The first set, consisting of those m variables x_j whose corresponding columns are in B, shall be called the basic variables. The second set consists of the remaining $(n - m)$ variables and we shall call these variables nonbasic. Note that some of these "nonbasic" variables x_j may be equal to u_j, while others are zero.

Suppose now that we have a basic feasible solution, and that $x_1, x_2, \ldots, x_m$ are the basic variables. Then, the equation corresponding to the ith row of the simplex tableau may be written

$$x_i + \sum_{j=m+1}^{n} y_{ij}x_j = \tilde{b}_i \qquad i = 1, 2, \ldots, m \tag{13–21}$$

Since it is possible that not all the nonbasic x_j are zero, the quantity $\tilde{b}_i$ is not necessarily equal to the current value of x_i, and in fact $\tilde{b}_i$ may be negative. The correct value of x_i is obtained by substituting the values of the nonbasic variables $x_{m+1}, \ldots, x_n$ into equation (13–21).

Suppose further that we have somehow decided that the nonbasic variable x_k is to enter the basis. Let us investigate how we determine the variable which leaves the basis. In the regular simplex method, we need only ensure that no basic variable becomes negative. Now, however, we must also make certain that no variable exceeds its upper bound.

Moreover, the entering variable x_k may currently be either 0 or u_k. These two cases must be considered separately.

Case 1: The Entering Variable $x_k = 0$

As x_k increases from zero, the values of the basic variables also change. Two situations are possible.

(a) x_k increases to its upper bound u_k, without any of the basic variables exceeding their lower bounds (0) or upper bounds (u_j). In this case, set $x_k = u_k$, but do not actually bring x_k into the basis, and choose another variable as a candidate for entry into the basis.

(b) One of the basic variables reaches its upper or lower bound before x_k does. Thus, x_k will enter the basis at some value between 0 and u_k.

For case (b), we must choose the variable which leaves the basis so that no variable exceeds its lower or upper bound.

If we let

$$U = \{j \mid x_j = u_j \text{ and } x_j \text{ is nonbasic}\}$$

then, the current value of the ith basic variable is, by (13–21),

$$x_i = \tilde{b}_i - \sum_{j \in U} y_{ij} u_j \equiv x_{Bi} \tag{13–22}$$

Since $x_k = 0$, currently, we can write (13–22) equivalently as

$$\begin{aligned} x_i &= \left(\tilde{b}_i - \sum_{j \in U} y_{ij} u_j\right) - y_{ik} x_k \\ &= x_{Bi} - y_{ik} x_k \end{aligned} \tag{13–23}$$

Now, if x_r leaves the basis, then no variable will become negative so long as

$$\theta_1 = \frac{x_{Br_1}}{y_{r_1 k}} = \operatorname*{Min}_i \left\{ \frac{x_{Bi}}{y_{ik}} \middle| y_{ik} > 0 \right\} \tag{13–24}$$

However, for those $y_{ik} < 0$, the corresponding x_{Bi} will increase. If a given x_{Br_2} $(= x_{r_2}$ here) leaves the basis by increasing to its upper bound u_{r_2}, then the new values of the basic variables are

$$\hat{x}_{Bi} = \hat{\tilde{b}}_i - \sum_{j \in U} \hat{y}_{ij} u_j - \hat{y}_{ir_2} u_{r_2} \tag{13–25}$$

where, according to the usual simplex transformation formulas,

$$\begin{aligned} \hat{y}_{ij} &= y_{ij} - \frac{y_{ik}}{y_{r_2 k}} y_{r_2 j}, \; i \neq r_2 \\ \hat{y}_{r_2 j} &= y_{r_2 j} / y_{r_2 k} \end{aligned} \tag{13–26}$$

$$\begin{aligned} \hat{\tilde{b}}_i &= \tilde{b}_i - \frac{\tilde{b}_{r_2}}{y_{r_2 k}} y_{ik}, \; i \neq r_2 \\ \hat{\tilde{b}}_{r_2} &= \frac{\tilde{b}_{r_2}}{y_{r_2 k}} \end{aligned} \tag{13–27}$$

Substitution of (13–26) and (13–27) into (13–25) yields

$$\begin{aligned} \hat{x}_{Bi} &= \left(\tilde{b}_i - \frac{\tilde{b}_{r_2}}{y_{r_2 k}} y_{ik}\right) - \sum_{j \in U} \left\{ y_{ij} - \frac{y_{ik}}{y_{r_2 k}} y_{r_2 j} \right\} u_j - \left(y_{ir_2} - \frac{y_{ik}}{y_{r_2 k}} y_{r_2 r_2} \right) u_{r_2} \\ &= \left(\tilde{b}_i - \sum_{j \in U} y_{ij} u_j\right) - \frac{y_{ik}}{y_{r_2 k}} \left(\tilde{b}_{r_2} - \sum_{j \in U} y_{r_2 j} u_j\right) - y_{ir_2} u_{r_2} + \frac{y_{ik}}{y_{r_2 k}} y_{r_2 r_2} u_{r_2} \\ &= x_{Bi} - \frac{y_{ik}}{y_{r_2 k}} x_{Br_2} - y_{ir_2} u_{r_2} + \frac{y_{ik}}{y_{r_2 k}} y_{r_2 r_2} u_{r_2} \\ &= (x_{Bi} - y_{ir_2} u_{r_2}) - \frac{y_{ik}}{y_{r_2 k}} (x_{Br_2} - y_{r_2 r_2} u_{r_2}), \; i \neq r_2 \end{aligned}$$

Moreover, since r_2 is the column corresponding to basic variable r_2, $y_{r_2r_2} = 1$ and $y_{ir_2} = 0$, $i \neq r_2$. Thus,

$$\hat{x}_{Bi} = x_{Bi} - \frac{y_{ik}}{y_{r_2k}}(x_{Br_2} - u_{r_2}),\ i \neq r_2 \tag{13–28}$$

For $i = r_2$, we have

$$\begin{aligned} \hat{x}_{Br_2} &= \frac{\tilde{b}_{r_2}}{y_{r_2k}} - \sum_{j\in U} \frac{y_{r_2j}}{y_{r_2k}} u_j - \frac{y_{r_2r_2}}{y_{r_2k}} u_{r_2} \\ &= \frac{1}{y_{r_2k}}\left(\tilde{b}_{r_2} - \sum_{j\in U} y_{r_2j}u_j\right) - \frac{1}{y_{r_2k}} u_{r_2}. \\ &= \frac{x_{Br_2} - u_{r_2}}{y_{r_2k}} \end{aligned} \tag{13–29}$$

Now, since $x_{Br_2} \leq u_{r_2}$, it is clear that $\hat{x}_{Br_2}$ will be positive if and only if $y_{r_2k} < 0$. And, since the other basic variables are not allowed to exceed their upper bounds, we require that $x_{Bi} \leq u_i$, $i \neq r$. From equation (13–28), therefore,

$$x_{Bi} - \frac{y_{ik}}{y_{r_2k}}(x_{Br_2} - u_{r_2}) \leq u_i$$

Or,

$$x_{Bi} - u_i \leq \frac{y_{ik}}{y_{r_2k}}(x_{Br_2} - u_{r_2})$$

Division by $y_{ik} < 0$ yields

$$\frac{x_{Br_2} - u_{r_2}}{y_{r_2k}} \leq \frac{x_{Bi} - u_i}{y_{ik}}$$

Thus, we see that, if x_{r_2} leaves the basis by increasing to its upper limit, and if none of the remaining basic variables is to exceed its upper bound, then r_2 must be chosen so that

$$\theta_2 \equiv \frac{x_{Br_2} - u_{r_2}}{y_{r_2k}} = \min_i \left\{\frac{x_{Bi} - u_i}{y_{ik}} \,\middle|\, y_{ik} < 0\right\} \tag{13–30}$$

Since $\hat{x}_{Br_2}$ ($= x_k$) must also be non-negative, we must choose the variable to leave according to min (θ_1, θ_2), for if $\theta_2 > \theta_1$ and x_{Br_2} leaves the basis, $\hat{x}_{Br_1} < 0$; conversely, if $\theta_2 < \theta_1$ and x_{Br_1} leaves the basis, then $\hat{x}_{Br_2} > u_{r_2}$.

Furthermore, if $\theta_2 > u_k$, then we cannot bring x_k into the basis, for $\hat{x}_k = \hat{x}_{Br_2} = \theta_2$, and hence x_k will exceed its upper bound. This latter is the case mentioned earlier, in which x_k is changed from its lower limit (0) to its upper limit (u_k), but is kept nonbasic.

To summarize this discussion, if the currently nonbasic—and zero—variable x_k is chosen as a candidate for entry into the basis, we compute θ_1 and θ_2 from equations (13–24) and (13–30), respectively, and then find

$$\theta = \min \{u_k, \theta_1, \theta_2\}$$

If $\theta = u_k$, then x_k is set equal to u_k, no transformation of the simplex tableau is performed, and a new candidate for entry into the basis is found; if $\theta = \theta_1$ (or θ_2), then x_{Br_1} (or x_{Br_2}) is chosen to leave the basis, and the tableau is transformed in the usual way.

Now, we must consider:

Case 2: The Entering Variable $x_k = u_k$

Arguments analogous to those in Case 1 lead to the following rules for determining the variable to leave the basis. The derivation of these rules is left as an exercise for the reader.

Compute

$$\theta_1 = \max_i \left\{ \frac{x_{Bi} - u_i}{y_{ik}} \,\middle|\, y_{ik} > 0 \right\} + u_k \tag{13–31a}$$

$$= \frac{x_{Br_1} - u_{r_1}}{y_{r_1 k}} + u_k$$

$$\theta_2 = \max_i \left\{ \frac{x_{Bi}}{y_{ik}} \,\middle|\, y_{ik} < 0 \right\} + u_k \tag{13–31b}$$

$$= \frac{x_{Br_2}}{y_{r_2 k}} + u_k$$

$$\theta = \max \{0, \theta_1, \theta_2\} \tag{13–32}$$

If $\theta = 0$, set $x_k = 0$ and perform no basis change; if $\theta = \theta_1$ or θ_2, remove x_{Br_1} or x_{Br_2}, respectively, from the basis, and bring x_k into the basis, and transform the tableau accordingly.

We have now completed our description of the procedure for determining the variable which leaves the basis, using the special upper bound algorithm. In order to complete our description of the algorithm, we must find a procedure for determining the variable which is to enter the basis. To do so, let us compute the quantities $z_j - c_j$, in the usual way, and write

$$z + \sum_{j=1}^{n} (z_j - c_j)x_j = \tilde{z}$$

where $\tilde{z}$ is the current value of the objective function. Since $z_j - c_j = 0$ for every basic variable, the above expression indicates clearly that:

(a) If the nonbasic variable x_j is currently equal to zero and its corresponding $z_j - c_j < 0$, then increasing x_j will increase z;

(b) If the nonbasic variable x_j is currently equal to u_k and its corresponding $z_j - c_j > 0$, then decreasing x_j will increase z;

(c) If all nonbasic variables currently equal to 0 have corresponding $z_j - c_j \geq 0$, and all nonbasic variables currently equal to u_k have corresponding $z_j - c_j \leq 0$, then the current solution is optimal.

Thus, one decision rule for choosing the variable to enter is to compute:

$$d_k = \operatorname{Min}_j \{d_j\} \tag{13–33}$$

where

$$d_j = \begin{cases} z_j - c_j, & \text{if } x_j = 0 \\ -(z_j - c_j), & \text{if } x_j = u_j \end{cases}$$

If $d_k \geq 0$, the solution is optimal; if $d_k < 0$, then x_k is the candidate for entry into the basis.

EXAMPLE 3

A small brewery, located in a remote section of a southwestern state, sells two brands of beer. For Brand A, the brewery makes a net profit of \$10 per barrel for the first 130 barrels sold, and \$8 for every barrel sold in excess of 130 barrels. For Brand B, the brewery makes a net profit of \$12 per barrel sold, up to 110 barrels, and a net profit of \$10 for every barrel sold beyond 110 barrels. The demands for Brands A and B are estimated to be 200 and 175 barrels, respectively, per week.

The process of combining all the ingredients for the beer requires 24 minutes per barrel for Brand A and 30 minutes per barrel for Brand B. A total of 100 hours per week is available for this activity.

Before the beer is finally ready for barreling, it must also be stored for a period of time in large vats. The brewery has the equivalent of only 225 barrels of beer in vat storage capacity. The brewery also must produce at least 50 barrels of beer per week, in order to meet a standing sales contract.

The brewery wishes to determine how many barrels of each brand of beer to produce, in order to maximize its net profit per week.

We begin our formulation of this problem by observing that the objective function is nonlinear (e.g., net profit for Brand A is not proportional to the number of barrels of Brand A produced) unless the proper choice of variables is made. With this in mind, we let

x_1 = no. of barrels of Brand A produced, up to the first 130 barrels

y_1 = no. of barrels Brand A produced in excess of 130 barrels

x_2 = no. of barrels of Brand B produced, up to the first 110 barrels

y_2 = no. of barrels of Brand B produced in excess of 110 barrels

Then the total net profit is

$$z = 10x_1 + 8y_1 + 12x_2 + 10y_2 \tag{13-34}$$

By the definitions of x_1, x_2, we have

$$\begin{aligned} x_1 &\leq 130 \\ x_2 &\leq 110 \end{aligned} \tag{13-35}$$

The demands on Brands A and B, of 200 and 175, respectively, give us the following two constraints:

$$\begin{aligned} x_1 + y_1 &\leq 200 \\ x_2 + y_2 &\leq 175 \end{aligned} \tag{13-36}$$

The ingredient processing constraint is:

$$\tfrac{24}{60}(x_1 + y_1) + \tfrac{30}{60}(x_2 + y_2) \leq 100$$

or

$$4(x_1 + y_1) + 5(x_2 + y_2) \leq 1000 \tag{13-37}$$

The vat storage capacity constraint is:

$$(x_1 + y_1) + (x_2 + y_2) \leq 225 \tag{13-38}$$

The minimum production requirement of 50 barrels yields:

$$(x_1 + y_1) + (x_2 + y_2) \geq 50 \tag{13–39}$$

Hence, the complete linear programming problem is to maximize the objective function (13–34) subject to the constraints (13–35) through (13–39), plus the usual non-negativity restrictions.

The addition of slack and surplus variables to the constraints would give us a problem with eight constraints and twelve variables to solve.

However, we can reduce the size of the problem considerably by being a bit more careful in formulating it and by using the upper bound constraint method.

First, the constraints (13–35) are already upper bound constraints, and these in turn imply, along with (13–36) that:

$$\begin{aligned} y_1 &\leq 70 \\ y_2 &\leq 65 \end{aligned} \tag{13–40}$$

Hence, we have replaced the two constraints (13–36) with the two upper bound constraints (13–40).

Furthermore, we can also replace the constraint (13–39) by an upper bound constraint, in the following way: Add a slack variable s_2 to the constraint (13–38). This yields

$$(x_1 + y_1) + (x_2 + y_2) + s_2 = 225$$

Or,

$$(x_1 + y_2) + (x_2 + y_2) = 225 - s_2 \tag{13–41}$$

Comparison of the left-hand sides of (13–39) and (13–41) suggests that

$$\begin{aligned} 225 - s_2 &\geq 50 \\ -s_2 &\geq -175 \\ s_2 &\leq 175 \end{aligned} \tag{13–42}$$

Hence, the constraint (13–39) can be replaced by the upper bound constraint (13–42).

Our linear programming problem has now become:

$$\left.\begin{aligned}
&\text{Maximize } z = 10x_1 + 8y_1 + 12x_2 + 10y_2 \\
&\text{subject to:} \\
&4(x_1 + y_1) + 5(x_2 + y_2) + s_1 = 1000 \\
&(x_1 + y_1) + (x_2 + y_2) + s_2 = 225 \\
&\text{and the upper bound constraints:} \\
&x_1 \leq 130 \\
&x_2 \leq 110 \\
&y_1 \leq 70 \\
&y_2 \leq 65 \\
&s_1 \leq \infty \\
&s_2 \leq 175
\end{aligned}\right\} \quad (13\text{–}43)$$

plus, of course, the usual non-negativity restrictions.

The linear programming problem (13–43) has only two regular constraints, six variables, and upper bound constraints on five of the variables.

Let us proceed to solve this problem by the upper bound method. Before doing so, we re-label the variables for convenience, letting $y_1 = x_3$, $y_2 = x_4$, $s_1 = x_5$, $s_2 = x_6$:

$$\begin{aligned}
&\text{Maximize } z = 10x_1 + 12x_2 + 8x_3 + 10x_4 \\
&4x_1 + 5x_2 + 4x_3 + 5x_4 + x_5 = 1000 \\
&x_1 + x_2 + x_3 + x_4 + x_6 = 225 \\
&0 \leq x_1 \leq 130 \\
&0 \leq x_2 \leq 110 \\
&0 \leq x_3 \leq 70 \\
&0 \leq x_4 \leq 65 \\
&0 \leq x_5 \leq \infty \\
&0 \leq x_6 \leq 175
\end{aligned}$$

Now, we observe that we cannot immediately find an initial basic feasible solution, because the usual initial solution with the two slack variables x_5 and x_6 basic violates the upper bound constraint on x_6. Thus, we must add an artificial variable x_7 to the second constraint, and use a Phase I procedure to drive x_7 from the basis.

Our initial simplex tableau is therefore:

Basis vectors	$\bar{x}_B$	$\tilde{b}$	$\bar{y}_1$	$\bar{y}_2$	$\bar{y}_3$	$\bar{y}_4$	$\bar{y}_5$	$\bar{y}_6$	$\bar{y}_7$
$\bar{a}_5$	1000	1000	4	5	4	5	1	0	0
$\bar{a}_7$	225	225	1	1	1	1	0	1	1
$z_j - c_j$	-225	-225	-1	-1	-1	-1	0	-1	0
		u_j	130	110	70	65	∞	175	∞
Current value			0	0	0	0	—	0	—

For record-keeping purposes and easy reference, we have added two rows to the simplex tableau, which are self-explanatory.

Since all nonbasic variables are currently 0, we see that $d_j = z_j - c_j = -1, j = 1, 2, 3, 4, 6$. We break the tie arbitrarily, and choose $\bar{a}_1$ to enter the basis. We now compute

$$\theta_1 = \text{Min}\left\{\frac{x_{B1}}{y_{11}}, \frac{x_{B2}}{y_{21}}\right\}$$

$$= \text{Min}\left\{\frac{1000}{4}, \frac{225}{1}\right\}$$

$$= 200$$

$$\theta_2 = +\infty, \quad \text{since no } y_{i1} < 0.$$

Thus,

$$\theta = \text{Min}\,\{u_1, \theta_1, \theta_2\}$$

$$= \text{Min}\,\{103, 200, \infty\}$$

$$= 130 = u_1$$

Therefore, we set $x_1 = u_1$, but do not bring it into the basis. The only changes in the tableau are the updating of the current values for $\bar{x}_B$ and $\bar{x}_1$:

Basis vectors	$\bar{x}_B$	$\tilde{b}$	$\bar{y}_1$	$\bar{y}_2$	$\bar{y}_3$	$\bar{y}_4$	$\bar{y}_5$	$\bar{y}_6$	$\bar{y}_7$
$\bar{a}_5$	480	1000	4	5	4	5	1	0	0
$\bar{a}_7$	95	225	1	1	1	1	1	0	0
$z_j - c_j$	-95	-225	-1	-1	-1	-1	0	-1	0
		u_j	130	110	70	65	∞	175	∞
Current value			130	0	0	0	—	0	—

Now, $d_1 = -(z_1 - c_1) = +1$, $d_j = z_j - c_j = -1$, $j = 2, 3, 4, 6$, and again we arbitrarily break the tie by choosing $\bar{a}_2$ to enter the basis. Next, we compute

$$\begin{aligned}\theta_1 &= \text{Min}\left\{\frac{x_{B1}}{y_{12}}, \frac{x_{B2}}{y_{22}}\right\}\\ &= \text{Min}\left\{\frac{480}{5}, \frac{95}{1}\right\}\\ &= 95\end{aligned}$$

$$\begin{aligned}\theta_2 &= +\infty\\ \theta &= \text{Min}\,\{u_2, \theta_1, \theta_2\}\\ &= \text{Min}\,\{110, 95, \infty\}\\ &= 95\end{aligned}$$

Hence, $\bar{a}_7$ leaves the basis. The transformed tableau is:

Basis vectors	$\bar{x}_B$	$\tilde{b}$	$\bar{y}_1$	$\bar{y}_2$	$\bar{y}_3$	$\bar{y}_4$	$\bar{y}_5$	$\bar{y}_6$
$\bar{a}_5$	5	−125	−1	0	−1	0	1	−5
$\bar{a}_2$	95	225	1	1	1	1	0	1
$z_j - c_j$	2440	2700	2	0	4	2	0	12
		u_1	130	110	70	65	∞	175
	Current value		130	—	0	0	—	0

We have driven the artificial variable from the basis, completing Phase I. We now recompute the $z_j - c_j$. These recomputed values are given in the above tableau, and the $\bar{a}_7$ column has been deleted.

Now,

$$\begin{aligned}d_1 &= -(z_1 - c_1) = -2\\ d_3 &= z_3 - c_3 = 4\\ d_4 &= z_4 - c_4 = 2\\ d_6 &= z_6 - c_6 = 12\end{aligned}$$

Hence, $\bar{a}_1$ is chosen to enter the basis. Since x_1 is currently at its upper bound, we use the formulas (13–31) and (13–32) to determine which variable leaves the basis:

$$\theta_1 = \frac{x_{B2} - u_{B2}}{y_{12}} + u_1$$
$$= \frac{95 - 110}{1} + 130$$
$$= 115$$
$$\theta_2 = \frac{x_{B1}}{y_{11}} + u_1$$
$$= \frac{5}{-1} + 130$$
$$= 125$$
$$\theta = \text{Max}\ \{0, \theta_1, \theta_2\}$$
$$= \text{Max}\ \{0, 115, 125\}$$
$$= 125$$

The variable which leaves the basis is therefore x_{B1}. The transformed tableau is:

Basis vectors	$\bar{x}_B$	$\bar{b}$	$\bar{y}_1$	$\bar{y}_2$	$\bar{y}_3$	$\bar{y}_4$	$\bar{y}_5$	$\bar{y}_6$
$\bar{a}_1$	125	125	1	0	1	0	−1	5
$\bar{a}_2$	100	100	0	1	0	1	1	−4
$z_j - c_j$	2450	2450	0	0	2	2	2	2
		u_j	130	110	70	65	∞	175
	Current values		—	—	0	0	0	0

Now, $d_j = z_j - c_j \geq 0$, all j, and the current solution is therefore optimal, with $x_1 = 125$, $x_2 = 100$, $z = \$2450$.

13-6 GENERALIZED UPPER BOUND TECHNIQUES

The decomposition and upper bound techniques described in the last three sections both deal with linear programming problems whose constraints have a particular special structure. In this section we shall consider still another special structure, and an efficient method for handling it. Specifically, we shall investigate linear programming problems of the following form:

$$\text{Maximize } z = \bar{c}'\bar{x} \tag{13–44a}$$

subject to:

$$A\bar{x} = \bar{b} \tag{13–44b}$$

and

$$\left.\begin{aligned} \sum_{j=n_0+1}^{n_1} x_j &= 1 \\ \sum_{j=n_1+1}^{n_2} x_j &= 1 \\ &\vdots \\ \sum_{j=n_{p-1}+1}^{n_p} x_j &= 1 \end{aligned}\right\} \tag{13–44c}$$

$$\bar{x} \geq \bar{0} \tag{13–44d}$$

where A is $m \times n$, and $n_p = n$.

Thus, in addition to the usual m constraints $A\bar{x} = \bar{b}$ and the non-negativity restrictions, the linear programming problem (13–44) has p constraints each of the form $\sum x_j = 1$. These latter constraints are the so-called generalized upper bound constraints. Each of the n variables x_j appears in at most† one of these constraints, and such a constraint may therefore be thought of as imposing an upper bound on a sum of some subset of the variables. Any constraint of the form

$$\sum a_j x_j \leq b,$$

where all $a_j > 0$, can be converted into the form $\sum \hat{x}_j = 1$, by first writing the constraint as $\sum (a_j/b)x_j \leq 1$, adding a slack variable, and then re-defining the variables, by $\hat{x}_j = (a_j/b)x_j$. (Negative a_j can also be handled by appropriately modifying the algorithm to be discussed. These modifications are described in Dantzig and Van Slyke[5] and in Lasdon.)[9]

The generalized upper bound constraints may obviously be handled by the decomposition principle, in which case each subproblem has only one constraint. However, such a procedure is not as efficient as the algorithm we shall develop in this section, which has proven to be quite effective in solving

† Note that the variables have been ordered so that the first n_0 variables do not appear in the generalized upper bound constraints.

some extremely large linear programming problems. Hirshfeld[8] has reported the successful solution of problems with as many as 50,000 constraints and over 600,000 variables (in which the use of the generalized upper bounding method reduced the effective basis size to approximately of order 650).

Many large-scale linear programming problems have the structure of (13–44), including transportation-type problems, and the problem obtained when the decomposition method is employed (see equation (13–12a)). Moreover, nonlinear programming problems solved by separable programming techniques also become linear programming problems with many generalized upper bound constraints (see, for example, Section 15–7, Beale,[1] or Dantzig[3]).

Let us begin our study of how to handle generalized upper bound constraints by re-writing the constraints (13–44b, c) as follows:

$$m \text{ rows } \{ \bar{a}_1 x_1 + \ldots + \bar{a}_{n_0} x_{n_0} + \bar{a}_{n_0+1} x_{n_0+1} + \ldots + \bar{a}_{n_1} x_{n_1} + \bar{a}_{n_1+1} x_{n_1+1} + \ldots + \bar{a}_{n_2} x_{n_2} + \ldots + \bar{a}_{n_{p-1}+1} x_{n_{p-1}+1} + \ldots + \bar{a}_n x_n = \bar{b} \tag{13–45}$$

$$p \text{ rows } \left\{ \begin{array}{lllr} x_{n_0+1} + \ldots + x_{n_1} & & & = 1 \\ & x_{n_1+1} + \ldots + x_{n_2} & & = 1 \\ & & & \cdot \\ & & & \cdot \\ & & & \cdot \\ & & x_{n_{p-1}} + \ldots + x_n & = 1 \end{array} \right.$$

Hence, $\bar{a}_j$ denotes the jth column of A. Let us denote by A_j the jth column of the coefficient matrix of the entire problem. Thus, A_j has $m + p$ components, and

$$A_j = \begin{bmatrix} \bar{a}_j \\ \bar{0} \end{bmatrix}, \; 1 \leq j \leq n_0$$

$$A_j = \begin{bmatrix} \bar{a}_j \\ \bar{e}_{m+i} \end{bmatrix}, \; (n_{i-1} + 1) \leq j \leq n_i, \; i = 1, 2, \ldots, p.$$

For simplicity, we let

$$S_0 = \{A_j \mid 1 \leq j \leq n_0\}$$
$$S_i = \{A_j \mid (n_{i-1} + 1) \leq j \leq n_i\}, \; i = 1, 2, \ldots, p.$$

We now observe that any feasible basis for (13–45) must contain at least one column from each of the sets $S_1, S_2, \ldots, S_p$, since at least one variable must be positive (hence, basic) in each of the generalized upper bound constraints.

Assume that we have an initial feasible basis for (13–45), denoted by

$$B = [A_{j_1} A_{j_2} \dots A_{j_{m+p}}]$$

Each set S_i must have at least one column in B; we shall choose one such column from each set and designate it as the *key column* for that set. Let A_{k_i} denote the key column corresponding to set S_i. If we order the basis vectors so that these p key columns appear first, then B can be partitioned as follows:

$$\left[\begin{array}{c|c} G_{m\times p} & H_{m\times m} \\ \hline I_p & F_{p\times m} \end{array}\right]$$

Key Columns   Nonkey Columns

$$G_{m\times p} = [\bar{a}_{k_1}\bar{a}_{k_2} \dots \bar{a}_{k_p}]$$

The rth column of the submatrix $F_{p\times m}$ is $\bar{0}$ if this column corresponds to a vector from S_0, and is $\bar{e}_i$ if this column corresponds to a vector from S_i, $i = 1, 2, \dots, p$. Hence, if we subtract the appropriate key columns from the nonkey columns, the resulting matrix will be of the form

$$\left[\begin{array}{c|c} G_{m\times p} & \tilde{B}_{m\times m} \\ \hline I_p & O_{p\times m} \end{array}\right]$$

It will prove to be very useful to obtain this form, so we shall proceed to do so. Observe that the above operations can be performed by performing a matrix multiplication on B; in particular, if

$$T = \left[\begin{array}{c|c} I_p & -F_{p\times m} \\ \hline O_{m\times p} & I_{m\times m} \end{array}\right] \tag{13–46}$$

Then BT has the desired form:

$$\begin{aligned} BT &= \left[\begin{array}{c|c} G_{m\times p} & H_{m\times m} \\ \hline I_p & F_{p\times m} \end{array}\right]\left[\begin{array}{c|c} I_p & -F_{p\times m} \\ \hline O_{m\times p} & I_m \end{array}\right] \\ &= \begin{bmatrix} (G_{m\times p}I_p + H_{m\times m}O_{m\times p}) & (-G_{m\times p}F_{p\times m} + H_{m\times m}I_m) \\ (I_pI_p + F_{p\times m}O_{m\times p}) & (-I_pF_{p\times m} + F_{p\times m}I_m) \end{bmatrix} \\ &= \left[\begin{array}{c|c} G_{m\times p} & \tilde{B}_{m\times m} \\ \hline I_p & O_{p\times m} \end{array}\right] \end{aligned} \tag{13–47}$$

where $\tilde{B} = -GF + H$.

However, since $B\bar{x}_B = \begin{bmatrix} \bar{b} \\ \bar{1} \end{bmatrix}$, (where $\bar{x}_B$ is the current basic feasible solution, and $\bar{1}$ is a p-component vector of 1's), we must transform the variables by letting

$$\bar{x}_B = T\bar{y}_B \tag{13–48}$$

so that

$$B\bar{x}_B = (BT)\bar{y}_B = \begin{bmatrix} \bar{b} \\ \hline \bar{1} \end{bmatrix} \tag{13–49}$$

In other words, (BT) is the basis matrix corresponding to the variables $\bar{y}_B$. (Since T and$B$ are both nonsingular, so is BT.)

Substitution of equation (4–47) into (4–49) yields

$$(BT)\bar{y}_B = \left[\begin{array}{c|c} G_{m\times p} & \tilde{B}_{m\times m} \\ \hline I_p & O_{p\times m} \end{array}\right] \begin{bmatrix} y_{B1} \\ \cdot \\ \cdot \\ \cdot \\ y_{Bp} \\ \hline y_{B,p+1} \\ \cdot \\ \cdot \\ \cdot \\ y_{B,p+m} \end{bmatrix} = \begin{bmatrix} \bar{b} \\ \hline \bar{1} \end{bmatrix}$$

Or,

$$G_{m\times p} \begin{bmatrix} y_{B1} \\ \cdot \\ \cdot \\ \cdot \\ y_{Bp} \end{bmatrix} + \tilde{B} \begin{bmatrix} y_{B,p+1} \\ \cdot \\ \cdot \\ \cdot \\ y_{B,p+m} \end{bmatrix} = \bar{b} \tag{13–50a}$$

$$I_p \begin{bmatrix} y_{B1} \\ \cdot \\ \cdot \\ \cdot \\ y_{Bp} \end{bmatrix} = \bar{1} \tag{13–50b}$$

Thus, $y_{Bi} = 1, i = 1, 2, \ldots, p$, and

$$\tilde{B}\begin{bmatrix} y_{B,p+1} \\ \cdot \\ \cdot \\ \cdot \\ y_{B,p+m} \end{bmatrix} = \bar{b} - \sum_{i=1}^{p} \bar{g}_i \qquad (13\text{–}51)$$

where $\bar{g}_i$ is the ith column of G. But, recall that $G = [\bar{a}_{k_1}\ \bar{a}_{k_2} \ldots \bar{a}_{k_p}]$. Therefore, equation (13–51) becomes

$$\tilde{B}\begin{bmatrix} y_{B,p+1} \\ \cdot \\ \cdot \\ \cdot \\ y_{B,p+m} \end{bmatrix} = \bar{b} - \sum_{i=1}^{p} \bar{a}_{k_i} \qquad (13\text{–}52)$$

Moreover, recall that BT was formed from B by subtracting key columns from the last m columns of B. If the $(p + r)^{\text{th}}$ column of B is A_{j_r}, then the $(p + r)^{\text{th}}$ column of BT is

$$A_{j_r} - A_{k_i}, \quad \text{where } A_{j_r} \in S_i$$

Thus, since $\tilde{B}$ is just the first m rows and the last m columns of (BT), the rth column of $\tilde{B}$ is $\bar{a}_{j_r} - \bar{a}_{k_i}$, where $A_{j_r} \in S_i$.

Now, let us return to the initial constraint set (13–45). The first m constraints can be expressed

$$\sum_{j=1}^{n_0} \bar{a}_j x_j + \sum_{j=n_0+1}^{n_1} \bar{a}_j x_j + \ldots + \sum_{j=n_{p-1}+1}^{n} \bar{a}_j x_j = \bar{b} \qquad (13\text{–}53)$$

If we solve each of the p generalized upper bound constraints for the key variable (each generalized upper bound constraint contains exactly one key variable) we obtain

$$x_{k_i} = 1 - \sum_{\substack{j=n_{i-1}+1 \\ j \neq k_i}}^{n_i} x_j \qquad i = 1, 2, \ldots, p \qquad (13\text{–}54)$$

Substitution of equations (13–54) into (13–53) yields

$$\sum_{j=1}^{n_0} \bar{a}_j x_j + \sum_{j=n_0+1}^{n_1} (\bar{a}_j - \bar{a}_{k_1}) x_j + \sum_{j=n_1+1}^{n_2} (\bar{a}_j - \bar{a}_{k_2}) x_j$$
$$+ \ldots + \sum_{j=n_{p-1}+1}^{n} (\bar{a}_j - \bar{a}_{k_p}) x_j + (\bar{a}_{k_1} + \bar{a}_{k_2} + \ldots + \bar{a}_{k_p}) = \bar{b} \qquad (13\text{–}55)$$

If we define

$$\bar{d}_j = \bar{a}_j - \bar{a}_{k_i}, \quad \text{if } A_j \in S_i \tag{13–56}$$

then equation (13–55) can be expressed

$$\sum_{j=1}^{n_0} \bar{a}_j x_j + \sum_{j=n_0+1}^{n_1} \bar{d}_j x_j + \sum_{j=n_1+1}^{n_2} \bar{d}_j x_j + \ldots + \sum_{j=n_{p-1}+1}^{n} \bar{d}_j x_j = \bar{b} - \sum_{i=1}^{p} \bar{a}_{k_i} \tag{13–57}$$

since in each sum, $\bar{d}_j = \bar{0}$ for $j = k_i$.

If we let $\bar{d}_j = \bar{a}_j$, $j = 1, 2, \ldots, n_0$, then (13–57) can be written more conveniently as

$$\sum_{j=1}^{n} \bar{d}_j x_j = \bar{b} - \sum_{i=1}^{p} a_{k_i} \tag{13–58}$$

We now show that the matrix $\tilde{B}$ is a basis for the system of equations (13–58). First of all, we observe that $\tilde{B}$ is nonsingular, for if its columns were linearly dependent, then so would be the last n columns of BT (see equation (13–47)); but we already know that BT is nonsingular. Hence, its columns are linearly independent, and $\tilde{B}$ is therefore nonsingular.

We have also already observed that the columns of $\tilde{B}$ are $\bar{a}_{j_r} - \bar{a}_{k_i}$, where $A_{j_r} \in S_i$, $r = 1, 2, \ldots, m$. From (13–56), we see then that the rth column of $\tilde{B}$ is

$$\bar{d}_{j_r} = \bar{a}_{j_r} - \bar{a}_{k_i}$$

Thus, $\tilde{B}$ consists of m linearly independent columns from the system (13–58), and hence constitutes a basis for this system. The system (13–58) is called the *reduced system*. We shall now show that each operation necessary to apply the revised simplex method to the original linear programming problem (13–44) may be performed by using quantities obtained from $\tilde{B}$. In this manner, only a basis of size m is required, instead of a basis of size $m + p$. The basis $\tilde{B}$ will therefore be called the *working basis*.

More specifically, in the revised simplex method the $(m + 1)$ order basis inverse B^{*-1} is available at each iteration, where

$$B^{*-1}\left[\begin{array}{c|c} 1 & \tilde{\bar{c}}'_B \tilde{B}^{-1} \\ \hline \bar{0} & \tilde{B}^{-1} \end{array}\right]$$

The first column corresponds to the objective function z, which is treated as a variable by the revised simplex method. Then, if $\bar{a}_j^* = \begin{bmatrix} -c_j \\ \bar{a}_j \end{bmatrix}$, we know that $z_j - c_j = [1, \tilde{\bar{c}}'_B \tilde{B}^{-1}]\bar{a}_j^*$. But, in our situation, these are the $z_j - c_j$ for the reduced system (13–58).

Suppose now, we proceed to calculate the $z_j - c_j$ for the original system (13–44). Let C_B be the $(m + p)$-component row vector defined by $C_B = [\bar{c}_B^{(1)}, \bar{c}_B^{(2)}]$, where $c_B^{(1)}$ is a p-component vector containing the c_j corresponding to the first p columns of the $(m + p)$ order basis matrix B, and $\bar{c}_B^{(2)}$ is an m-component vector containing the c_j corresponding to the last m columns of B.

If we let $\overline{w} = C'_B B^{-1}$, then, by definition,

$$\begin{aligned} z_j - c_j &= C'_B B^{-1} A_j - c_j \\ &= \overline{w} A_j - c_j, j = 1, 2, \ldots, n \end{aligned} \tag{13–59}$$

In order to compute these $z_j - c_j$, we need $\overline{w}$.

Suppose we partition $\overline{w}$ into a p-component vector $\overline{w}^{(1)}$ and an n-component vector $\overline{w}^{(2)}$, $\overline{w} = [\overline{w}^{(1)}, \overline{w}^{(2)}]$; then

$$\begin{aligned} C_B &= \overline{w} B \\ C_B T &= \overline{w} B T \\ &= [\overline{w}^{(1)}, \overline{w}^{(2)}] \left[\begin{array}{c|c} G_{m\times p} & \tilde{B} \\ \hline I_p & O_{p\times m} \end{array}\right] \\ &= [(\overline{w}^{(1)} G + \overline{w}^{(2)}), \overline{w}^{(1)} \tilde{B}] \end{aligned} \tag{13–60}$$

Moreover,

$$\begin{aligned} C_B T &= [\bar{c}_B^{(1)}, \bar{c}_B^{(2)}] \left[\begin{array}{c|c} I_p & -F_{p\times m} \\ \hline O_{m\times p} & I_m \end{array}\right] \\ &= [\bar{c}_B^{(1)}, -\bar{c}_B^{(1)} F + \bar{c}_B^{(2)}] \end{aligned} \tag{13–61}$$

Hence, comparing the right-hand sides of equations (13–60) and (13–61) yields

$$\overline{w}^{(1)} G + w^{(2)} = \bar{c}_B^{(2)} \tag{13–62}$$

$$\overline{w}^{(1)} \tilde{B} = -\bar{c}_B^{(1)} F + \bar{c}_B^{(2)} \tag{13–63}$$

Postmultiplication of (13–63) by $\tilde{B}^{-1}$ yields

$$\overline{w}^{(1)} = (-\bar{c}_B^{(1)} F + \bar{c}_B^{(2)}) \tilde{B}^{-1} \tag{13–64}$$

But the row vector $(-\bar{c}_B^{(1)} F + \bar{c}_B^{(2)})$ is precisely the vector of c_j's corresponding to the basic variables of the reduced system (13–58), which we have called $\tilde{\bar{c}}_B$ (the reader is asked to prove this statement in the exercises). Thus, with $\overline{w}^{(1)} = \tilde{\bar{c}}_B \tilde{B}^{-1}$, we obtain from equation (13–62) that

$$\begin{aligned} \overline{w}^{(2)} &= \bar{c}_B^{(2)} - \overline{w}^{(1)} G \\ &= \bar{c}_B^{(2)} - \tilde{\bar{c}}'_B \tilde{B}^{-1} G \end{aligned}$$

Furthermore, since $G = [\bar{a}_{k_1}\ \bar{a}_{k_2} \ldots \bar{a}_{k_p}]$, the ith component of $\overline{w}^{(2)}$ is

$$w_i^{(2)} = c_{Bi}^{(2)} - (\tilde{\bar{c}}'_B \tilde{B}^{-1} \bar{a}_{k_i}) \tag{13–65}$$

Thus, equations (13–64) and (13–65) show us that the vector $\overline{w}$ can be computed directly from $B^{*-1} = \begin{bmatrix} 1 & \tilde{\bar{c}}'_B \tilde{B}^{-1} \\ \bar{0} & \tilde{B}^{-1} \end{bmatrix}$ and the key columns $a_{k_1}, \bar{a}_{k_2}, \ldots, \bar{a}_{k_p}$.

From equation (13–59) we see that

$$z_j - c_j = \overline{w}A_j - c_j$$

$$= \begin{cases} [\overline{w}^{(1)}, \overline{w}^{(2)}] \begin{bmatrix} \bar{a}_j \\ \bar{0} \end{bmatrix} - c_j, & 1 \le j \le n_0 \\ [\overline{w}^{(1)}, \overline{w}^{(2)}] \begin{bmatrix} \bar{a}_j \\ \bar{e}_j \end{bmatrix} - c_j, & n_{i-1}+1 \le j \le n_i,\ i = 1, 2, \ldots, p \end{cases}$$

$$= \begin{cases} \overline{w}^{(1)}\bar{a}_j - c_j, & 1 \le j \le n_0 \\ \overline{w}^{(1)}\bar{a}_j + w_i^{(2)} - c_j & n_{i-1}+1 \le j \le n_i,\ i = 1, 2, \ldots, p \end{cases}$$

$$z_j - c_j = \begin{cases} \tilde{\bar{c}}'_B \tilde{B}^{-1} \bar{a}_j - c_j, & 1 \le j \le n_0 \\ \bar{c}_B \tilde{B}^{-1}(\bar{a}_j - \bar{a}_{k_i}) + (c_{Bi}^{(2)} - c_j), & n_{i-1}+1 \le j \le n_i,\ i = 1, 2, \ldots, p \end{cases} \tag{13–66}$$

Hence, at each iteration we compute $z_j - c_j$ for all variables which are not in the basis $\tilde{B}$ or which are not key variables (which are actually basic variables and whose $z_j - c_j = 0$). If all $z_j - c_j \ge 0$, we have the optimal solution. If not all $z_j - c_j \ge 0$, we bring x_s into the basic solution, where $z_s - c_s = \text{Min}\,\{z_j - c_j\}$.

Next, we must determine which variable leaves the basis. To do so, we must first compute

$$Y_s = B^{-1}A_s \tag{13–67}$$

To derive an expression for Y_s in terms of known quantities, we let $Y_s = TV_s$, so that

$$\begin{aligned} A_s &= BY_s \\ &= BTV_s \end{aligned} \tag{13–68}$$

Let us partition V_s into a p-component vector $\bar{v}_s^{(1)}$ and an m-component vector $\bar{v}_s^{(2)}$. For simplicity, we shall consider the two cases $1 \le s \le n_0$ and $n_{i-1} + 1 \le s \le n_i$ separately.

Case 1: $1 \leq s \leq n_0$

Equation (13–68) may be written in partitioned form:

$$\begin{bmatrix} \bar{a}_s \\ \bar{0} \end{bmatrix} = \left[\begin{array}{c|c} G & \tilde{B} \\ \hline I & 0 \end{array}\right] \begin{bmatrix} \bar{v}_s^{(1)} \\ \bar{v}_s^{(2)} \end{bmatrix}$$
$$= [G\bar{v}_s^{(1)} + \tilde{B}\bar{v}_s^{(2)}]$$

Hence, $\bar{v}_s^{(1)} = \bar{0}$, $\tilde{B}\bar{v}_s^{(2)} = \bar{a}_s$, and therefore

$$\bar{v}_s^{(2)} = \tilde{B}^{-1}\bar{a}_s \tag{13–69}$$

Partitioning Y_s into a p-component vector $\bar{y}_s^{(1)}$ and an m-component vector $\bar{y}_s^{(2)}$ yields, from $Y_s = TV_s$ and the above,

$$Y_s = \begin{bmatrix} \bar{y}_s^{(1)} \\ \bar{y}_s^{(2)} \end{bmatrix} = \left[\begin{array}{c|c} I & -F \\ \hline 0 & I \end{array}\right] \begin{bmatrix} \bar{0} \\ \tilde{B}^{-1}\bar{a}_s \end{bmatrix}$$
$$= \begin{bmatrix} -F\tilde{B}^{-1}\bar{a}_s \\ \tilde{B}^{-1}\bar{a}_s \end{bmatrix} \tag{13–70}$$

Since the rth column of F is either $\bar{0}$ or $\bar{e}_i$ (depending on whether this column corresponds to a vector from S_0 or S_i, respectively) it is clear that Y_s is easily computed if $\tilde{B}^{-1}$ is known.

Case 2: $n_{i-1} + 1 \leq s \leq n_i$

In partitioned form, equation (13–68) now becomes

$$\begin{bmatrix} \bar{a}_s \\ \bar{e}_i \end{bmatrix} = \left[\begin{array}{c|c} G & \tilde{B} \\ \hline I & 0 \end{array}\right] \begin{bmatrix} \bar{v}_s^{(1)} \\ \bar{v}_s^{(2)} \end{bmatrix}$$
$$= \begin{pmatrix} G\bar{v}_s^{(1)} + \tilde{B}\bar{v}_s^{(2)} \\ \bar{v}_s^{(1)} \end{pmatrix}$$

Hence, $\bar{v}_s^{(1)} = \bar{e}_i$, and

$$\begin{aligned} \bar{v}_s^{(2)} &= \tilde{B}^{-1}[\bar{a}_s - G\bar{e}_i] \\ &= \tilde{B}^{-1}[\bar{a}_s - \bar{a}_{k_i}] \\ &= \tilde{B}^{-1}\bar{d}_s \quad \text{(by equation (13–56))} \end{aligned}$$

We now have, then that

$$Y_s = \begin{bmatrix} \bar{y}_s^{(1)} \\ \bar{y}_s^{(2)} \end{bmatrix} = \begin{bmatrix} I & -F \\ 0 & I \end{bmatrix} \begin{bmatrix} \bar{e}_i \\ \tilde{B}^{-1}\bar{d}_s \end{bmatrix} = \begin{bmatrix} \bar{e}_i - F\tilde{B}^{-1}\bar{d}_s \\ \tilde{B}^{-1}\bar{d}_s \end{bmatrix} \tag{13–71}$$

The vector to leave the basis is chosen in the usual way, from

$$\theta = \frac{x_{Br}}{y_{rs}} = \underset{i}{\text{Min}} \left\{ \frac{x_{Bi}}{y_{is}} \,\middle|\, y_{is} > 0 \right\}$$

(where y_{is} is the ith component of Y_s).

If all $y_{is} \leq 0$, the problem has an unbounded solution. Otherwise, the rth basic variable leaves the basis. The new basic variables are easily computed, according to the usual transformation formulas:

$$\hat{x}_{Br} = \theta$$
$$\hat{x}_{Bi} = x_{Bi} - \theta y_{is},\ i = 1, 2, \ldots, m + p;\ i \neq r$$

We have left only to determine how to compute the new $\tilde{B}^{-1}$ for our development of the generalized upper bounding method to be complete.

The transformation formulas for $\tilde{B}^{-1}$ are somewhat more complicated, for the following reason. We have assumed throughout that the full $(m + p)$ order basis B contains p key columns, one from each set S_i, $i = 1, 2, \ldots, p$. If we remove one of these key columns from the basis, another basic column from the same set must be designated as the key column for that set. But, this entails reordering the columns of B, and also means that the columns corresponding to $\tilde{B}$ may therefore be changed.

Suppose that the entering vector $A_s \in S_q$ and the leaving vector $A_{j_r} \in S_t$. There are three cases we must consider.

Case 1: A_{j_r} Is Not a Key Column

In this case, the column leaving corresponds to a column of $\tilde{B}$. Since the key columns remain the same, $\tilde{B}^{-1}$ may be transformed in the usual way; namely, append the column $\bar{y}_s^{(2)}$ to the tableau and perform a pivot, with r^{TH} component of $\bar{y}_s^{(2)}$ as the pivot element, where $r = p + \hat{r}$. This operation may be illustrated schematically as follows:

$$[\tilde{B}^{-1},\ y_s] \rightarrow [\hat{\tilde{B}}^{-1}, \bar{e}_{\hat{r}}]$$

Since we are using the revised simplex method, we are actually interested in transforming:

$$\left[\begin{array}{c|c|c} 1 & \tilde{c}'_B\tilde{B}^{-1} & z_s - c_s \\ \hline \bar{0} & \tilde{B}^{-1} & \bar{y}_s^{(2)} \end{array}\right] \rightarrow \left[\begin{array}{c|c|c} 1 & \hat{\tilde{c}}'_B\hat{\tilde{B}}^{-1} & 0 \\ \hline \bar{0} & \hat{\tilde{B}}^{-1} & \bar{e}_{\hat{r}} \end{array}\right]$$

Case 2: A_{j_r} *Is a Key Column and* $q \neq t$

In this case, the key column $A_{k_t} = A_{j_r}$ is leaving the basis. Hence, another basic column from S_t must be designated as a key column. (Recall that each set S_i, $i = 1, 2, \ldots, p$, always contains at least one basic column.) Denote this new key column by $A_{k^*_t}$. Since we have ordered the basic columns so that the key columns are the first p columns of B, we must now reorder the basic columns to maintain this property, before performing the transformation.

The columns $\bar{d}^*_j$ of the reduced system corresponding to this interchange of the key columns are given by

$$\bar{d}^*_j = \begin{cases} \bar{d}_j, j \notin S_t \\ \bar{a}_j - \bar{a}_{k^*_t}, \quad j \in S_t \quad \text{and} \quad j \neq k^*_t \\ \bar{a}_{k_t} - \bar{a}_{k^*_t} \end{cases} \tag{13–72}$$

Observe that, by equation (13–56), $\bar{d}^*_{k^*_t} = -\bar{d}_{k^*_t}$, and

$$\begin{aligned} \bar{d}^*_j &= \bar{a}_j - \bar{a}_{k^*_t} \\ &= (\bar{a}_j - \bar{a}_{k_t}) + (\bar{a}_{k_t} - \bar{a}_{k^*_t}) \\ &= \bar{d}_j - \bar{d}^*_{k^*_t}, \quad j \in S_t, \quad j \neq k^*_t \end{aligned} \tag{13–73}$$

Since the columns of $\tilde{B}$ are merely a subset of these $\bar{d}_j$, the columns of the working basis $\tilde{B}^*$, corresponding to the new reduced system, are just the corresponding $\bar{d}^*_j$. Hence, the jth column of $\tilde{B}^*$ is the jth column of $\tilde{B}$ if $j \notin S_t$; the jth column of $\tilde{B}^*$ is minus one times the $k^{*\text{th}}_t$ column of $\tilde{B}$ added to the jth column of $\tilde{B}$, for $j \in S_t$ and $j \neq k^*_t$; and the $k^{*\text{th}}_t$ column of $\tilde{B}^*$ is the negative of the $k^{*\text{th}}_t$ column of $\tilde{B}$. It is easy to show that

$$\tilde{B}^* = \tilde{B}E \tag{13–74}$$

where

$$E = \begin{bmatrix} 1 & 0 & 0 & \cdots & & & & & 0 \\ 0 & 1 & 0 & \cdots & & & & & 0 \\ \cdot & & & & & & & & \\ \cdot & & & & & & & & \\ \cdot & & & & & & & & \\ 0 & \cdots -1 & \cdots & 0 \quad 0 & \cdots -1 & -1 & \cdots 0 & \cdots & -1 \quad \cdots \\ \cdot & & & & & \cdot & & & \\ \cdot & & & & & & \cdot & & \\ \cdot & & & & & & & \cdot & \\ 0 & & \cdots & & & & & & 1 \end{bmatrix} \leftarrow \text{row } i^*$$

$(m \times m)$

and where $k_t^* = p + i^*$. The ith row of E is simply the ith row of the m^{th}order identity matrix, $i \neq i^*$. The -1's in row i^* occur in columns corresponding to $A_j \in S_t$. (Thus, E is obtained from I_m by adding minus one times the $k_t^{*\text{th}}$ column of I_m to the jth column, for $j \in S_t$, and then multiplying the $k_t^{*\text{th}}$ column by minus one.) The reader is asked to verify the validity of equation (13–74) in the exercises, and also to prove that $EE = I_m$, which implies that $E = E^{-1}$. With these results, we find that

$$\begin{aligned} (\tilde{B}^*)^{-1} &= E^{-1}\tilde{B}^{-1} \\ &= E\tilde{B}^{-1} \end{aligned} \tag{13–75}$$

Hence, when the key column k_t is replaced by the column k_t^*, the inverse of the working basis $\tilde{B}$ is modified by premultiplying it by the matrix E. This amounts to adding minus one times row i^* of $\tilde{B}^{-1}$ to each of the rows corresponding to vectors in S_t, and then multiplying row i^* by minus one.

Once this change in $\tilde{B}^{-1}$ is made, the situation is that of Case 1: The vector leaving the basis is no longer a key column. Hence, the transformation of $\tilde{B}^{-1}$ is accomplished as described in Case 1.

Case 3: A_{j_r} Is a Key Column and $q = t$. In this case, the key column $A_{k_t} = A_{j_r}$ is leaving the basis and is to be replaced by A_s, another column from S_t. Hence, it is possible that no other column (besides A_{k_t}) from S_t is currently basic. If this is so, then A_s must become the new key column for S_t. However, it will also be true that no change is necessary in the working basis, since none of the columns in $\tilde{B}$ are affected by this change. Therefore, no transformation is necessary.

If, however, there are other basic columns from S_t, we merely choose one of these to become the new key column, and treat this case the same as Case 2.

We have now completed our development of the generalized upper bounding method. We have shown how a linear programming problem with

m general constraints and p generalized upper bound constraints can be solved with a working basis of order m. At each iteration, a small amount of extra computation may be necessary, to appropriately modify the inverse of the working basis, as described in Cases 2 and 3, above. Moreover, some extra bookkeeping is necessary to keep track of which columns are key columns and to which set S_j each of the columns in the working basis belongs. However, it is obvious that this extra computation and bookkeeping effort is extremely small compared to the extra storage and computation required to solve the same problem with a basis of order $m + p$, if there are more than just a few generalized upper bound constraints.

PROBLEMS

1. Solve the problem of exercise 6, Chapter 5, by the composite simplex method.
2. Solve the problem of exercise 7, Chapter 5, by the composite simplex method.
3. Solve the problem of exercise 8, Chapter 5, by the composite simplex method.
4. As noted in the text, the quantities d_j, defined by equation (13–3), cannot be transformed at each iteration of the composite simplex method according to the usual simplex transformation formulas. Develop a set of formulas for transforming these quantities without re-computing each d_j from its definition at each iteration.
5. Show how to use the composite simplex algorithm in conjunction with the revised simplex method.
6. Solve by the decomposition method:

$$\begin{aligned}
\text{Maximize } z = 3x_1 + 4x_2 + x_3 + 2x_4 + 2x_5 & \\
x_1 + 2x_2 + x_3 \qquad\qquad\qquad & = 8 \\
2x_1 + x_2 + 2x_3 \qquad\qquad\qquad & = 6 \\
x_4 + 3x_5 & = 9 \\
x_1 + x_2 + x_3 + x_4 + x_5 & = 13 \\
x_1, x_2, x_3, x_4, x_5 \geq 0 \qquad &
\end{aligned}$$

7. Discuss how to modify the decomposition principle so that the restriction that each subproblem be strictly bounded can be removed.

8. Prove that the set of vectors $\bar{p}_j, \bar{q}_j, j = 1, 2, \ldots, m$, defined on page 319, is linearly independent, if and only if, $\bar{a}_1, \bar{a}_2, \ldots, \bar{a}_m$ are linearly independent.
9. In the upper bounding method discussed in Section 13–5, if the variable which enters the basis is currently at its upper bound, then the variable which leaves the basis is determined according to equations (13–31) and (13–32). Derive these equations.
10. Solve the following problem by the upper bound method of Section 13–5:

$$\begin{aligned}
\text{Maximize } z = x_1 + 3x_2 & \\
x_1 + x_2 & \leq 8 \\
4x_1 + x_2 & \leq 26 \\
-x_1 + x_2 & \leq 4 \\
0 \leq x_1 & \leq 5 \\
0 \leq x_2 & \leq 7
\end{aligned}$$

Illustrate each iteration graphically.

11. Solve the following problem by the upper bound method of Section 13–5:

$$\begin{aligned} \text{Maximize } z &= x_1 + x_2 \\ x_1 - x_2 &\leq 2 \\ x_1 - x_2 &\geq -4 \\ 0 \leq x_1 &\leq 4 \\ 0 \leq x_2 &\leq 6 \end{aligned}$$

Illustrate each iteration graphically.

12. Verify the validity of equation (13–34), which is used in the generalized upper bounding method of Section 13–6. Also, show that $EE = I_n$, where E is defined immediately below equation (13–74).

13. Solve by the generalized upper bound method of Section 13–6:

$$\text{Maximize } z = x_1$$

$$\begin{array}{rcl} 5x_1 + 10x_2 + 5x_3 + 15x_4 + 10x_5 + 15x_6 + 5x_7 + 10x_8 &=& 34 \\ 5x_1 + 10x_2 + 10x_3 + 5x_4 + 20x_5 + 15x_6 + 10x_7 + 10x_8 &=& 38 \\ x_2 + x_3 &=& 1 \\ x_4 + x_5 + x_6 &=& 1 \\ x_7 + x_8 &=& 1 \end{array}$$

$$x_1, x_2, x_3, x_4, x_5, x_6, x_7, x_8 \geq 0$$

14. In the discussion of the generalized upper bounding technique in Section 13-6, we stated that the row vector $-(\bar{c}_B^{(1)}F + \bar{c}_B^{(2)})$ of equation (13-64) is precisely the vector of c_j's corresponding to the basic variables of the reduced system (13-58). Discuss the validity of this statement.

REFERENCES

1. Beale, E. M. L.: *Mathematical Programming in Practice.* Sir Isaac Pitman and Sons, Ltd., London, 1968.
2. Beale, E. M. L., P. A. B. Hughes, and R. E. Small: Experience in using a decomposition program. *Computer Journal*, 8: 13–18, 1965.
3. Dantzig, G. B.: *Linear Programming and Extensions.* Princeton University Press, Princeton, N.J. 1963.
4. Dantzig, G. B.: Upper bounds, secondary restraints, and block triangularity. *Econometrica*, 23: 53–72, April 1965.
5. Dantzig, G. B., and R. M. Van Slyke: Generalized upper bounding techniques, *J. Computer System Sci.*, 1: 213–226, 1967.
6. Dantzig, G. B., and P. Wolfe: Decomposition principle for linear programs. *Operations Research*, 8: 101–111, Jan. 1960.
7. Dantzig, G. B., and P. Wolfe: The decomposition principle for linear programming. *Econometrica*, 29: 767–778, 1961.
8. Hirshfeld, David S.: Very large linear programming models and how to solve them professionally. Paper presented at 41st National ORSA Meeting, New Orleans, April 1972.
9. Lasdon, Leon S.: *Optimization Theory for Large Systems*, The Macmillan Company, New York, 1970.
10. Wismer, David A. (ed.): *Optimization Methods for Large Scale Systems with Applications.* McGraw-Hill Book Company, New York, 1971.
11. Wolfe, Philip: The Composite Simplex Algorithm. *SIAM Review*, 7 (No. 1): 42–54 January 1965.

14

Network Flow Problems

14–1 INTRODUCTION

In this chapter we shall deal with some problems relating to "networks," in particular maximal flow problems and a network flow approach to solving transportation problems. We have chosen these two problems for several reasons. First, the maximal flow problem and the transportation problem are important types of linear programming problems and hence deserve mention.

The maximal flow problem is a linear programming problem, but because of its highly specific structure we shall develop an algorithm unrelated to the simplex algorithm which is extremely efficient. For the transportation problem, we shall apply the primal-dual algorithm of Chapter 9, which, because of the special structure of the transportation problem, results in a computational algorithm generally regarded as more efficient than the method described in Chapter 10.

There are many other network problems, such as minimum length problems, generalizations of maximal flow problems, etc. The interested reader may consult references 1 and 2.

It is sufficient for the problems we wish to consider to use the simplest terminology. A network is a collection of points or *nodes* some or all of which are connected by lines or *arcs* and through which the flow of some "material" occurs. A collection of a set of nodes and arcs is called a *graph*. If a flow of some kind occurs in the graph, it is generally referred to as a *network*. Mathematically, a specification of nodes and arcs defines a graph. However, for convenience we often pictorialize a graph as in Figure 14–1.

Let us consider some examples of situations that can be thought of as networks:

1. A system of chemical processing units in which some product is undergoing change. The nodes correspond to the various units such as heat

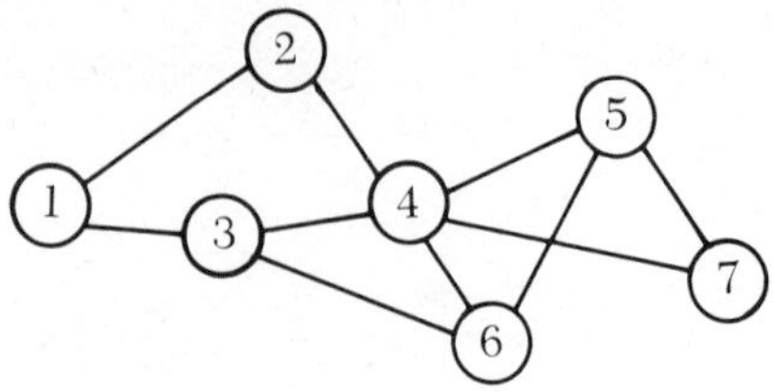

Figure 14–1

exchanges, reactors, and so forth. The arcs correspond to pipelines between units.

2. A computer network in which the nodes correspond to computer systems or terminals, and the telephone lines or microwave transmission lines correspond to the arcs of the network.
3. A transportation system in which the nodes are cities, and the arcs are roads over which some material is transported.

There are many real world situations similar to the above which can be profitably regarded as networks.

The graph depicted in Figure 14–1 is such that there is no indicated orientation to the arcs, i.e., the arcs do not have an indication of a sense of direction. If we are interested in networks, it is often the case that if a flow occurs it can or should occur along certain directions. It is, of course, possible in many situations to have flow in both directions along an arc. A graph with a sense of direction imparted to the arcs is called an *oriented graph.* Figure 14–2 is an example of an oriented graph and Figure 14–1 is an example of an *unoriented graph.*

For the sections that follow we shall require a few definitions that we now give. The *capacity* of an arc in one (or both) directions is the maximum allowable flow that the arc can accommodate. A flow capacity is a nonnegative number. In the case in which there is no practical finite upper bound on an allowable flow, we shall consider the flow capacity to be infinite. For a number of reasons, which will be explained later, it is convenient to indicate flow capacities between nodes in both directions. For example, if x_{ij} represents the flow from node i to node j over the arc connecting node i and node j, then, if the arc was oriented from i to j, $x_{ij} = 5$ would be an allowable flow, assuming the capacity of the arc to be at least 5, but x_{ji} would always be zero. On the other hand, if the arc was not oriented then x_{ji} could be different from zero.

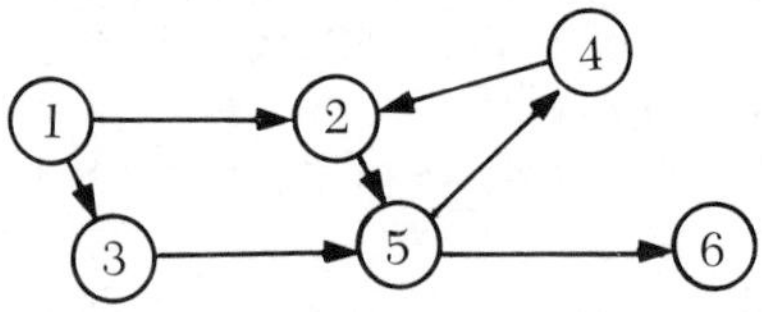

Figure 14–2

In most networks there are one or more nodes which are distinguished as *sources*. A source is a node such that all arcs connected to it are oriented so that the flows are away from the node. Similarly, a *sink* is a node such that the flows in the arcs connected to it are all directed into the node. In brief, flows originate at sources and terminate at sinks.

14–2 THE MAXIMAL FLOW PROBLEM

Let us consider an oriented network with a single source, a single sink and a set of intermediate connected nodes. For each node, with the exception of the source and sink, we shall assume that conservation of flow is maintained, i.e., flow into a given node equals flow out of the node. Suppose the network has N nodes, where node 1 is the source and node N is the sink. We will designate the arc connecting node i and node j by (i, j). We will also designate the flow along arc (i, j) by x_{ij} and the capacity of the arc (i, j) by c_{ij}.

Let us now consider the formulation of the *maximal flow problem*, i.e., how to maximize the flow of whatever is "flowing" from source to sink subject to constraints of network arc capacities and conservation of flow. It is clear that no flow in a given arc may exceed the capacity of that arc. Hence, we require constraints of the form:

$$0 \leq x_{ij} \leq c_{ij} \qquad \text{all} \quad i, j \tag{14–1}$$

In addition, as we have noted, there is conservation of flow at each node, i.e., the flow into each node equals the flow out of the node. Hence we have:

$$\sum_i x_{ik} = \sum_j x_{kj} \qquad k = 2, 3, \ldots, N - 1 \tag{14–2}$$

Our objective is to maximize flow. Hence, the objective function may be stated as maximizing the sum of flows originating at the source or flowing into the sink, i.e., we may write:

$$\text{Max } z = \sum_j x_{1j}$$

or

$$\text{Max } z = \sum_i x_{iN} \tag{14–3}$$

To summarize, the maximal flow problem is given by:

$$\begin{aligned} &\text{Max } z = \sum_j x_{1j} \\ &\sum_i x_{ik} - \sum_j x_{kj} = 0 \qquad k = 2, 3, \ldots, N - 1 \\ &0 \leq x_{ij} \leq c_{ij} \qquad \text{all } i, j \end{aligned} \tag{14–4}$$

It is easily seen that the problem stated in (14–4) is a linear programming problem. However, for a network with many nodes and arcs, the problem can be quite large. For example, a network with 100 nodes and 200 arcs would have either a basis of size 98 + 200 or a basis of size 98 plus a special technique for handling the upper bounds if the simplex method were used. Instead of this, we shall present a *labeling* method devised by Ford and Fulkerson[1] which is quite efficient for solving our problem.

In order to make our description of this method perfectly general let us assume that each arc (i, j) has associated with it capacities c_{ij} and c_{ji}. In the event that flows are allowed in one direction only, one of these capacities would be zero, e.g., $c_{ij} > 0$, $c_{ji} = 0$. Let us now define the term *excess capacity*, which we designate d_{ij}. Given any set of flows x_{ij} and x_{ji} in arc (i, j) then the d_{ij} are defined as:

$$\begin{aligned} d_{ij} &= c_{ij} - x_{ij} + x_{ji} \\ d_{ji} &= c_{ji} - x_{ji} + x_{ij} \end{aligned} \tag{14–5}$$

We will now describe the labeling algorithm for solving the maximal flow problem. Following the description, we shall prove that the method does in fact yield the optimal solution.

Initially, all flows are set at zero. We also calculate for each arc its excess capacities, d_{ij} and d_{ji}. The basic aim of the labeling method is to start at the source and attempt to reach or "label" the sink by finding a path consisting of arcs with positive excessive capacity. We do this as follows.

We begin at the source, node 1, and consider all nodes which are connected to the source by arcs of positive excess capacity. We designate the nodes which are connected to the source this way with the index p. Call this set of nodes P. We then label each node $p \in P$ on our network with two numbers or labels, (α_p, β_p). These labels are defined as follows:

$$\begin{aligned} \alpha_p &= \text{excess capacity from source to node } p \\ \beta_p &= \text{the node from which we came (the source)} \end{aligned}$$

Therefore, we have:

$$(\alpha_p, \beta_p) = (d_{1p}, 1) \qquad p \in P \tag{14–6}$$

If we had labeled the sink in the first step, which is theoretically possible but highly unlikely in any practical situation, we move to the final part of this algorithm. In this part, we give a method for increasing the flow.

Generally, the sink is not labeled in the first step. Hence, from the set of nodes $p \in P$, we choose $p = \bar{p}_1$. We then seek out nodes, not yet labeled, which are joined to the node $p = p_1$, $p_1 \in P$ by arcs of positive excess capacity. If there are none, we go to $p = p_2$, $p_2 \in P$ and repeat. If we can reach any unlabeled nodes in this fashion, let us designate this set of nodes with the index q and this set of nodes will be called Q. We now proceed to label the nodes $q \in Q$ as follows:

$$\begin{aligned} \alpha_q &= \min (d_{pq}, \alpha_p) \\ \beta_q &= p \end{aligned} \tag{14–7}$$

The label α_q represents the minimum excess capacity of the two arcs $(1, p)$ and (p, q). The label β_q tells us the node from which we came to reach node q. We carry out this labeling process for all nodes $p \in P$. Again, if we have labeled the sink during this step, we proceed to the latter part of this algorithm.

We now give the general step of the algorithm, which is basically the same as the previous step. If we have a set of labeled nodes $r \in R$, we look for nodes, not yet labeled, which are connected to nodes r by arcs which have positive excess capacity. We do this for each $r \in R$. If we find a set of unlabeled nodes $t \in T$, we label these nodes as follows:

$$\begin{aligned} \alpha_t &= \min (d_{rt}, \alpha_r) \\ \beta_t &= r \end{aligned} \tag{14–8}$$

We repeat this general step. Since the network has a finite number of nodes and arcs we must reach one of the two following conditions in a finite number of steps: (1) the sink is labeled; or (2) we cannot label the sink and no other nodes can be labeled. If we reach condition (1), we can increase the flow. If we reach condition (2), the existing flow is the maximal flow and we have solved the problem.

Let us now consider how to increase the flow if condition (1) holds, i.e., the sink has been labeled. The label on the sink, α_N, indicates how much positive excess capacity exists from source to sink over the particular path traversed, and therefore, how much the flow can be increased. Furthermore, it is a simple matter to traverse this path, since the second label on the nodes, β_j, indicates the preceding node leading to node j. Hence, we can trace the path backwards from the sink.

Let us designate by d_{uv} the excess capacities of the arcs in this particular path from source to sink, which enabled us to label the sink. We can increase the flow by α_N and we do this, in effect, by recalculating the excess capacities for each arc. These new excess capacities are given by:

$$\begin{aligned} \hat{d}_{uv} &= d_{uv} - \alpha_N \\ \hat{d}_{vu} &= d_{vu} + \alpha_N \end{aligned} \tag{14–9}$$

$\hat{d}_{ij} = d_{ij}$ for branches not in the sink labeling path. We now repeat the entire labeling process for the network with the new excess capacities, starting again at the source. It is clear that since we are dealing with finite maximal flows in a finite network, we must reach condition (2) in a finite number of steps.

At each stage we can calculate the *net flows* in each arc using the definition of excess capacities given in (14–5). The net flow in the arc (i, j) is:

$$x_{ij} = c_{ij} - d_{ij} \quad \text{if} \quad c_{ji} = 0 \tag{14–10}$$

If both $c_{ij} \neq 0$, $c_{ji} \neq 0$ then either:

$$x_{ij} = c_{ij} - d_{ij} \quad \text{and} \quad x_{ji} = 0$$

or (14–11)

$$x_{ji} = c_{ji} - d_{ji} \quad \text{and} \quad x_{ij} = 0$$

depending upon which of the quantities $c_{ij} - d_{ij}$ and $c_{ji} - d_{ji}$ is positive.

Let us now prove that the above algorithm, when it terminates under the conditions stated, does indeed yield the maximal flow. In order to do so let us first define the concept of a *cut* in a network. A cut in a network is defined as a set of oriented arcs of the network such that every path from the source to the sink contains at least one arc of the set. In a non-trivial network containing many nodes and arcs, it is obvious that there are many cuts for any network, but the number is finite. The concept of a cut is of central importance because we can see that the maximal flow cannot be greater than the sum of the capacities of the arcs in any cut. This is true because the flow in every path is limited by the arc of *least* capacity and every path must contain an arc of the cut.

It is obvious that at each stage the constraints are satisfied since the capacity restrictions are satisfied by our definition of excess capacity and calculation of flows. Further, we are obviously satisfying the conservation of flow at each node. Hence, what we need to show is that the final flow is the maximal flow. Now, let us consider the situation when we terminate our calculation. We see that we have two disjoint sets of nodes: nodes that we have been able to label and those we have not. Let us designate the nodes we have been able to reach in the final labeling process as S_L and those we have not been able to label as S_U. If we now consider the set of oriented arcs that connect members of S_L to members of S_U, this set of arcs, say J_c, is clearly a *cut* of the network because unless each path from source to sink contains at least one of these arcs, we cannot reach the sink. What we will now show is that the sum of the capacities in J_c is equal to the flow in the network.

If we combine the definition of excess capacity which is given by equation (14–5) and the constraints on flow conservation (14–2), we see that for all nodes except the source and sink we can write:

$$\sum_{\substack{j=1 \\ j \neq i}}^{N} (c_{ij} - d_{ij}) = 0 \qquad i = 2, \ldots, N - 1 \tag{14–12}$$

For the source, $i = 1$, we have:

$$\sum_{\substack{j=1 \\ j \neq 1}}^{N} (c_{1j} - d_{1j}) = \sum_{j=2}^{N} x_{1j} = \text{total flow} \tag{14–13}$$

If we now sum the left hand side of (14–12) over *labeled nodes*, i.e., over nodes in S_L, and recall that the source is labeled (by definition), we see from (14–12) and (14–13) that:

$$\sum_{i \in S_L} \sum_{\substack{j=1 \\ j \neq i}}^{N} (c_{ij} - d_{ij}) = \sum_{j=2}^{N} x_{1j} = \text{total flow} \tag{14–14}$$

We now note the following about equation (14–14). If both $j \in S_L$ and $i \in S_L$, i.e., if nodes i and j are both labeled, then we have terms for both $c_{ij} - d_{ij}$ and $c_{ji} - d_{ji}$ present in the summation and they cancel each other in (14–14). Hence, all terms cancel for $j \in S_L$, and only terms for $j \in S_U$ are present. However, if $i \in S_L$ and $j \in S_U$, then, because we know that S_L and S_U are disjoint sets, we know that $d_{ij} = 0$. Hence, (14–14) reduces to:

$$\sum_{i \in S_L} \sum_{j \in S_U} c_{ij} = \text{total flow} \tag{14–15}$$

From the definition of cut, we see that a set of arcs such as that given by (14–15) is clearly a cut. However, (14–15) says that this particular cut *equals* the total flow. We have already noted that the total flow cannot *exceed* the sum of the capacities in any cut. However, in (14–15) we have found a cut in which the total flow *equals* the capacities of the arcs in the cut. Hence, for this particular cut, the flow must be maximal and this proves that the labeling algorithm we have given provides the maximal flow.

It so happens that the particular cut that gives the maximal flow is the one for which the sum of cut capacities is minimal. This is often stated as the *max flow–min cut theorem.* Our proof showed this in finding the cut whose cut capacities equaled the total flow. All other cuts would have provided sums of cut capacities that exceeded the total flow.

14–3 MAXIMAL FLOW COMPUTATIONAL ALGORITHM

It is a relatively simple matter to organize the maximal flow algorithm into a compact tableau format. For a network with N nodes we merely require an Nth order matrix of the excess capacities d_{ij}. We also require two additional columns for the labels. We will also note the order of labeling in a

Nodes	1	2	...	j	...	N	Labels		Step
1	—	d_{12}	...	d_{1j}	...	d_{1N}	α	β	
2	d_{21}	—	...	d_{2j}	...	d_{2N}	α_2	β_2	
⋮									
i	d_{i1}	d_{i2}	...	d_{ij}	...	d_{iN}	α_i	β_i	
⋮									
N	d_{N1}	d_{N2}	...	d_{Nj}	...	—	α_N	β_N	

Figure 14–3 Tableau for maximal flow calculation.

third additional column. The tableau format is more convenient than performing the calculation on a network diagram. The maximal flow tableau format is shown in Figure 14–3.

In order to use the tableau shown in Figure 14–3, we begin with the matrix $||d_{ij}||$ of excess capacities. Initially, $d_{ij} = c_{ij}$ we start in row 1 and look for columns with positive excess capacity, i.e., $d_{ij} > 0$. For all such columns we set $\alpha_j = d_{ij}$ *and* $\beta_j = 1$. Assuming that we have not labeled the sink, i.e., a value has not been given to α_N, we look at the first of the rows i that have been labeled at the previous step. We keep track of these in the final column marked "step." In row i we look for $d_{ij} > 0$ for which row j has not yet been labeled. For all such rows, we set $\alpha_j = \min(d_{ij}, \alpha_i)$, $\beta_i = i$. We continue this for each of the rows labeled at the first step, as long as α_N is not computed, i.e., the sink is not labeled. If all rows are labeled and we still haven't labeled the sink, we now examine the rows labeled in the second step. We repeat the entire procedure. Finally, since the algorithm is finite, we must either label the sink or find that we cannot reach the sink, i.e., assign a value to α_N. If α_N is not labeled, the flow is maximal. Otherwise, we know we can increase the flow.

Suppose we have labeled the sink and thus have some value α_N and $\beta_N = r$. We need to construct a new tableau. In order to do so we first calculate the d_{uv} in the path from the source to the sink. Since $\beta_N = r$, we know that the final row was reached from row r. Therefore, we need to recalculate d_{rN} and:

$$\hat{d}_{rN} = d_{rN} - \alpha_N \tag{14–16}$$

Now suppose for row r, $\beta_r = s$. Hence, we next recalculate d_{sr} in row s as:

$$\hat{d}_{sr} = d_{sr} - \alpha_N \tag{14–17}$$

If $\beta_s = t$, then we next change d_{ts}, etc. Next we recalculate d_{vu}; i.e., for each element d_{uv} that was changed by subtracting α_N, we recalculate the corresponding d_{vu} by adding α_N. For example, with respect to (14–16) and (14–17) we have:

$$\begin{aligned} \hat{d}_{Nr} &= d_{Nr} + \alpha_N \\ \hat{d}_{rs} &= d_{rs} + \alpha_N \\ &\text{etc.} \end{aligned} \tag{14–18}$$

All the other d_{ij} are unchanged in the new tableau. Once the new tableau is constructed we repeat the entire process. When we reach a tableau in which α_N cannot be given a value we have found the maximal flow.

In order to find the total flow for any tableau we need merely add all the entries in column 1 or all the entries in row N. This follows since $c_{i1} = c_{Nj} = 0$ since all flow is away from node 1 and all flow is into node N, by definition. Therefore, we have that since:

$$d_{ij} = c_{ij} - x_{ij} + x_{ji}$$

then

$$d_{i1} = 0 - 0 + x_{1i}$$

and

$$d_{Nj} = 0 - 0 + x_{jN}$$

Therefore:

$$\text{Total flow} = \sum_{i=1}^{N} d_{i1} = \sum_{j=1}^{N} d_{Nj} \tag{14–19}$$

14-4 MAXIMAL FLOW COMPUTATIONAL EXAMPLE

Consider the problem for the network depicted in Figure 14–4. The entries shown on the oriented arcs are the c_{ij} in the direction of the arrows. For example, $c_{36} = 5$ and $c_{63} = 0$. Since all initial flows, $x_{ij} = 0$, it can be

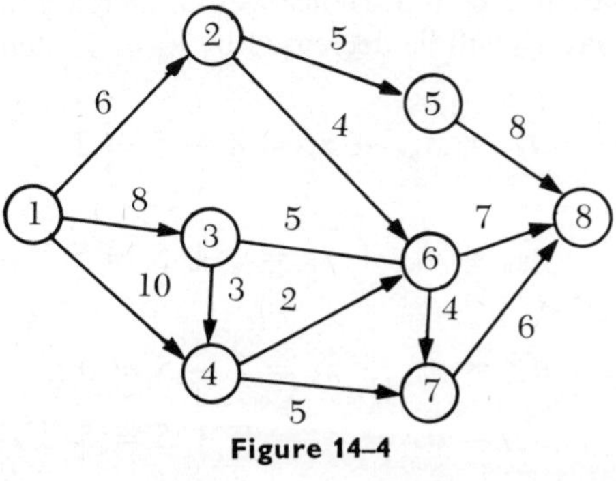

Figure 14–4

Nodes	1	2	3	4	5	6	7	8	Labels		Step
									α	β	
1	—	6	8	10	0	0	0	0			
2	0	—	0	0	5	4	0	0	6	1	1
3	0	0	—	3	0	5	0	0	8	1	1
4	0	0	0	—	0	2	5	0	10	1	1
5	0	0	0	0	—	0	0	8	5	2	2
6	0	0	0	0	0	—	4	7	4	2	2
7	0	0	0	0	0	0	—	6	5	4	2
8	0	0	0	0	0	0	0	—	5	5	3

Flow = 0

Figure 14–5 Tableau 1.

seen from (14–5) that, initially, $d_{ij} = c_{ij}$. Hence the initial tableau is as shown in Figure 14–5. In step 1 we can label α_2, α_3, α_4 with 6, 8, 10, respectively. Then, we examine row 2 and find that we can label nodes 5 and 6. We calculate:

$$\alpha_5 = \min(d_{25} = 5, \alpha_2 = 6) = 5, \qquad \beta_5 = 2$$

$$\alpha_6 = \min(d_{26} = 4, \alpha_2 = 6) = 4, \qquad \beta_6 = 2$$

Next, we examine row 3 and find that we can label no new rows. From row 4 we can label row 7

$$\alpha_7 = \min(5, 10) = 5, \qquad \beta_7 = 4$$

Since we have not yet labeled the sink, we next examine row 5 and find that

$$\alpha_8 = \min(8, 5) = 5, \qquad \beta_8 = 5$$

We have labeled the sink, $\alpha_8 = 5$. Hence we can increase the flow by 5. Since $\beta_8 = 5$ we know that d_{58} will be decreased by $\alpha_8 = 5$. Hence,

$$\hat{d}_{58} = d_{58} - \alpha_8 = 8 - 5 = 3$$

and

$$\hat{d}_{85} = d_{85} + \alpha_8 = 0 + 5 = 5$$

Since $\beta_5 = 2$

$$\hat{d}_{25} = d_{25} - \alpha_8 = 5 - 5 = 0$$

$$\hat{d}_{52} = d_{52} + \alpha_8 = 0 + 5 = 5$$

Nodes	1	2	3	4	5	6	7	8	Labels		Step
1	—	1	8	10	0	0	0	0	α	β	
2	5	—	0	0	0	4	0	0	1	1	1
3	0	0	—	3	0	5	0	0	8	1	1
4	0	0	0	—	0	2	5	0	10	1	1
5	0	5	0	0	—	0	0	3			
6	0	0	0	0	0	—	4	7	1	2	2
7	0	0	0	0	0	0	—	6	5	2	2
8	0	0	0	0	5	0	0	—	1	6	3

Flow = 5

Figure 14–6 Tableau 2.

Since $\beta_2 = 1$

$$\hat{d}_{12} = d_{12} - \alpha_8 = 6 - 5 = 1$$
$$\hat{d}_{21} = d_{21} + \alpha_8 = 0 + 5 = 5$$

All other values of excess capacity d_{ij} are unchanged. The new tableau is shown in Figure 14–6. We now repeat the labeling in Tableau 2 with the results shown. We see we have labeled the sink and that $\alpha_8 = 1$. We note further that

$$\beta_8 = 6, \qquad \beta_6 = 2, \qquad \beta_2 = 1$$

$$\hat{d}_{68} = 7 - 1 = 6$$

Therefore:

$$\hat{d}_{86} = 0 + 1 = 1$$
$$\hat{d}_{26} = 4 - 1 = 3$$
$$\hat{d}_{62} = 0 + 1 = 1$$
$$\hat{d}_{12} = 1 - 1 = 0$$
$$\hat{d}_{21} = 5 + 1 = 6$$

Tableau 3 is given in Figure 14–7. We again carry out the labeling process as shown. From Tableau 3 we see that $\alpha_8 = 5$. We also have that

$$\beta_8 = 6, \qquad \beta_6 = 3, \qquad \beta_3 = 1$$

Nodes	1	2	3	4	5	6	7	8	Labels		Step
1	—	0	8	10	0	0	0	0	α	β	
2	6	—	0	0	0	3	0	0	1	6	3
3	0	0	—	3	0	5	0	0	8	1	1
4	0	0	0	—	0	2	5	0	10	1	1
5	0	5	0	0	—	0	0	3			
6	0	1	0	0	0	—	4	6	5	3	2
7	0	0	0	0	0	0	—	6	5	4	2
8	0	0	0	0	5	1	0	—	5	6	3

Flow = 6

Figure 14–7 Tableau 3.

Therefore:

$$\hat{d}_{68} = 6 - 5 = 1$$
$$\hat{d}_{86} = 1 + 5 = 6$$
$$\hat{d}_{36} = 5 - 5 = 0$$
$$\hat{d}_{63} = 0 + 5 = 5$$
$$\hat{d}_{13} = 8 - 5 = 3$$
$$\hat{d}_{31} = 0 + 5 = 5$$

The new tableau is given in Figure 14–8 with the labeling as shown.

Nodes	1	2	3	4	5	6	7	8	Labels		Step
1	—	0	3	10	0	0	0	0	α	β	
2	6	—	0	0	0	3	0	0			
3	5	0	—	3	0	0	0	0	3	1	1
4	0	0	0	—	0	2	5	0	10	1	1
5	0	5	0	0	—	0	0	3			
6	0	1	5	0	0	—	4	1	2	4	2
7	0	0	0	0	0	0	—	6	5	4	2
8	0	0	0	0	5	6	0	—	1	6	3

Flow = 11

Figure 14–8 Tableau 4.

Nodes	1	2	3	4	5	6	7	8	Labels		Step
1	—	0	3	9	0	0	0	0	α	β	
2	6	—	0	0	0	3	0	0	1	6	3
3	5	0	—	3	0	0	0	0	3	1	1
4	1	0	0	—	0	1	5	0	9	1	1
5	0	5	0	0	—	0	0	3			
6	0	1	5	1	0	—	4	0	1	4	2
7	0	0	0	0	0	0	—	6	5	4	2
8	0	0	0	0	5	7	0	—	5	7	3

Flow = 12

Figure 14–9 Tableau 5.

Again in Tableau 4 we have labeled the sink with $\alpha_8 = 1$ and $\beta_8 = 6$, $\beta_6 = 4$, $\beta_4 = 1$. Therefore,

$$\hat{d}_{68} = 1 - 1 = 0$$
$$\hat{d}_{86} = 6 + 1 = 7$$
$$\hat{d}_{46} = 2 - 1 = 1$$
$$\hat{d}_{64} = 0 + 1 = 1$$
$$\hat{d}_{14} = 10 - 1 = 9$$
$$\hat{d}_{41} = 0 + 1 = 1$$

Tableau 5 is shown in Figure 14–9 and the labeling that results. In Tableau 5, $\alpha_8 = 5$, $\beta_8 = 7$, $\beta_7 = 4$, $\beta_4 = 1$. Therefore,

$$\hat{d}_{78} = 6 - 5 = 1$$
$$\hat{d}_{87} = 0 + 5 = 5$$
$$\hat{d}_{47} = 5 - 5 = 0$$
$$\hat{d}_{74} = 0 + 5 = 5$$
$$\hat{d}_{14} = 9 - 5 = 4$$
$$\hat{d}_{41} = 1 + 5 = 6$$

The new tableau is shown in Figure 14–10. The labeling process is again carried out.

Nodes	1	2	3	4	5	6	7	8	Labels		Step
1	—	0	3	4	0	0	0	0	α	β	
2	6	—	0	0	0	3	0	0	1	6	3
3	5	0	—	3	0	0	0	0	3	1	1
4	6	0	0	—	0	1	0	0	4	1	1
5	0	5	0	0	—	0	0	3			
6	0	1	5	1	0	—	4	0	1	4	2
7	0	0	0	5	0	0	—	1	1	6	3
8	0	0	0	0	5	7	5	—	1	7	4

Flow = 17

Figure 14–10 Tableau 6.

In Tableau 6, $\alpha_8 = 1$, $\beta_8 = 7$, $\beta_7 = 6$, $\beta_6 = 4$, $\beta_4 = 1$. Therefore,

$$\begin{aligned} \hat{d}_{78} &= 0, & \hat{d}_{87} &= 6 \\ \hat{d}_{67} &= 3, & \hat{d}_{76} &= 1 \\ \hat{d}_{46} &= 0, & \hat{d}_{64} &= 2 \\ \hat{d}_{14} &= 3, & \hat{d}_{41} &= 7 \end{aligned}$$

Tableau 7 is given in Figure 14–11. Once more, we carry out the labeling process.

Nodes	1	2	3	4	5	6	7	8	Labels		Step
1	—	0	3	3	0	0	0	0	α	β	
2	6	—	0	0	0	3	0	0			
3	5	0	—	3	0	0	0	0	3	1	1
4	7	0	0	—	0	0	0	0	3	1	1
5	0	5	0	0	—	0	0	3			
6	0	1	5	2	0	—	3	0			
7	0	0	0	5	0	1	—	0			
8	0	0	0	0	5	7	6	—			

Flow = 18

Figure 14–11 Tableau 7.

In Tableau 7, we can only label rows 3 and 4 and from rows 3 and 4 we can label no further. We cannot reach the sink. Hence, we have found the maximal flow.

The flow in each arc (i, j) of Figure 14–4 is obviously given by

$$x_{ij} = \max(c_{ij} - d_{ij}, 0) \tag{14–20}$$

For example

$$x_{13} = \max(8 - 3, 0) = 5$$

$$x_{34} = \max(3 - 3, 0) = 0$$

14–5 THE TRANSPORTATION PROBLEM— A PRIMAL-DUAL ALGORITHM

In Chapter 9 we discussed the primal-dual algorithm for solving linear programming problems. Since a transportation problem is a linear programming problem, it is obvious that we could use this algorithm for solving transportation problems. However, since the transportation problem has a fixed and known structure for a problem of any size, it turns out that the primal-dual algorithm can be simplified when it is adapted to the transportation problem. It will be recalled from Chapter 9 that only vectors that satisfied a complementary slackness condition could enter the basis. It turns out that this portion of the calculation can be handled by the application of our maximal flow algorithm as part of the general primal-dual scheme. The algorithm described here is essentially that of Ford and Fulkerson.[1]

Let us consider that our primal transportation problem is given as

$$\text{Min } z = \sum_{i=1}^{m} \sum_{j=1}^{n} c_{ij}x_{ij}$$

$$\begin{aligned} \sum_{j=1}^{n} x_{ij} &= a_i \qquad i = 1, 2, \ldots, m \\ \sum_{i=1}^{m} x_{ij} &= b_j \qquad j = 1, 2, \ldots, n \\ x_{ij} &\geq 0 \qquad i = 1, 2, \ldots, m \\ & \qquad\quad\; j = 1, 2, \ldots, n \end{aligned} \tag{14–21}$$

For this primal it is a relatively simple matter to find an initial solution to the dual without the complications we encountered in Chapter 9. The dual of

(14–21) can be written as

$$\begin{aligned} \text{Max } w &= \sum_{i=1}^{m} a_i u_i + \sum_{j=1}^{n} b_j v_j \\ u_i + v_j \le c_{ij} \qquad & i = 1, 2, \ldots, m \\ & j = 1, 2, \ldots, n \end{aligned} \tag{14–22}$$

It is easy to see that a solution that satisfies the dual constraints of (14–22) can be chosen as

$$\begin{aligned} u_i &= \min_j c_{ij} \qquad i = 1, 2, \ldots, m \\ v_j &= \min_i (c_{ij} - u_i) \qquad j = 1, 2, \ldots, n \end{aligned} \tag{14–23}$$

In the primal-dual algorithm we added a set of artificial variables, and the initial basis consisted entirely of the corresponding artificial vectors. The primal-dual algorithm drives all these vectors out of the basis while maintaining complementary slackness and, when the Phase I objective function is zero, we obtain the optimal solution to the original primal problem. We shall do the same here except that it is not necessary to explicitly add the artificial variables. This is a consequence of the simple structure of transportation problems. We see this as follows. If any artificial variables are positive, then, since the primal constraints of (14–21) are equalities, these constraints cannot be satisfied exactly by the legitimate variables x_{ij} that are basic. However, since the artificial variables (see Chapter 9) are non-negative, the x_{ij} must satisfy constraints of the form

$$\begin{aligned} \sum_{j=1}^{n} x_{ij} &\le a_i \qquad i = 1, 2, \ldots, m \\ \sum_{i=1}^{m} x_{ij} &\le b_j \qquad j = 1, 2, \ldots, n \end{aligned} \tag{14–24}$$

It will be recalled that the objective function of (9–50) in Chapter 9 is to maximize the negative sum of the artificial variables or, equivalently, to minimize the sum of the artificial variables. From (14–24) it is clear that the sum of the artificial variables is equal to

$$\sum_{i=1}^{m} a_i + \sum_{j=1}^{n} b_j - 2 \sum_{i=1}^{m} \sum_{j=1}^{n} x_{ij} \tag{14–25}$$

Hence, since the first two terms are constant, minimizing the sum of the artificial variables is equivalent to

$$\text{Max } z_a = \sum_{i=1}^{m} \sum_{j=1}^{n} x_{ij} \tag{14–26}$$

We recall from Chapter 9 that not all x_{ij} are allowed to enter the primal basis. Using a notation similar to Chapter 9, we define

$$V = \{v_{sij} \mid v_{sij} = 0\} \tag{14–27}$$

where v_{sij} is the slack variable of the $(i + j)^{\text{th}}$ dual constraint of (14–22). Then the set of primal variables x_{ij} which is allowed to enter the basis is given by:

$$W = \{x_{ij} \mid v_{sij} \in V\} \tag{14–28}$$

We see, therefore, that in this fashion we maintain complementary slackness, since $x_{ij} = 0$ if $v_{sij} \notin V$; i.e., the dual constraint is a strict inequality.

We may now note that the problem:

$$\begin{aligned} \text{Max } z_a &= \sum_{i=1}^{m} \sum_{j=1}^{n} x_{ij} \\ \sum_{j=1}^{n} x_{ij} &\le a_i \qquad i = 1, 2, \ldots, m \\ \sum_{i=1}^{m} x_{ij} &\le b_j \qquad j = 1, 2, \ldots, n \\ x_{ij} &\ge 0 \qquad i = 1, 2, \ldots, n \\ & \qquad\qquad j = 1, 2, \ldots, n \end{aligned} \tag{14–29}$$

can be solved as a network flow problem. To see this we need to construct an appropriate network with a single source and sink. Such a network is shown in Figure 14–12. All arcs (i, j) are directed arcs and either have no capacity restriction or have a capacity of zero if x_{ij} may not be positive. No flows

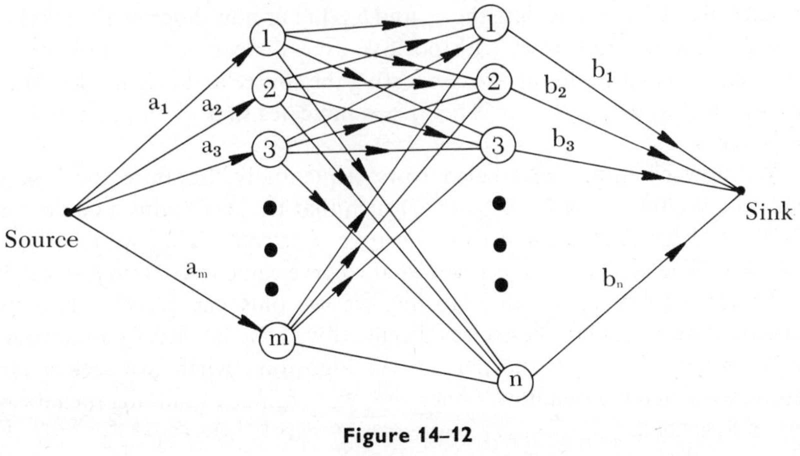

Figure 14–12

from j to i (destinations to origins) are allowed. When we have solved the maximal flow problem for this network, we have indeed maximized z_a in (14–29).

Since we are dealing with a transportation problem it is more convenient to use the tableau form of Chapter 10 with some additional entries, as well as rows and columns for our labeling process. Excess capacities are simply handled in this problem. Recall from equations (14–5) that since the arc (i, j) has no capacity restriction if it can enter the basis, there is no need to record d_{ij}. In this case, d_{ji} will equal x_{ij} so we need merely record x_{ij}, which we would do in any case in a transportation tableau.

Unlike the labeling process we described in the preceding sections of this chapter, we have *two* sets of labels to contend with. One set is for the origins and the other is for the destinations. Hence, we will add two rows and two columns to the transportation tableau for this purpose. We will call the origin nodes α_i, β_i and the destination nodes γ_j, δ_j.

It is also necessary to record the excess capacity of the arcs leading from the source to the origin nodes and from the destination nodes to the sink. For this purpose we define

$$
\begin{aligned}
a_{si} &= a_i - \sum_{j=1}^{n} x_{ij} \qquad i = 1, 2, \ldots, m \\
b_{sj} &= b_j - \sum_{i=1}^{m} x_{ij} \qquad j = 1, 2, \ldots, n
\end{aligned}
\tag{14–30}
$$

where a_{si} and b_{sj} are the source and sink excess capacities, respectively. We add another row and column for these quantities. The tableau format is given in Figure 14–13.

Let us now consider how a typical step of this algorithm is carried out. Suppose those cells of the transportation tableau which are allowed to have positive x_{ij}, i.e., $x_{ij} \in W$, have been noted by placing circles in them. We assume that the network has some flow and the x_{ij} are indicated in their allowed cells. We also calculate the a_{si} and b_{sj}. Let us now describe the labeling process. We seek those rows with positive excess capacities a_{si}. This means there is excess capacity in the arc connecting the source and origin i. For these rows we label $\alpha_i = a_{si}$, $\beta_i = s$. The $\beta_i = s$ indicates that we have come from the source.

We now examine the labeled rows sequentially. Suppose the first of these is i_1. We then look for cells (i_1, j) such that (i_1, j) contains a circle, i.e., $x_{ij} \in W$, and hence is allowed to be positive. For each such j we label $\gamma_j = a_{si}$, $\delta_j = i_1$. The label δ_j tells us from what row we came to column j. It will be noted that in carrying out this labeling we use only circled cells. It is not necessary that $x_{ij} > 0$ in these circled cells. If we have labeled a column such that $b_{sj} > 0$ we move to the part of the algorithm where we seek a new solution with an increased flow. Otherwise, we continue examining the labeled rows as described.

	1	2	...	n				
$u_i \backslash v_j$	v_1	v_2	...	v_n	a_i	a_{si}	α_i	β_i
u_1	c_{11} x_{11}	c_{12} x_{12}	...	c_{1n} x_{1n}	a_1	a_{s1}	α_1	β_1
u_2	c_{21} x_{21}	c_{22} x_{22}	...	c_{2n} x_{2n}	a_2	a_{s2}	α_2	β_2
⋮	⋮	⋮	...	⋮	⋮	⋮	⋮	⋮
u_m	c_{m1} x_{m1}	c_{m2} x_{m2}	...	c_{mn} x_{mn}	a_m	a_{sm}	α_m	β_m
b_j	b_1	b_2	...	b_n		$\Sigma\, a_{si}$		
b_{sj}	b_{s1}	b_{s2}	...	b_{sn}				
γ_j	γ_1	γ_2	...	γ_n				
δ_j	δ_1	δ_2	...	δ_n				

Figure 14–13 Primal-dual transportation problem tableau.

If no column with $b_{sj} > 0$ has been labeled, we examine the labeled columns sequentially. Now we look for cells (i, j_1), if we are in column j_1 for which $x_{ij_1} > 0$. If there are some rows i not yet labeled, then for each such row i, we label

$$\alpha_i = \min(\gamma_{j_1}, x_{ij})$$

$$\beta_i = j_1$$

Again β_i tells us we came from column j_1. The reason x_{ij} must be positive in this step is that, since we are moving in the wrong direction, no *net* flow is permitted in this direction.

We continue this process, alternately labeling rows and columns until, since this is a finite process, we reach one of the three following conditions:

1. No column with $b_{sj} > 0$ has been labeled and

$$\sum_{i=1}^{m} a_{si} = 0$$

This implies that

$$\sum_{i=1}^{m} \sum_{j=1}^{n} x_{ij} = \sum_{i=1}^{m} a_i$$

Hence, we have a feasible and therefore optimal solution to the problem and we are done.

2. No column with $b_{sj} > 0$ has been labeled and $\sum_{i=1}^{m} a_{si} > 0$. This means we need a new solution to the dual so that at least one new x_{ij} may become positive. We describe below how this is done.

3. A column with $b_{sj} > 0$ has been labeled. This means the flow in the network can be increased by a certain amount. Suppose column l was such that $b_{sl} > 0$. Then the flow can be increased by:

$$\Delta = \min (b_{sl}, \gamma_l) \tag{14–31}$$

If $\delta_l = k$, we increase x_k by Δ, i.e.,

$$\hat{x}_{kl} = x_{kl} + \Delta$$

If $\beta_k = p$, we decrease x_{kp} by Δ, i.e.,

$$\hat{x}_{kp} = x_{kp} - \Delta$$

Now, if $\delta_p = s$, we increase x_{sp} by Δ, etc., until we come to a label $\beta_t = s$, i.e., we are back at the source. With these new values of x_{ij} we construct a new flow problem to label again. We repeat this process until we reach conditions 1 or 2.

If we reach condition 2, we know we need a new solution to the dual of our transportation problem. Let L_R denote the set of labeled rows and L_C denote the set of labeled columns that existed in the tableau when we reached condition 2. We now define a quantity

$$\rho = \min_{\substack{i \in L_R \\ j \notin L_C}} (c_{ij} - u_i - v_j) > 0 \tag{14–32}$$

where u_i, v_j are the current values of the dual variables (available from the tableau). ρ as defined by (14–32) is positive because $j \notin L_C$, which means j is the index of an unlabeled column and hence cell $(i, \ j)$ does not contain a circle, so that the dual constraint must be:

$$u_i + v_j < c_{ij} \tag{14–33}$$

We now construct a new solution to the dual as follows:

$$\begin{aligned} \hat{u}_i &= u_i + \rho & i &\in L_R \\ \hat{u}_i &= u_i & i &\notin L_R \\ \hat{v}_j &= v_j - \rho & j &\in L_C \\ \hat{v}_j &= v_j & j &\notin L_C \end{aligned} \tag{14–34}$$

To prove that (14–34) is a new solution to the dual constraints we note the following cases:

1. $$i \notin L_R, \quad j \notin L_C$$

For this case $\hat{u}_i = u_i$ and $\hat{v}_j = v_j$. Then, by the previous solution

$$\hat{u}_i + \hat{v}_j \leq c_{ij}$$

since

$$u_i + v_j \leq c_{ij}$$

2. $$i \in L_R, \quad j \in L_C$$

For this case, $\hat{u}_i = u_i + \rho$ and $\hat{v}_j = v_j - \rho$. Then we have

$$\hat{u}_i + \hat{v}_j = u_i + \rho + v_j - \rho \leq c_{ij}$$

3. $$i \notin L_R, \quad j \in L_C$$

For this case, $\hat{u}_i = u_i$ and $\hat{v}_j = v_j - \rho$. Then we have

$$\hat{u}_i + \hat{v}_j = u_i + v_j - \rho \leq c_{ij}$$

4. $$i \in L_R, \quad j \notin L_C$$

Here we have $\hat{u}_i = u_i + \rho$ and $\hat{v}_j = v_j$. Because of (14–32) and how ρ is chosen we see that

$$\hat{u}_i + \hat{v}_j = u_i + v_j + \rho \leq c_{ij}$$

Hence, we have proved that (14–34) is a new solution to the dual.

Because of the way ρ is chosen in (14–32) it is clear that $\hat{u}_i + \hat{v}_j = c_{ij}$ for at least one $v_{sij} \in V$; i.e., at least one new x_{ij} can be positive. We can use the previous tableau as the new starting solution because all previous x_{ij} which were allowed to be positive are still allowed to be positive. We prove this as follows by looking at all cases:

1. $$x_{ij} > 0, \quad i \notin L_R, \quad j \notin L_C$$

$$\hat{u}_i + \hat{v}_j = u_i + v_j = c_{ij}$$

since $v_{sij} = 0$ and $\hat{u}_i = u_i$, $\hat{v}_j = v_j$

2. $$x_{ij} > 0, \qquad i \in L_R, \qquad j \in L_C$$
$$\hat{u}_i + \hat{v}_j = u_i + \rho + v_j - \rho = c_{ij}$$

since $v_{sij} = 0$

3. $$i \in L_R, \qquad j \in L_C$$

There can be no $x_{ij} > 0$ under these conditions since it contradicts our assumptions.

4. $$i \notin L_R, \qquad j \in L_C$$

There can be no $x_{ij} > 0$, otherwise we would label row i. Hence, our assertion is proven.

To show that we have a new solution to the dual which improves (is greater than) the dual objective function w, we note that

$$\begin{aligned} \hat{w} &= \sum_{i \in L_R} a_i(u_i + \rho) + \sum_{j \in L_C} b_j(v_j - \rho) + \sum_{i \notin L_R} a_i u_i + \sum_{j \notin L_C} b_j v_j \\ &= w + \rho\left(\sum_{i \in L_R} a_i - \sum_{j \in L_C} b_j\right) \end{aligned} \tag{14–35}$$

It will be recalled that for $a_{si} > 0$, $i \in L_R$ and that for $j \in L_C$ we have that:

$$\sum_{i \in L_R} x_{ij} = b_j \tag{14–36}$$

If (14–36) were not the case we could increase the flow. For $j \in L_C$, either there is no circle in cell (i, j) or if there is, $x_{ij} = 0$ if $i \notin L_R$. This must be true or row i could be labeled. Therefore,

$$\sum_{j \in L_C} \sum_{i \in L_R} x_{ij} = \sum_{j \in L_C} b_j \tag{14–37}$$

Now we know $a_{si} > 0$ for at least one $i \in L_R$. Therefore, we have

$$\sum_{i \in L_R} a_i > \sum_{j \in L_C} \sum_{i \in L_R} x_{ij} \tag{14–38}$$

From (14–37) and (14–38) we see that

$$\sum_{i \in L_R} a_i - \sum_{j \in L_C} b_j > 0 \tag{14–39}$$

From (14–39) and (14–35) we see that

$$\hat{w} > w$$

Therefore we have increased the dual objective function as we claimed.

The new solution to the dual gives us a new set of x_{ij} which are allowed to be positive. We then solve the new primal problem using the labeling algorithm. We continue the entire iterative process until we end in condition 1 with the optimal solution.

14–6 THE PRIMAL-DUAL TRANSPORTATION ALGORITHM—AN EXAMPLE

Consider the following problem:

		Destinations						
		1	2	3	4	5	6	a_i
Origins	1	2	1	3	1	5	2	10
	2	3	2	4	4	2	3	11
	3	4	3	6	5	1	7	21
	4	1	4	1	2	1	1	17
	b_j	12	15	8	7	12	5	

Figure 14–14 Data for example.

The entries in the matrix of Figure 14–14 are the c_{ij} of the objective function. The availabilities a_i and requirements b_j are also given. The initial tableau is shown in Figure 14–15. Using equations (14–23) we calculate the u_i, which are the minimum costs in each row. Whenever the minimum occurs we place a circle, since for that cell x_{ij} can be positive. We next calculate the v_j. For each column containing a circle, it is clear that $v_j = 0$. In our tableau $v_j = 0$ for all j.

In order to find an initial flow, we use a variant of the "Northwest Corner Rule," involving only the circled cells, since only those x_{ij} are allowed to be positive. In row 1 we set $x_{12} = \min(10, 15) = 10$. Since 5 units of b_2 are still available we calculate $x_{22} = \min(5, 11) = 5$. This leaves 6 units available from a_2, so we calculate $x_{25} = \min(6, 12) = 6$. We now have 6 units left from b_5 so we calculate $x_{35} = \min(6, 21) = 6$. There are a number of ways we can satisfy a_4. We choose $x_{41} = 12$ and $x_{46} = 5$. We now calculate the a_{si} and b_{sj} by difference. For example,

$$a_{s3} = a_3 - x_{35} = 21 - 6 = 15$$
$$b_{s1} = b_1 - x_{41} = 12 - 12 = 0$$

	$u_i \backslash v_j$	1: 0	2: 0	3: 0	4: 0	5: 0	6: 0	a_i	a_{si}	α_i	β_i
1	1	2	1 (10)	3	1 (○)	5	2	10	0	6	2
2	2	3	2 (5)	4	4	2 (6)	3	11	0	6	5
3	1	4	3	6	5	1 (6)	7	21	15	15	S
4	1	1 (12)	4	1 (○)	2	1 (○)	1 (5)	17	0		
	b_j	12	15	8	7	12	5				
	b_{sj}	0	0	8	7	0	0		15		
	γ_j		6		6	15					
	δ_j		2		1	3					

Figure 14–15 Tableau 1.

Since we now have an initial flow we begin the labeling process. Only one row has $a_{si} > 0$, namely row 3. Therefore,

$$\alpha_3 = a_{s3} = 15, \qquad \beta_3 = s$$

In row 3 there is one circled cell (3, 5). Therefore column 5 is the only column to be labeled as:

$$\gamma_5 = 15, \qquad \delta_5 = 3$$

In column 5, row 2 has $x_{25} > 0$. Therefore, we label row 2 as:

$$\alpha_2 = \min(\gamma_5, x_{25}) = \min(15, 6) = 6, \qquad \beta_2 = 5$$

In row 2, we see that one can label column 2 from:

$$\gamma_2 = \alpha_2 = 6, \qquad \delta_2 = 2$$

From column 2 we can label row 1 to obtain

$$\alpha_1 = \min(6, 10) = 6, \qquad \beta_2 = 2$$

Row 1 contains a circle in column 4. Hence we label column 4 as

$$\gamma_4 = \alpha_1 = 6, \qquad \delta_4 = 1$$

		1	2	3	4	5	6				
	u_i \ v_j	0	0	0	0	0	0	a_i	a_{si}	α_i	β_i
1	1	2	1 (4)	3	1 (6)	5	2	10	0		
2	2	3	2 (11)	4	4	2 ○	3	11	0		
3	1	4	3	6	5	1 (12)	7	21	9	9	S
4	1	1 (12)	4	1 ○	2	1 ○	1 (5)	17	0		
	b_j	12	15	8	7	12	5				
	b_{sj}	0	0	8	1	0	0		9		
	γ_j					9					
	δ_j					3					

Figure 14–16 Tableau 2.

Since we have labeled a column with $b_{sj} > 0$, i.e., $b_{s4} = 7$, we must modify the flow. We calculate Δ as

$$\Delta = \min(b_{s4}, \gamma_4) = \min(7, 6) = 6$$

Therefore, since

$$\begin{aligned} \delta_4 &= 1, & \hat{x}_{14} &= 0 + 6 = 6 \\ \beta_1 &= 2, & \hat{x}_{12} &= 10 - 6 = 4 \\ \delta_2 &= 2, & \hat{x}_{22} &= 5 + 6 = 11 \\ \beta_2 &= 5, & \hat{x}_{25} &= 6 - 6 = 0 \\ \delta_5 &= 3, & \hat{x}_{35} &= 6 + 6 = 12 \\ \beta_3 &= S \end{aligned}$$

Figure 14–16 contains Tableau 2 with the new flows x_{ij}. We now relabel Tableau 2 with the results shown. After labeling column 5 we see that there are no other circles with $x_{ij} > 0$, and $\sum_i a_{si} > 0$. Hence we require a new solution to the dual so that at least one new x_{ij} can be positive. We use equations (14–32) and (14–34) to construct this solution. We see that $L_R = \{3\}$, $L_C = \{5\}$. Therefore,

$$\rho = \min_{\substack{i=3 \\ j \neq 5}} (c_{3j} - u_3 - v_j)$$

$$= \min (4 - 1 - 0, 3 - 1 - 0, 6 - 1 - 0, 5 - 1 - 0, 7 - 1 - 0)$$

$$= \min (3, 2, 5, 4, 6) = 2$$

Therefore,

$$\hat{u}_3 = u_3 + \rho = 1 + 2 = 3$$

$$\hat{v}_5 = v_5 - \rho = 0 - 2 = -2$$

The new tableau is given in Figure 14–17. All circled cells are those for which $u_i + v_j = c_{ij}$ and, hence, x_{ij} is allowed to be positive. The new set of x_{ij} is calculated as we did initially. We again carry out the labeling procedure. We see that we have labeled column 4 and $b_{s4} = 7$. Hence, we can increase the flow

$$\Delta = \min (7, 6) = 6$$

$$\delta_4 = 1, \qquad \hat{x}_{14} = 0 + 6 = 6$$

$$\beta_1 = 2, \qquad \hat{x}_{12} = 10 - 6 = 4$$

$$\delta_2 = 2, \qquad \hat{x}_{22} = 5 + 6 = 11$$

$$\beta_2 = S$$

		1	2	3	4	5	6				
	u_i \ v_j	0	0	0	0	−2	0	a_i	a_{si}	α_i	β_i
1	1	2	1 (10)	3	1 ○	5	2	10	0	6	2
2	2	3	2 (5)	4	4	2	3	11	6	6	S
3	3	4	3 ○	6	5	1 (12)	7	21	9	9	S
4	1	1 (12)	4	1 ○	2	1	1 (5)	17	0		
	b_j	12	15	8	7	12	5				
	b_{sj}	0	0	8	7	0	0		15		
	γ_j		6		6	9					
	δ_j		2		1	3					

Figure 14–17 Tableau 3.

	$u_i \backslash v_j$	1: 0	2: 0	3: 0	4: 0	5: −2	6: 0	a_i	a_{si}	α_i	β_i
1	1	2	1 (4)	3	1 (6)	5	2	10	0	4	2
2	2	3	2 (11)	4	4	2	3	11	0	9	2
3	3	4	3 ( )	6	5	1 (12)	7	21	9	9	S
4	1	1 (12)	4	1 ( )	2	1	1 (5)	17	0		
	b_j	12	15	8	7	12	5				
	b_{sj}	0	0	8	1	0	0		9		
	γ_j		9		4	9					
	δ_j		3		1	3					

Figure 14–18 Tableau 4.

The new tableau is given in Figure 14–18. We again carry out the labeling procedure as shown.

In Tableau 4, we have labeled column 4 and $b_{s4} = 1$. Hence we can again increase the flow.

$$\Delta = \min(1, 4) = 1$$

$$\delta_4 = 1, \qquad \hat{x}_{14} = 6 + 1 = 7$$
$$\beta_1 = 2, \qquad \hat{x}_{12} = 4 - 1 = 3$$
$$\delta_2 = 3, \qquad \hat{x}_{32} = 0 + 1 = 1$$
$$\beta_3 = S$$

The new tableau is given in Figure 14–19. We again carry out the entire labeling procedure.

In Tableau 5, Figure 14–19, we have not been able to label a column with $b_{sj} > 0$ and $\sum_i a_{si} > 0$. Hence we must find a new solution to the dual. We note that:

$$L_R = \{1, 2, 3\} \quad \text{and} \quad L_C = \{2, 4, 5\}$$
$$\rho = \min_{\substack{i=1,2,3 \\ j=1,3,6}} (c_{ij} - u_i - v_j) = 1$$

$$\hat{u}_1 = 1 + 1 = 2 \qquad \hat{v}_2 = 0 - 1 = -1$$
$$\hat{u}_2 = 2 + 1 = 3 \qquad \hat{v}_4 = 0 - 1 = -1$$
$$\hat{u}_3 = 3 + 1 = 4 \qquad \hat{v}_5 = -2 - 1 = -3$$

		1	2	3	4	5	6				
	u_i \ v_j	0	0	0	0	−2	0	a_i	a_{si}	α_i	β_i
1	1	2	1 (3)	3	1 (7)	5	2	10	0	3	2
2	2	3	2 (11)	4	4	2	3	11	0	8	2
3	3	4	3 (1)	6	5	1 (12)	7	21	8	8	S
4	1	1 (12)	4	1 ○	2	1	1 (5)	17	0		
	b_j	12	15	8	7	12	5				
	b_{sj}	0	0	8	0	0	0		8		
	γ_j		8		3	8					
	δ_j		3		1	3					

Figure 14–19 Tableau 5.

		1	2	3	4	5	6				
	u_i \ v_j	0	−1	0	−1	−3	0	a_i	a_{si}	α_i	β_i
1	2	2 (10)	1 ○	3	1 ○	5	2 ○	10	0	3	1
2	3	3 (2)	2 (9)	4	4	2	3 ○	11	0	2	1
3	4	4	3 (6)	6	5	1 (12)	7	21	3	3	S
4	1	1	4	1 (8)	2	1	1 (5)	17	4	4	S
	b_j	12	15	8	7	12	5				
	b_{sj}	0	0	0	7	0	0		7		
	γ_j	3	3	4	3	3	4				
	δ_j	3	3	4	1	3	4				

Figure 14–20 Tableau 6.

	$u_i \backslash v_j$	1: 0	2: −1	3: 0	4: −1	5: −3	6: 0	a_i	a_{si}	α_i	β_i
1	2	2 (7)	1 ○	3	1 (3)	5	2 ○	10	0	4	1
2	3	3 (2)	2 (9)	4	4	2	3 ○	11	0	2	1
3	4	4 (3)	3 (6)	6	5	1 (12)	7	21	0	3	1
4	1	1 ○	4	1 (8)	2	1	1 (5)	17	4	4	S
	b_j	12	15	8	7	12	5				
	b_{sj}	0	0	0	4	0	0		4		
	γ_j	4	4	4	4		4				
	δ_j	4	1	4	1		4				

Figure 14–21 Tableau 7.

	$u_i \backslash v_j$	1: 0	2: −1	3: 0	4: −1	5: −3	6: 0	a_i	a_{si}	α_i	β_i
1	2	2 (3)	1 ○	3	1 (7)	5	2 ○	10	0		
2	3	3 (2)	2 (9)	4	4	2	3 ○	11	0		
3	4	4 (3)	3 (6)	6	5	1 (12)	7	21	0		
4	1	1 (4)	4	1 (8)	2	1	1 (5)	17	0		
	b_j	12	15	8	7	12	5				
	b_{sj}	0	0	0	0	0	0		0		
	γ_j										
	δ_j										

Figure 14–22 Tableau 8.

The new tableau with the new flows is given in Figure 14–20. The subsequent tableaus are given in Figures 14–21 and 14–22. In Tableau 8 we see that $\sum_{i=1}^{m} a_{si} = 0$. Hence we have obtained the optimal solution, which is shown in the tableau.

PROBLEMS

1. Devise a method for converting a flow problem with several sources and several sinks into an equivalent problem with a single source and a single sink.

2. Find all the cuts for the following network:

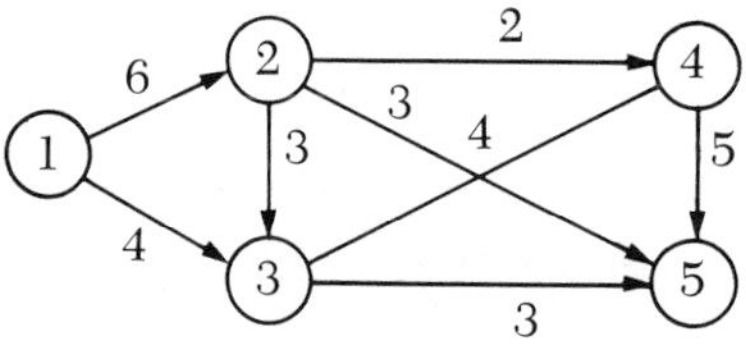

3. Prove that for the networks discussed in this chapter, if the capacity of each arc is a non-negative integer, then the maximal flow will be a non-negative integer.

4. Find a maximal flow for the following network. The numbers associated with the arcs are capacities in the orientation indicated.

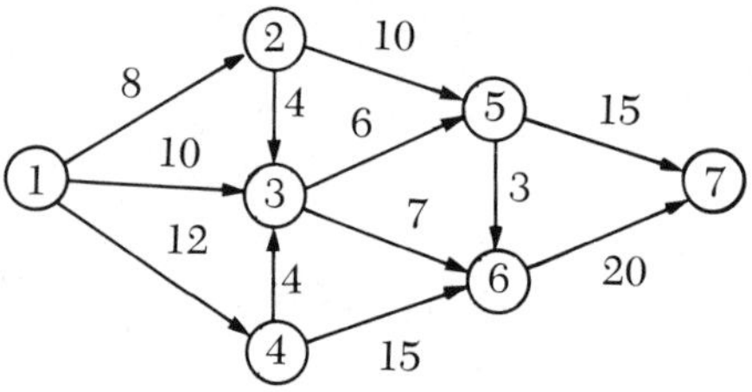

5. Find a maximal flow for the following network:

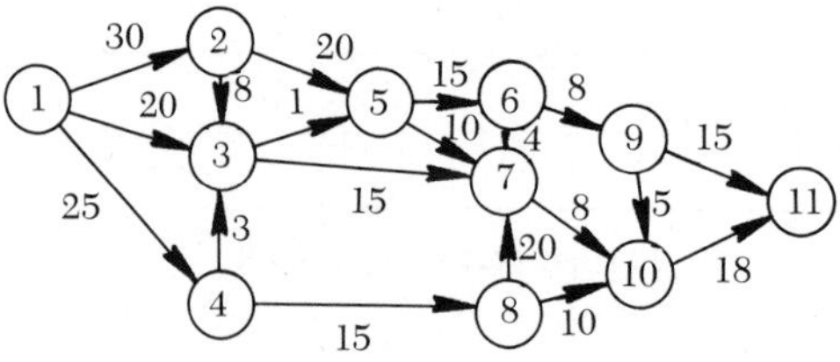

6. Solve the following transportation problem using the primal-dual method:

Destinations

Origins	1	2	3	4	5	6	a_i
1	5	1	3	1	4	6	50
2	6	2	2	1	2	2	25
3	2	4	8	2	1	5	20
4	1	3	4	3	4	5	10
b_j	20	10	5	15	25	30	

7. Solve the following transportation problem using the primal dual method.

Destinations

Origins	1	2	3	4	5	6	7	8	a_i
1	1	2	6	2	4	3	7	3	25
2	4	3	1	1	7	4	2	6	25
3	6	1	7	3	6	6	3	3	30
4	3	4	4	3	5	4	5	5	20
5	2	5	5	2	4	2	1	4	50
b_j	10	5	15	20	30	40	10	20	

REFERENCES

1. Ford, L. R., and Fulkerson, D. R.: *Flows in Networks*, Princeton, N.J., Princeton University Press, 1962.
2. Hu, T. C.: *Integer Programming and Network Flows*, Reading, Mass., Addison-Wesley, 1969.

15

Integer Linear Programming: Models

15–1 INTRODUCTION

The models we have considered in this text thus far have all been formulated as linear programming problems. In such problems the variables are allowed to be continuous; that is, the variables can assume any values in the feasible region. However, there are many situations in which it would not make sense for these variables to assume other than integer values. For example, if the variables in a given problem represent numbers of machines to be purchased, or numbers of men to be assigned to a particular job, fractional solutions would be of little value.

Although in a few of the models we have studied (e.g., transportation, assignment, and network flow problems) integer solutions can be obtained automatically, such is not usually the case: The extreme points of a general linear programming problem are not ordinarily restricted to having only integer-valued components.

Moreover, as we shall see in the example below, one cannot merely round the fractional values of an optimal solution of a linear programming problem and expect to thereby produce the optimal integer-valued solution.

EXAMPLE 1

Consider the following problem, illustrated in Figure 15–1:

$$\text{Maximize } z = 38x_1 + 81x_2$$

subject to:

$$18x_1 + 40x_2 \le 237$$

$$8x_1 - 8x_2 \le 23$$

$$x_1, x_2 \ge 0$$

The optimal solution is found to be $z^* = 489.336$ at extreme point C: $x_1 = 6.069$, $x_2 = 3.194$. However, suppose we are interested in the optimal integer solution. If we round the optimal continuous solution to $x_1 = 6$, $x_2 = 3$, we find that this solution is not feasible, as the second constraint is violated. Moreover, all of the integer points resulting from the various combinations of rounding x_1 to 6 or 7 and x_2 to 3 or 4 are infeasible. The nearest feasible integer point to (6.069, 3.194) is (5, 3), with $z = 434$. But, as the table below indicates, there are several other feasible integer points which are better solutions than (5, 3):

Point	z
(1, 5)	443
(2, 5)	481
(4, 4)	476

In fact, the optimal integer solution is (2, 5). This example illustrates that rounding the optimal continuous solution of a linear programming problem not only may not yield the optimal integer solution, but rounding may not even yield a feasible integer solution. Hence, if integer solutions are desired, the simplex method cannot normally be expected to find them, and different methods must be employed. A wide variety of algorithms have been devised to solve such problems. Some of these methods are presented in Chapter 16.

A linear programming problem in which some or all of the variables are required to be integer-valued is called an *integer linear programming problem*, or simply an *integer programming problem*. If all the variables are required to be integer-valued, the problem is sometimes called a *pure integer programming problem*; if some of the variables in an integer programming problem are continuous, the problem is sometimes called a *mixed integer programming problem*, or a *mixed integer-continuous variable programming problem*.

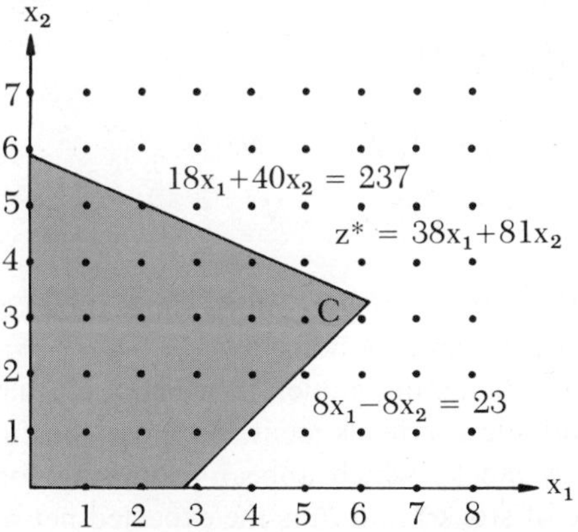

Extreme Point	z
A (0,0)	0
B(3.875,0)	109.925
C(6.069,3.194)	489.336
D(0,5.925)	479.925

Figure 15–1 Linear programming problem of Example 1.

There are many situations which can be modeled as integer programming problems but which cannot be described in terms of an ordinary linear programming model. These include certain types of nonlinear functions, sequencing models, and combinatorial models. In order to give the reader a better feel for these types of applications of integer programming problems, the rest of this chapter is devoted to a presentation of several of these applications and models.

15-2 CAPITAL BUDGETING MODELS

Suppose a construction company, which specializes in speculative homebuilding, has a total of D dollars available for immediate use. The company has identified N different potential homesites upon which it could build a house. The cost of constructing a house on homesite j is d_j dollars; because of the different types of neighborhoods in which these homesites are located, the type of house which can be built—and hence the cost—varies from homesite to homesite. The expected profit to be realized from the sale of a house on homesite j is P_j dollars. The company cannot build all N houses, because $\sum_{j=1}^{N} d_j > D$. Hence, it must select from among all N homesites that subset which yields the largest total expected profit.

The problem may be formulated as follows:

$$\text{Let } x_j = \begin{cases} 1, & \text{if homesite } j \text{ is selected} \\ 0, & \text{if homesite } j \text{ is not selected} \end{cases}$$

Then the total profit is $\sum_{j=1}^{N} P_j x_j$, and the total amount spent is $\sum_{j=1}^{N} d_j x_j$. Hence,

$$\sum_{j=1}^{N} d_j x_j \leq D$$

$$0 \leq x_j \leq 1, \qquad j = 1, 2, \ldots, N$$

$$x_j \text{ integer}, \qquad j = 1, 2, \ldots, N$$

Observe that we must require the x_j to be integer-valued, since it makes no sense to allocate money to build a fraction of a house.

Consider, now, another capital budgeting model, in which a mutual investment fund has D dollars available for investment. Suppose the fund's investment counselor has found N stocks which appear promising for investment. If d_j is the current price of stock j and P_j is the expected net profit per share of stock, then we obtain the following model:

Let x_j = No. of shares of stock j purchased.

$$\text{Maximize } z = \sum_{j=1}^{N} P_j x_j$$

subject to

$$\sum_{j=1}^{N} d_j x_j \leq D$$

$$x_j \geq 0 \qquad j = 1, 2, \ldots, N$$

$$x_j \text{ integer}, \qquad j = 1, 2, \ldots, N$$

Again, the variables must be restricted to integer values, since fractional shares cannot ordinarily be purchased (of course, if a given company does allow fractional shares of its stock to be purchased, the corresponding variable can be allowed to be continuous).

For more discussion of capital budgeting problems, see Weingartner.[9]

15-3 THE TRAVELING SALESMAN PROBLEM

The traveling salesman problem, which is briefly discussed in Chapter 11, is a problem for which efficient solution procedures are still being sought. As mentioned in Chapter 11, special algorithms have been devised to solve traveling salesman problems,[1,5,7] but thus far no method has been found which is practical for problems with more than several hundred cities.

The traveling salesman problem can also be formulated in terms of integer programming, and thus can be solved by any general integer programming algorithm, such as any of those discussed in Chapter 16. However, the current state of the art with respect to integer programming algorithms is such that it is even more impractical to try to solve a traveling salesman problem (except perhaps a very small problem) by such an algorithm than by one of the specialized algorithms.

In this section, we wish to present an integer programming formulation of the traveling salesman problem solely for instructional purposes, to illustrate the use of integer variables in modeling; the reader should not infer that traveling saleman problems are actually solved as integer programming problems.

Let us begin by reviewing the discussion in Chapter 11 of the traveling salesman problem. The problem may be stated as follows: Given a set of n cities, numbered from 1 to n, a salesman wishes to determine a route (called a "tour") which begins in City 1 and then goes through each of the other $(n - 1)$ cities exactly once, and then returns to City 1, in such a way that the total distance traveled is a minimum.

In Chapter 11 we partially formulated the traveling salesman problem in terms of integer programming by defining the integer variables x_{ij}:

$$x_{ij} = \begin{cases} 1, & \text{if the salesman goes from city } i \text{ to city } j \\ 0, & \text{otherwise} \end{cases}$$

Then, if c_{ij} denotes the distance between city i and city j, the objective function is

$$\text{minimize } z = \sum_{i=1}^{n} \sum_{j=1}^{n} c_{ij} x_{ij}$$

The following $2n$ constraints ensure that each city will be included in the tour once and only once:

$$\sum_{j=1}^{n} x_{ij} = 1, \qquad i = 1, 2, \ldots, n \tag{15–1}$$

$$\sum_{i=1}^{n} x_{ij} = 1, \qquad j = 1, 2, \ldots, n \tag{15–2}$$

However, as we observed in Chapter 11, these constraints do not eliminate the possibility of subtours occurring. Recall that a subtour is a tour through a subset of the n cities. A subtour which goes through k cities is called a subtour of length k. We shall now show how to add constraints to the above formulation which exclude all subtours of length $(n - 1)$ or less.

Consider the following $(n - 1)^2 - (n - 1)$ constraints:

$$y_i - y_j + n x_{ij} \leq n - 1, \qquad \begin{array}{l} i = 2, 3, \ldots, n \\ j = 2, 3, \ldots, n \\ i \neq j \end{array} \tag{15–3}$$

We have also introduced $(n - 1)$ additional variables $y_2, y_3, \ldots, y_n$.

Let us see why the constraints (15–3) exclude all subtours of length $(n - 1)$ or less. If a solution contains a subtour, then it must contain at least two subtours, since each city must be visited once, by the constraints (15–1) and (15–2). Hence, there is always a subtour which does not contain City 1. Assume that such a subtour of length k exists, $k \leq (n - 1)$. Then, corresponding to this subtour are k variables x_{ij} equal to one.† Adding up the k

† We are assuming that all subtours of length 1 will be excluded from consideration, so that all $x_{ii} = 0$. This is easily accomplished by setting all $c_{ii} = +\infty$.

constraints from (15–3) which contain these k variables yields the inequality $kn \leq k(n - 1)$, which is not valid for any $k > 0$. For example, in the subtour of length three illustrated in Figure 11–2, we have $x_{34} = x_{45} = x_{53} = 1$. The corresponding constraints of (15–3) are

$$y_3 - y_4 + nx_{34} \leq n - 1$$
$$y_4 - y_5 + nx_{45} \leq n - 1$$
$$y_5 - y_3 + nx_{53} \leq n - 1$$

Adding these inequalities yields

$$y_3 - y_4 + y_4 - y_5 + y_5 - y_3 + n(x_{34} + x_{45} + x_{53}) \leq 3(n - 1)$$
$$3n \leq 3(n - 1)$$

Hence, the variables x_{34}, x_{45} and x_{53} cannot simultaneously assume the value one, and this subtour is eliminated as a feasible solution.

In order to show that our formulation of the traveling salesman problem is complete, we must also show that the constraints (15–3) do not exclude any feasible tours. To do so, suppose we have a feasible tour in which city i is the m_i-th visited, $i = 1, 2, \ldots, n$. Then, suppose we let $y_i = m_i$, $i = 1, 2, \ldots, n$.

Now, if $x_{ij} = 1$ in this tour, then city j is visited immediately after city i, so that $y_j = m_i + 1$, and the constraint (15–3) containing x_{ij} is satisfied, since

$$y_i - y_j + nx_{ij} = m_i - (m_i + 1) + n(1) = n - 1.$$

Also, if $x_{ij} = 0$ in this tour, the constraint containing x_{ij} is again satisfied since $(y_i - y_j)$ is at most equal to $(n - 1)$: The largest y_i is n and the smallest is one.

Hence, we have formulated the traveling salesman problem as an integer programming problem, defined by (15–1) through (15–3). There are n^2 variables x_{ij}, and $(n - 1)$ variables y_i, for a total of $n^2 + n - 1$ variables. Note that the y_i need not be constrained to be integer-valued. The number of constraints in this formulation is $n^2 - n + 2$ constraints.

Another, perhaps more straightforward, formulation of the traveling salesman problem in terms of integer programming involves the use of a triply subscripted variable x_{ijk}, defined as follows:

$$x_{ijk} = \begin{cases} 1, \text{ if the salesman goes from city } i \text{ to city } j \\ \quad \text{on the } k\text{-th leg of the tour} \\ 0, \text{ otherwise} \end{cases} \qquad (15\text{–}4)$$

We leave the details of this formulation as an exercise for the reader. For another example of an application of the traveling salesman model, the reader may wish to review Example 11 of Chapter 11.

15–4 JOB-SHOP SCHEDULING MODELS†

Consider the situation in which a company manufactures different items on a made-to-order basis. Suppose the company has essentially one major piece of machinery which performs all the various types of operations necessary to manufacture the ordered products. However, the machine can only perform one operation at a time. Each such operation shall be called a "task," and we shall assume that it is known how long each task requires. Any one of the ordered products ("end products") requires the performance of some of the tasks, perhaps in a specified sequence.

The scheduling problem consists of determining an order in which the tasks may be performed which satisfies the following constraints:

1. Sequencing requirements—certain tasks may have to be performed in a particular order.
2. Machine non-interference restrictions—no two tasks can be performed simultaneously.
3. Specific delivery requirements—certain end products may have specified delivery dates which must be met.

To formulate these constraints in terms of integer programming, let us suppose there are a total of n tasks to be performed, and define the integer-valued variable x_j as the day on which task j is to be begun. Suppose, also, that it is known that task j requires a_j days to complete. Moreover, we assume that it is known that all tasks can be begun by the Tth day (for T chosen sufficiently large, this assumption is not really restrictive). Thus, each x_j can assume only the values $0, 1, \ldots, T$.

Now, the sequencing requirements can easily be formulated. If, for example, task j must be performed before task k, then task j must be begun at least a_j days before task k can begin. Hence,

$$x_j + a_j \leq x_k \tag{15–5}$$

One such constraint would be included for each sequencing restriction.

The machine non-interference restrictions are a bit more tricky to formulate. If, for example, there is no sequencing requirement between task i and task j, then in order that tasks i and j do not occupy the machine at the same

---

† The material in this section is from Manne.[6]

time we must require that either

$$x_i + a_i \leq x_j \tag{15–6}$$

or

$$x_j + a_j \leq x_i \tag{15–7}$$

Constraint (15–6) implies that task j cannot begin until task i is completed, while constraint (15–7) does not allow task i to begin until task j is completed. Obviously, both constraints cannot be satisfied simultaneously. If our problem only contained two such constraints, we could easily resolve this difficulty by solving the problem with only one of these constraints included, and then solving it with only the other constraint included, and choose the best of the resulting two optimal solutions.

However, we must include machine non-interference constraints for every pair of tasks for which there is no sequencing restriction. If there are m tasks without sequencing restrictions, then there are $m(m - 1)/2$ machine non-interference restrictions. Instead of attempting to solve a large number of separate problems, consider the following:

$$\begin{Bmatrix} x_i + a_i \leq x_j + (T + a_i)y_{ij} \\ x_j + a_j \leq x_i + (T + a_j)(1 - y_{ij}) \\ y_{ij} = 0 \text{ or } 1 \end{Bmatrix} \tag{15–8}$$

Observe that these two constraints can be satisfied simultaneously; moreover, if $y_{ij} = 1$, then the first constraint becomes redundant, since it only constrains x_i to be less than or equal to T days (recall that it is already known that all tasks will be begun by day T). The second constraint, with $y_{ij} = 1$, requires task j to be completed before task i is begun. On the other hand, if $y_{ij} = 0$, then the second constraint is redundant, and the first constraint requires task i to be completed before task j is begun.

Thus, we have seen that the machine non-interference restrictions can indeed be formulated in terms of integer programming, with the introduction of the integer variables y_{ij}.

Let us now turn to the specific delivery requirements. If task k is the final task for an end product which is supposed to be completed by day d_k, then we can include the constraint

$$x_k + a_k \leq d_k \tag{15–9}$$

to ensure this delivery date is met.

We have not yet discussed an objective function for this model. There are many possibilities, depending on the particular situation. Manne[6] suggests that the total time to complete all the tasks be minimized. To do so, we denote this time by t, and add the following (n = no. of tasks) constraints:

$$x_j + a_j \leq t \qquad j = 1, 2, \ldots, n \tag{15-10}$$

The problem is then to minimize t, subject to the constraints (15–5), (15–8), (15–9), and (15–10).

15-5 THE FIXED CHARGE PROBLEM

Consider a company which manufactures n products and wishes to minimize its total production costs. Suppose that for each unit of product j produced, there is a cost of c_j dollars. Suppose also that, for each product j, there is a fixed cost of k_j dollars incurred, if any units of product j are produced at all, regardless of the quantity involved. This fixed cost is the cost of setting up the manufacturing equipment for the product.

Thus, if x_j denotes the number of units of product j produced, then the quantity we wish to minimize is

$$z = \sum_{j=1}^{n} (c_j x_j + k_j y_j) \tag{15-11}$$

where

$$y_j = \begin{cases} 1, & \text{if } x_j > 0 \\ 0, & \text{if } x_j = 0 \end{cases} \tag{15-12}$$

Suppose, further, that the constraints on production can be formulated as linear constraints, which we shall denote by $A\bar{x} = \bar{b}$, $\bar{x} \geq \bar{0}$, in the usual way. Thus, if all the k_j were equal to zero (and if the x_j are not integer-constrained) the problem would be an ordinary linear programming problem. However, the inclusion of the k_j makes the problem not only nonlinear, but also extremely difficult to solve.

The above problem, called the fixed charge problem, can be formulated as a mixed integer programming problem, by introducing certain constraints to represent the situation described by (15–12). In particular, if an *a priori* upper bound u_j is known for each variable x_j, then we can write

$$\begin{aligned} &y_j \geq x_j/u_j, \qquad j = 1, 2, \ldots, n \\ &y_j = 0 \text{ or } 1 \end{aligned} \tag{15-13}$$

In this way, if x_j is positive, then y_j cannot be 0, and since $x_j/u_j \leq 1$, y_j must be equal to 1, as stipulated in (15–12). Moreover, if x_j is zero, then the inequality (15–13) merely requires $y_j \geq 0$. Hence, y_j can assume either of the values 0 or 1; however, since the objective function in this case will be smaller if $y_j = 0$, and we are minimizing, we do not have to construct any additional constraints to force $y_j = 0$ if $x_j = 0$.

Several special algorithms have been devised for solving fixed charge problems, but these do not appear to be computationally feasible for problems in which $n \geq 100$ (e.g., see Steinberg[8]). However, there are some very good "heuristic" methods available which usually find nearly optimal solutions for the fixed charge problem (see Cooper and Drebes[2] and Steinberg[8]). These methods frequently do obtain the exact optimal solution, but do not guarantee to do so. Moreover, since they are basically just variations of the simplex method, solutions for very large fixed charge problems can be obtained by them. In most practical applications these heurestic methods are more effective than trying to solve a fixed charge problem by a mixed integer programming algorithm or by a special exact algorithm for the fixed charge problem.

15–6 DISCRETE ALTERNATIVES

Frequently, models arise in which it is desired that the solution must satisfy not an entire set of constraints but rather, a subset of the set of constraints. Which particular subset, however, may not be known in advance of solving the problem. We have already seen one such situation, in section 15–4, involving the machine non-interference constraints (15–6) and (15–7), in which the solution was required to satisfy either (15–6) or (15–7).

Let us restate our resolution of this "either/or" situation, in more general terms. Suppose it is required that either the constraint $\sum_j a_{1j}x_j \leq b_1$ or the constraint $\sum_j a_{2j}x_j \leq b_2$ be satisfied. Suppose further that numbers K_1, K_2 are known such that for all feasible $\bar{x}$,

$$\sum_j a_{1j}x_j - b_1 \leq K_1$$

$$\sum_j a_{2j}x_j - b_2 \leq K_2$$

Then, by introducing the integer variable y, we can represent this situation by

$$\sum_j a_{1j}x_j - b_1 \leq K_1 y$$

$$\sum_j a_{2j}x_j - b_2 \leq K_2(1 - y)$$

$$y = 0 \text{ or } 1$$

Thus, if $y = 1$, the first constraint above becomes redundant while the second constraint $\left(\sum_j a_{2j}x_j \leq b_2\right)$ must be satisfied. If $y = 0$, the second constraint is redundant and the first must be satisfied.

A variation of the above formulation is obtained by introducing two integer variables y_1 and y_2 as follows:

$$\sum_j a_{1j}x_j - b_1 \leq K_1y_1$$

$$\sum_j a_{2j}x_j - b_2 \leq K_2y_2$$

$$y_1 + y_2 = 1$$

$$y_1, y_2 \text{ integers}$$

Although the latter formulation is not as efficient as the first formulation, since it contains an additional integer variable and an additional constraint, it is suggestive of how the technique can be generalized to the situation in which it is required that at least p constraints out of a given set of q constraints must be satisfied, but it is not stipulated which particular p constraints must be satisfied.

In particular, if we have the following q constraints

$$\sum_j a_{ij}x_j \leq b_i, \qquad i = 1, 2, \ldots, q \tag{15–14}$$

and we have somehow determined quantities $K_1, K_2, \ldots, K_q$ such that for all feasible $\bar{x}$,

$$\sum_j a_{ij}x_j - b_i \leq K_i \qquad i = 1, 2, \ldots, q$$

then, the requirement that at least p of the constraints (15–14) be satisfied can be represented by

$$\sum_j a_{ij}x_j - b_i \leq K_iy_i \qquad i = 1, 2, \ldots, q \tag{15–15}$$

$$\sum_{i=1}^{q} y_i = q - p \tag{15–16}$$

$$y_i = 0, \text{ or } 1, \qquad i = 1, 2, \ldots, q \tag{15–17}$$

The constraints (15–16) and (15–17) can only be satisfied if exactly $(q - p)$ of the y_i are equal to 1 and the remaining p $y_i = 0$. The p constraints (15–14) corresponding to these latter y_i must therefore be satisfied.

EXAMPLE 2

A farm equipment manufacturer has two types of harvesters, and wishes to set up his equipment for a production run. The manufacturer has decided that he will make at least 10 harvesters of Model *A* or at least 10 of Model *B*. Furthermore, to take advantage of certain economies of scale, he will make either at least twice as many harvesters of Model *A* as of Model *B* or twice as many of Model *B* as of Model *A*, unless the number of harvesters of each model exceeds 30, in which case these restrictions are unnecessary. Because of his current sales forecasts, the manufacturer has determined not to produce more than a total of 75 harvesters.

The net profit per harvester is $150 for Model *A* and $120 for Model *B*.

If we let x_1 and x_2 denote the number of harvesters produced of Model *A* and Model *B*, respectively, and assume for the moment that x_1 and x_2 are not required to be integers, then the feasible region for this problem is as indicated by the shaded portion of Figure 15–2.

Observe that the feasible region (which is composed of three disjoint regions) can be described by requiring

$$x_1 + x_2 \leq 75$$

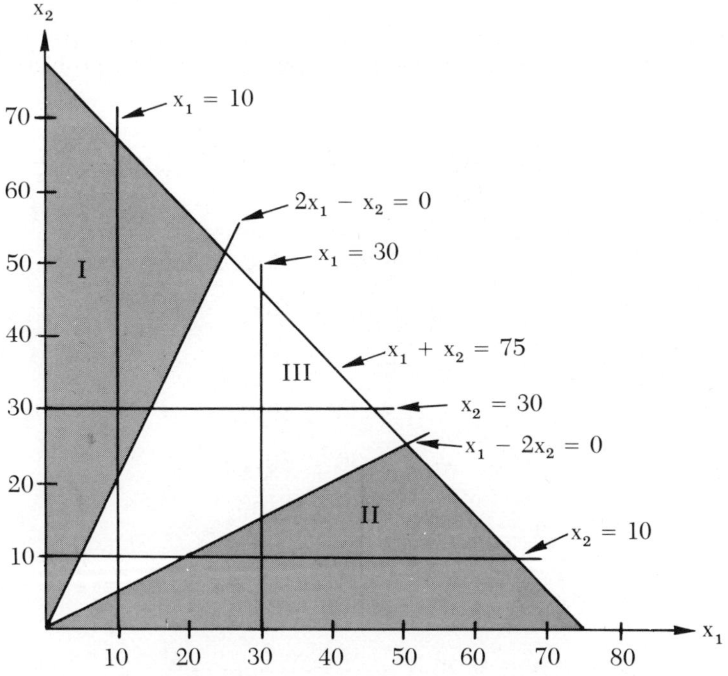

Figure 15–2 Feasible region for Example 2.

and that at least two of the following constraints be met:

$$
\begin{aligned}
x_1 &\geq 10 \\
2x_1 - x_2 &\geq 0 \\
x_1 &\geq 30 \\
x_2 &\geq 30 \\
x_1 - 2x_2 &\leq 0 \\
x_2 &\geq 10
\end{aligned}
$$

For example, Region I is described by

$$
\begin{aligned}
x_1 + x_2 &\leq 75 \\
x_2 &\geq 10 \\
2x_1 - x_2 &\geq 0
\end{aligned}
$$

Adding the six integer variables y_i, $i = 1, 2, \ldots, 6$, as per (15–15), (15–16), and (15–17), yields the desired formulation of this problem:

$$\text{Maximize } z = 150x_1 + 120x_2$$

Subject to

$$
\begin{aligned}
x_1 + x_2 &\leq 75 \\
-x_1 + 10 &\leq 10y_1 \\
-2x_1 + x_2 &\leq 150y_2 \\
-x_1 + 30 &\leq 30y_3 \\
-x_2 + 30 &\leq 30y_4 \\
x_1 - 2x_2 &\leq 150y_5 \\
-x_2 + 10 &\leq 10y_6 \\
y_1 + y_2 + y_3 + y_4 + y_5 + y_6 &= 4 \\
y_i = 0 \text{ or } 1, \qquad i &= 1, 2, \ldots, 6 \\
x_1, x_2 &\geq 0
\end{aligned}
$$

The reader should verify that each constraint is appropriately redundant when its corresponding $y_i = 1$.

If we wish to require x_1 and x_2 to be integer-valued, as well, the above formulation remains essentially unchanged, except, of course, we would add in the restriction, "x_1, x_2 integer."

An alternate formulation of the region of Figure 15–2 may be obtained as follows: A point (x_1, x_2) is feasible if and only if it lies in one of the three regions I, II, III. These regions are described by the following three constraint sets:

$$\text{Region I} \quad \begin{cases} x_1 + x_2 \leq 75 \\ x_2 \qquad\quad \geq 10 \\ 2x_1 - x_2 \geq 0 \\ \qquad x_1, x_2 \geq 0 \end{cases}$$

$$\text{Region II} \quad \begin{cases} x_1 + x_2 \leq 75 \\ x_1 \qquad\quad \geq 10 \\ x_1 - 2x_2 \leq 0 \\ \qquad x_1, x_2 \geq 0 \end{cases}$$

$$\text{Region III} \quad \begin{cases} x_1 + x_2 \leq 75 \\ x_1 \qquad\quad \geq 30 \\ \qquad\quad x_2 \geq 30 \\ \qquad x_1, x_2 \geq 0 \end{cases}$$

Suppose we define three integer variables:

$$y_1, y_2, y_3 = \begin{cases} 0, \text{ if } (x_1, x_2) \text{ is in Region I, II, III,} \\ \quad \text{respectively} \\ 1, \text{ otherwise} \end{cases}$$

Then, the following constraints also correctly represent the feasible region:

$$
\begin{cases}
x_1 + x_2 - 75 \le 75y_1 \\
-x_2 + 10 \le 10y_1 \\
-2x_1 + x_2 \le 150y_1
\end{cases}
$$

$$
\begin{cases}
x_1 + x_2 - 75 \le 75y_2 \\
-x_1 + 10 \le 10y_2 \\
x_1 - 2x_2 \le 150y_2
\end{cases}
$$

$$
\begin{cases}
x_1 + x_2 - 75 \le 75y_3 \\
-x_1 + 30 \le 30y_3 \\
-x_2 + 30 \le 30y_3
\end{cases}
$$

$$
\begin{aligned}
y_1 + y_2 + y_3 &= 1 \\
y_i = 0 \text{ or } 1, &\quad i = 1, 2, 3 \\
x_1, x_2 &\ge 0
\end{aligned}
$$

Although the above formulation has more constraints than the previous one, it has the advantage of having fewer integer variables. Moreover, since the constraint $x_1 + x_2 \le 75$ is common to all three regions, we can also describe the whole feasible region by

$$
\begin{aligned}
x_1 + x_2 &\le 75 \\
-x_2 + 10 &\le 10y_1 \\
-2x_1 + x_2 &\le 150y_1 \\
-x_1 + 10 &\le 10y_2 \\
x_1 - 2x_2 &\le 150y_2 \\
-x_1 + 30 &\le 30y_3 \\
-x_2 + 30 &\le 30y_3
\end{aligned}
$$

$$
\begin{aligned}
y_1 + y_2 + y_3 &= 1 \\
y_i = 0 \text{ or } 1, &\quad i = 1, 2, 3 \\
x_1, x_2 &\ge 0
\end{aligned}
$$

The latter two formulations may suggest to the reader how we can extend the idea of requiring any p of q constraints to be satisfied, to the case of requiring at least one of q convex regions to be satisfied. Another application of this idea is given in Example 3.

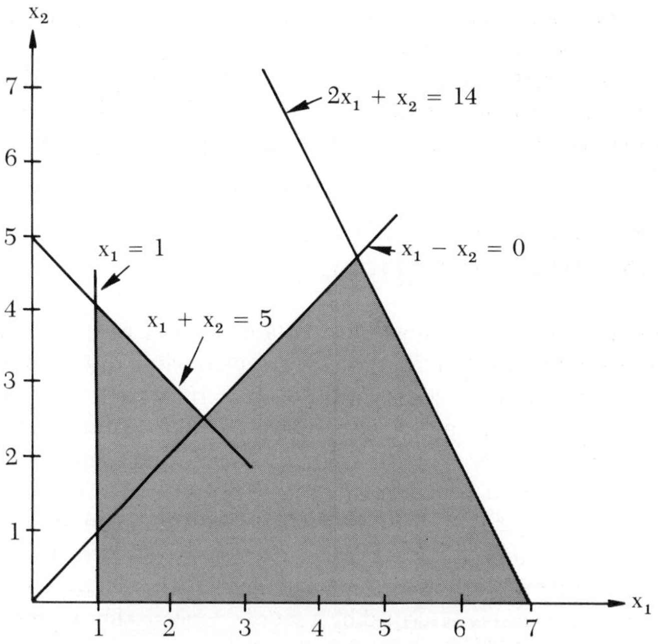

Figure 15–3 Feasible region for Example 3.

EXAMPLE 3

Suppose the feasible region for a particular problem is as indicated by the shaded portion of Figure 15–3. Observe that this region is not convex. However, any point in the region lies in one of the following two convex regions:

$$\begin{Bmatrix} x_1 + x_2 \le 5 \\ x_1 \qquad \ge 1 \\ \qquad x_2 \ge 0 \end{Bmatrix} \quad \text{and} \quad \begin{Bmatrix} 2x_1 + x_2 \le 14 \\ x_1 - x_2 \ge 0 \\ x_1, x_2 \ \ge 0 \end{Bmatrix}$$

Hence, we can represent the entire nonconvex region as follows:

$$\begin{aligned} x_1 + x_2 - 5 &\le 9y_1 \\ -x_1 \qquad + 1 &\le 1y_1 \\ 2x_1 + x_2 - 14 &\le 1y_2 \\ -x_1 + x_2 \qquad &\le 7y_2 \\ y_1 + y_2 \qquad &= 1 \\ y_1, y_2 &= 0 \text{ or } 1 \\ x_1, x_2 &\ge 0 \end{aligned}$$

Again, the reader should verify that each constraint does actually become redundant when its corresponding $y_i = 1$.

The fact that certain types of nonconvex regions can be represented in terms of integer linear programming is a very useful application of integer programming. We shall continue developing these ideas in the next section.

15-7 APPROXIMATION OF SEPARABLE NONLINEAR FUNCTIONS

In this section we shall show how certain types of nonlinear functions can be approximated by functions which are piecewise linear; that is, functions $\hat{f}(x)$ which are linear over various subintervals of x. We will then discuss how this type of approximation can be used to formulate certain nonlinear programming problems as "almost linear" programming problems and also as integer linear programming problems.

We shall begin our discussion by considering the function $f(x)$ shown in Figure 15–4. An example of a piecewise linear approximation to $f(x)$ is the function $\hat{f}(x)$, illustrated in Figure 15–4 with dashed lines. Thus, we see that a piecewise linear function is described by a set of connected line segments. This is also referred to as a "polygonal line," and the approximation $\hat{f}(x)$ as a "polygonal approximation."

Suppose we wish to approximate $f(x)$ in the interval $a_1 \leq x \leq a_4$. In our polygonal approximation in Figure 15–4, we have chosen to subdivide the interval into three subintervals and thus to approximate $f(x)$ by a set of three connected line segments; the widths of the subintervals need not be equal. At the endpoints of each subinterval $\hat{f}(x) = f(x)$. Hence, the equation which

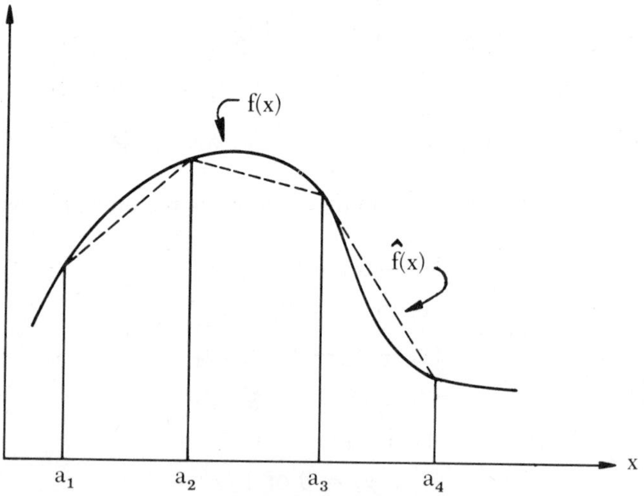

Figure 15–4 A polygonal approximation of a nonlinear function.

represents $\hat{f}(x)$ in the interval $[a_1, a_2]$ is

$$\hat{f}(x) = f(a_1) + \frac{f(a_2) - f(a_1)}{a_2 - a_1}(x - a_1) \qquad a_1 \le x \le a_2 \tag{15-18}$$

Another way of representing the line segment $\hat{f}(x)$ in $[a_1, a_2]$ is as follows: For any x in the interval $[a_1, a_2]$ we can determine a value of λ, $0 \le \lambda \le 1$, such that

$$x = \lambda a_1 + (1 - \lambda)a_2 \tag{15-19}$$

If we denote this value of λ by λ_1, and let $\lambda_2 = 1 - \lambda_1$, then equation (15–19) becomes

$$x = \lambda_1 a_1 + \lambda_2 a_2 \tag{15-20}$$

where

$$\lambda_1 + \lambda_2 = 1, \qquad \lambda_1 \ge 0, \qquad \lambda_2 \ge 0 \tag{15-21}$$

Upon subtracting a_1 from both sides of equation (15–20), we find that

$$\begin{aligned} x - a_1 &= \lambda_1 a_1 + \lambda_2 a_2 - a_1 \\ &= \lambda_2 a_2 - (1 - \lambda_1)a_1 \\ &= \lambda_2 a_2 - \lambda_2 a_1 \\ &= \lambda_2(a_2 - a_1) \end{aligned}$$

Substitution of this result into equation (15–18) yields

$$\begin{aligned} \hat{f}(x) &= f(a_1) + \frac{f(a_2) - f(a_1)}{a_2 - a_1}\lambda_2(a_2 - a_1) \\ &= f(a_1) + \lambda_2 f(a_2) - \lambda_2 f(a_1) \\ &= (1 - \lambda_2)f(a_1) + \lambda_2 f(a_2) \\ &= \lambda_1 f(a_1) + \lambda_2 f(a_2) \end{aligned} \tag{15-22}$$

In a like manner we can represent $\hat{f}(x)$ in the remaining two subintervals:

$$\left\{\begin{aligned} &\hat{f}(x) = \lambda_2 f(a_2) + \lambda_3 f(a_3), \qquad a_2 \le x \le a_3 \\ &x = \lambda_2 a_2 + \lambda_3 a_3 \\ &\lambda_2 + \lambda_3 = 1, \qquad \lambda_2 \ge 0, \qquad \lambda_3 \ge 0 \end{aligned}\right\} \tag{15-23}$$

$$\left\{\begin{aligned} &\hat{f}(x) = \lambda_3 f(a_3) + \lambda_4 f(a_4), \qquad a_3 \le x \le a_4 \\ &x = \lambda_3 a_3 + \lambda_4 a_4 \\ &\lambda_3 + \lambda_4 = 1, \qquad \lambda_3 \ge 0, \qquad \lambda_4 \ge 0 \end{aligned}\right\} \tag{15-24}$$

However, in order to actually use the above representations of $\hat{f}(x)$, we need to know in which interval x lies. If we are attempting to use $\hat{f}(x)$ in a problem in which x is the unknown variable, we will obviously not know this before solving the problem.

Consider, instead, the following. Suppose we write

$$\left.\begin{aligned} \hat{f}(x) &= \lambda_1 f(a_1) + \lambda_2 f(a_2) + \lambda_3 f(a_3) + \lambda_4 f(a_4) \\ x &= \lambda_1 a_1 + \lambda_2 a_2 + \lambda_3 a_3 + \lambda_4 a_4 \\ & \lambda_1 + \lambda_2 + \lambda_3 + \lambda_4 = 1 \\ & \lambda_1, \lambda_2, \lambda_3, \lambda_4 \geq 0 \end{aligned}\right\} \tag{15–25}$$

Observe that (15–25) does not require *a priori* knowledge of the particular subinterval in which x lies. This interval—and the value of x—is determined by the values assigned to the λ_i's. Observe also that the relations in (15–25) are all linear in the variables λ_i, $i = 1, 2, 3, 4$.

However, (15–25) by itself is not completely equivalent to the representations of $\hat{f}(x)$ given by (15–20) through (15–24), because, for example, if $\lambda_1 = \lambda_3 = 0$, then according to (15–25) we have

$$\begin{aligned} \hat{f}(x) &= \lambda_2 f(a_1) + \lambda_4 f(a_4) \\ x &= \lambda_2 a_2 + \lambda_4 a_4 \\ \lambda_2 + \lambda_4 &= 1, \qquad \lambda_2 \geq 0, \qquad \lambda_4 \geq 0 \end{aligned}$$

These relations imply that x is in the interval $[a_2, a_4]$, and that $\hat{f}(x)$ is the line segment joining the points $(a_2, f(a_2))$ and $(a_4, f(a_4))$, as shown in Figure 15–5. This is obviously not the intended approximation. Fortunately, this situation is easily remedied, by imposing the additional restrictions on the

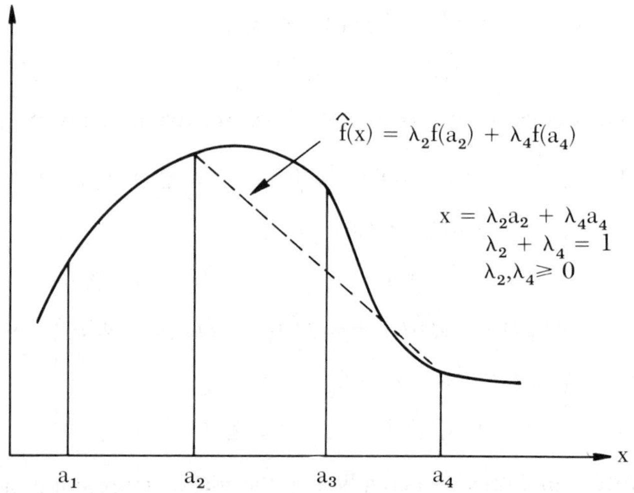

Figure 15–5 Incorrect polygonal approximation.

λ_i's that: (1) At most two λ_i can be positive, with the rest zero; and (2) Only "adjacent" λ_i are allowed to be positive.

Requirement (1) is self-explanatory; requirement (2) means that if, for example, λ_2 is positive, then either λ_1 or λ_3 can also be positive, but not both and not λ_4. The reader should be certain he is convinced that (15–20) through (15–24) is equivalent to (15–25), subject to the requirements above.

As we noted earlier, the relations in (15–25) are linear, in the continuous variables λ_i, $i = 1, 2, 3, 4$. Let us now see how we can incorporate the two additional requirements on the λ_i, given above, within an integer programming framework. Suppose we introduce three variables y_1, y_2, y_3, each constrained to be 0 or 1, as follows:

$$\lambda_1 \le y_1 \tag{15–26a}$$
$$\lambda_2 \le y_1 + y_2 \tag{15–26b}$$
$$\lambda_3 \le y_2 + y_3 \tag{15–26c}$$
$$\lambda_4 \le y_3 \tag{15–26d}$$
$$y_1 + y_2 + y_3 = 1 \tag{15–26e}$$
$$\lambda_1 + \lambda_2 + \lambda_3 + \lambda_4 = 1 \tag{15–26f}$$
$$y_i = 0 \text{ or } 1, \qquad i = 1, 2, 3 \tag{15–26g}$$
$$\lambda_i \ge 0 \qquad i = 1, 2, 3, 4 \tag{15–26h}$$

The constraints (15–26e, g) are satisfied if and only if exactly one $y_i = 1$, the others being zero. For the one y_i which is equal to one, the corresponding two constraints from (15–26a, b, c, d) allow exactly two adjacent λ_i's to be positive; also, the other λ_i's must be zero.

To summarize, we have shown how the polygonal approximation $\hat{f}(x)$ of a nonlinear function $f(x)$ can be represented by the following linear relations, which involve the continuous variables λ_i and the 0–1 integer variables y_i:

$$\left.\begin{aligned}
\hat{f}(x) &= \lambda_1 f(a_1) + \lambda_2 f(a_2) + \lambda_3 f(a_3) + \lambda_4 f(a_4) \\
&\lambda_1 \le y_1 \\
&\lambda_2 \le y_1 + y_2 \\
&\lambda_3 \le y_2 + y_3 \\
&\lambda_4 \le y_3 \\
&y_1 + y_2 + y_3 = 1 \\
&\lambda_1 + \lambda_2 + \lambda_3 + \lambda_4 = 1 \\
&\lambda_i \ge 0 \qquad i = 1, 2, 3, 4 \\
&y_i = 0 \text{ or } 1, \qquad i = 1, 2, 3
\end{aligned}\right\} \tag{15–27}$$

where

$$x = \lambda_1 a_1 + \lambda_2 a_2 + \lambda_3 a_3 + \lambda_4 a_4$$

Recall that the number of λ_i's in this representation of $\hat{f}(x)$ depends on the number of subintervals we have employed. In general, if $\hat{f}(x)$ consists of

P connected line segments (i.e., there are P subintervals: $a_1 < a_2 < \ldots < a_{P+1}$) then the system (15–27) becomes

$$\left.\begin{aligned} \hat{f}(x) &= \sum_{k=1}^{P+1} \lambda_k f(a_k) \\ \lambda_1 &\leq y_1 \\ \lambda_k &\leq y_{k-1} + y_k \qquad k = 2, 3, \ldots, P \\ \lambda_{P+1} &\leq y_P \\ \sum_{k=1}^{P} y_k &= 1 \\ \sum_{k=1}^{P+1} \lambda_k &= 1 \\ \lambda_k &\geq 0 \qquad k = 1, 2, \ldots, P+1 \\ y_k &= 0 \text{ or } 1, \qquad k = 1, 2, \ldots, P \end{aligned}\right\} \tag{15–28}$$

where

$$x = \sum_{k=1}^{P+1} \lambda_k a_k \tag{15–29}$$

Thus far in this section we have restricted our attention to a nonlinear function of a single variable. Let us now turn to the problem of dealing with a nonlinear function of more than one variable. Suppose we have a function $f(x_1, x_2, \ldots, x_n)$ which can be expressed as the sum of n functions $f_j(x_j)$, $j = 1, 2, \ldots, n$, each of which is a function of a single variable. Thus,

$$f(x_1, x_2, \ldots, x_n) = \sum_{j=1}^{n} f_j(x_j)$$

Such a function is said to be *separable.* A great many nonlinear programming problems can be formulated in terms of separable functions. The general separable nonlinear programming problem may be expressed as follows:

$$\left.\begin{aligned} &\text{Maximize } z = f(x_1, x_2, \ldots, x_n) = \sum_{j=1}^{n} f_j(x_j) \\ &\text{Subject to} \\ &\sum_{j=1}^{n} g_{ij}(x_j) \leq b_i \qquad i = 1, 2, \ldots, m \\ &x_j \geq 0 \qquad j = 1, 2, \ldots, n \end{aligned}\right\} \tag{15–30}$$

Thus, the objective function and each of the constraints consists entirely of separable functions. Each of the functions $f_j(x_j)$, $g_{ij}(x_j)$ can then be approximated by a polygonal line, $\hat{f}(x_j)$, $\hat{g}_{ij}(x_j)$. If the variable x_j ranges over the interval† $L_j \leq x_j \leq U_j$, and if we subdivide this interval into P_j

† We have assumed that some upper bound U_j is known for each variable x_j. Usually, the lower bound L_j is simply 0, from the non-negativity restriction $x_j \geq 0$.

subintervals, with end points $L_j = a_{1j}, a_{2j}, \ldots, a_{P_j j}, a_{P_j+1,j} = U_j$, then from equations (15–28) and (15–29) we obtain

$$\left.\begin{aligned}
\hat{f}_j(x_j) &= \sum_{k=1}^{P_j+1} \lambda_{kj} f_j(a_{kj}) \\
\hat{g}_{ij}(x_j) &= \sum_{k=1}^{P_j+1} \lambda_{kj} g_{ij}(a_{kj}) \\
\lambda_{1j} &\le y_{1j} \\
\lambda_{kj} &\le y_{k-1,j} + y_{kj}, \qquad k = 2, 3, \ldots, P_j \\
\lambda_{P_j+1,j} &\le y_{P_j j} \\
\sum_{k=1}^{P_j} y_{kj} &= 1 \\
\sum_{k=1}^{P_j+1} \lambda_{kj} &= 1 \\
\lambda_{kj} &\ge 0, \qquad k = 1, 2, \ldots, P_j + 1 \\
y_{kj} &= 0 \text{ or } 1, \qquad k = 1, 2, \ldots, P_j \\
\text{where} \qquad x_j &= \sum_{k=1}^{P_j+1} \lambda_{kj} a_{kj}
\end{aligned}\right\} \qquad (15\text{–}31)$$

Substitution of (15–31) into (15–30) yields the following integer linear programming problem:

$$\left.\begin{aligned}
&\text{Maximize } z = \sum_{j=1}^{n} \sum_{k=1}^{P_j+1} \lambda_{kj} f_j(a_{kj}) \\
&\text{Subject to} \\
&\sum_{j=1}^{n} \sum_{k=1}^{P_j+1} \lambda_{kj} g_{ij}(a_{kj}) \le b_i, \quad i = 1, 2, \ldots, m \\
&\lambda_{1j} \le y_{1j} \qquad j = 1, 2, \ldots, n \\
&\lambda_{kj} \le y_{k-1,j} + y_{kj} \qquad j = 1, 2, \ldots, n; \quad k = 1, 2, \ldots, P_j \\
&\lambda_{P_j+1,j} \le y_{P,j} \qquad j = 1, 2, \ldots, n \\
&\sum_{k=1}^{P_j} y_{kj} = 1 \qquad j = 1, 2, \ldots, n \\
&\sum_{k=1}^{P_j+1} \lambda_{kj} = 1 \qquad j = 1, 2, \ldots, n \\
&\lambda_{kj} \ge 0 \qquad j = 1, 2, \ldots, n; \quad k = 1, 2, \ldots, P_j + 1 \\
&y_{kj} = 0 \text{ or } 1, \qquad j = 1, 2, \ldots, n; \quad k = 1, 2, \ldots, P_j
\end{aligned}\right\} \qquad (15\text{–}32)$$

where†

$$x_j = \sum_{k=1}^{P_j+1} \lambda_{kj} a_{kj}, \qquad j = 1, 2, \ldots, n \tag{15–33}$$

Note that the problem (15–32) is only an approximation of the original nonlinear programming problem (15–30). Observe also the large number of additional variables and constraints needed to convert the nonlinear programming problem (15–30) into the integer linear programming problem (15–32). The latter contains $\left(m + 3n + \sum_{j=1}^{n} P_j\right)$ constraints (excluding the non-negativity restrictions on the λ_{kj} and the 0, 1 restrictions on the y_{kj}). There are also $\left(\sum_{j=1}^{n} P_j + n\right)$ continuous variables and $\sum_{j=1}^{n} P_j$ integer variables in (15–32). Contrast this with the original problem (15–30), which has m constraints and n continuous variables.

At this point we wish to make one final observation: The separable nonlinear programming problem (15–30) may also be approximated by an "almost" linear programming problem, without the use of the integer variables y_{kj}. This can be accomplished by generalizing the representation of the polygonal approximation given by (15–25), along with the two additional requirements stated shortly after (15–25), and then applying this generalization to (15–30). The resulting problem is:

$$\text{Maximize } z = \sum_{j=1}^{n} \sum_{k=1}^{P_j+1} \lambda_{kj} f_j(a_{kj})$$

Subject to

$$\sum_{j=1}^{n} \sum_{k=1}^{P_j+1} \lambda_{kj} g_{ij}(a_{kj}) \leq b_i, \qquad i = 1, 2, \ldots, m$$

$$\sum_{k=1}^{P_j+1} \lambda_{kj} = 1, \qquad j = 1, 2, \ldots, n$$

$$\lambda_{kj} \geq 0, \qquad j = 1, 2, \ldots, n; \quad k = 1, 2, \ldots, P_j + 1 \tag{15–34}$$

and

(1) For each j: At most two λ_{kj} can be positive with the rest 0; and,

(2) For each j: Only adjacent λ_{kj} are allowed to be positive.

† The equations (15–33) are not used to solve the problem. Once the problem is solved the optimal values of the λ_{kj} may be inserted into (15–33) to obtain the optimal values of the original variables x_j).

Except for the two requirements (1) and (2) above, (15–34) is an ordinary linear programming problem with $(m + n)$ constraints and $\left(\sum_{j=1}^{n} P_j + n\right)$ variables. If we modify the simplex method appropriately, it can be used to solve (15–34). The modification needed is merely that no variable may enter the basis if either of the requirements (1) and (2) cannot be satisfied. This modification is sometimes referred to as the "restricted basis entry simplex method." One might at first suppose that it would be far preferable to solve the latter than to solve the integer programming formulation, because (15–34) contains fewer variables and constraints and has no integer variables. However, unless the functions $f_j(x_j)$ and $g_{ij}(x_j)$ have certain very special properties, the restricted basis entry simplex method is not guaranteed to find the optimal solution, whereas if the integer programming formulation is solved by an exact integer programming algorithm, the optimal solution will be obtained. For a further discussion of separable nonlinear programming and other integer programming formulations of separable nonlinear programming problems, see Cooper and Steinberg[3] or Hadley.[4]

PROBLEMS

1. Formulate as an integer programming problem:

A bakery sells six varieties of doughnuts. The preparation of varieties 1, 2, and 3 involves a rather complicated process, and so the bakery has decided not to make these varieties unless it should prove desirable to make at least 10 dozen doughnuts of varieties 1, 2, and 3 combined. The capacity of the bakery is such that no more than 40 dozen doughnuts in total can be made. Moreover, past sales data indicate that no more than 15 dozen, 12 dozen, and 10 dozen doughnuts of varieties 4, 5, and 6, respectively, should be made. If p_j dollars denotes the profit per dozen of variety j, $j = 1, 2, 3, 4, 5, 6$, how many dozens of each variety of doughnut should the bakery make to maximize its profit?

2. Formulate as an integer programming problem:

A family of four is moving its residence from New York to San Francisco. In order to reduce its costs, the family has decided to make the journey by train and, to reduce the tedium of such a long trip, the family has also decided to divide the trip into four segments. In the first segment, the family will travel to either Washington, Chicago or St. Louis. In the second segment, the family will travel to either Omaha or Kansas City, and in the third segment the family will travel to either Denver, Los Angeles or Las Vegas. The fare for each of these trips is given on page 400 (in dollars).

As the family is attempting to save money, determine the route which will minimize its total travel cost.

To / From	Washington	Chicago	St. Louis
New York	20	40	50

To / From	Omaha	Kansas City
Washington	65	50
Chicago	25	35
St. Louis	30	15

To / From	Denver	Los Angeles	Las Vegas
Omaha	40	75	60
Kansas City	30	65	70

To / From	San Francisco
Denver	20
Los Angeles	30
Las Vegas	15

3. Formulate as an integer programming problem:

A vacuum cleaner manufacturer currently has m factories and p regional warehouses. All sales are made directly from the warehouses, and costs of shipments of merchandise from the warehouses are borne by the customer. However, to reduce the latter, thereby making the product more competitive on a cost basis, and also to reduce the shipping costs from the factories to the warehouses, the company is considering building r additional warehouses. The manufacturer has already selected r potential sites for these warehouses, and also has determined that the annual demands for each warehouse will be b_j vacuum cleaners, $j = p + 1, p + 2, \ldots, p + r$; the demands at the existing warehouses are b_j vacuum cleaners, $j = 1, 2, \ldots, p$. The capacities of the m factories are a_i vacuum cleaners per year, $i = 1, 2, \ldots, m$.

For each of the new warehouses, the cost of building the warehouse, distributed over its estimated lifetime, is D_j dollars per year, $j = p + 1, p + 2, \ldots, p + r$.

The shipping costs are c_{ij} dollars per vacuum cleaner shipped from factory i to warehouse j, $i = 1, 2, \ldots, m$ and $j = 1, 2, \ldots, p, \ldots, p + r$.

Determine which, if any, of the new warehouses should be built, and the new shipping schedule (how many vacuum cleaners should be shipped from each factory to each warehouse) so that the total annual cost is a minimum.

4. Graph the feasible region for the following constraints, and represent it as a set of linear constraints in continuous and integer variables.

$$x_1 + x_2 \le 4 \quad \text{or} \quad x_2 \ge 6$$
$$x_1 - x_2 \le 1$$
$$x_1 \le 2 \qquad \text{or} \quad x_1 \ge 3$$

5. Graph the feasible region for the following constraints, and represent it as a set of linear constraints in continuous and integer variables.

$$|x_1 - 2x_2| \ge 4$$
$$|x_1 + x_2| \ge 3$$
$$|x_1 - 2| \le 4$$
$$|x_2 - 3| \le 4$$

6. Formulate the feasible region indicated by the shaded portion of Figure 15–6 as a set of linear constraints in continuous and integer variables, with as few of the latter as possible.

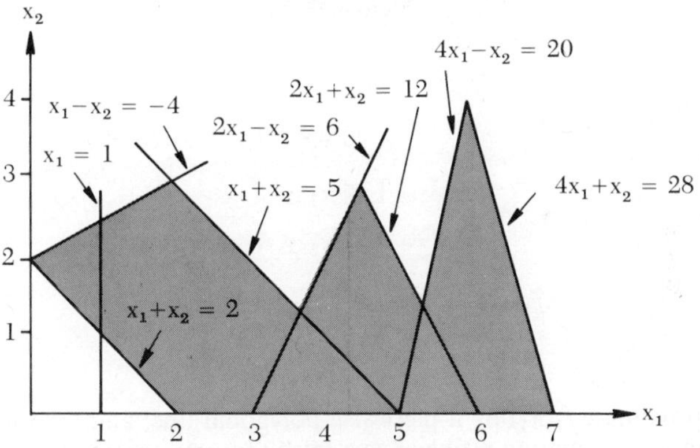

Figure 15–6

7. Formulate the set $S = \{(x_1, x_2) \mid 2x_1 + x_2 \le 4, \text{ or } x_1 + 2x_2 \le 4, \text{ or both}; \text{ and } x_1, x_2 \text{ non-negative}\}$ as a set of linear constraints in continuous and integer variables.

8. Formulate the set $S = \{(x_1, x_2) \mid 2x_1 + x_2 \le 4, \text{ or } x_1 + 2x_2 \le 4, \text{ but not both}; \text{ and } x_1, x_2 \text{ non-negative}\}$ as a set of linear constraints in continuous and integer variables.

9. Formulate the traveling salesman problem with the triply subscripted variable x_{ijk} defined by equation (15–4). Compare your formulation (in terms of the numbers of variables and constraints) with that presented in Section 15–3. [Hint: If any given $x_{ijk} = 1$, then $x_{jp,k+1} = 1$ for exactly one p, $p \neq i, j$. Why?].

10. Consider the following separable nonlinear programming problem:

$$\text{Maximize } z = f_1(x_1) + f_2(x_2)$$

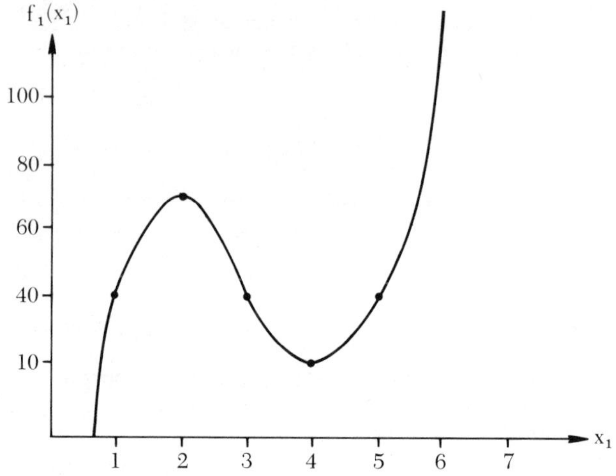

Figure 15–7

subject to:

$$x_1 + x_2 \leq 10$$
$$1 \leq x_1 \leq 5$$
$$0 \leq x_2 \leq 7$$

where

$$f_1(x_1) = 10x_1^3 - 90x_1^2 + 230x_1 - 110$$
$$f_2(x_2) = 75x_2$$

(a) Approximate $f_1(x_1)$ by a piecewise polygonal line, and then formulate the resulting approximating problem as a mixed integer-continuous variable linear programming problem. A graph of $f_1(x_1)$ is given in Figure 15–7.
(b) Formulate the approximating problem found in part (a) into one involving only continuous variables, as per equation (15–34).

REFERENCES

1. Bellmore, M., and Nemhauser, G. L.: The traveling salesman problem: A survey, *Oper. Res.*, 16:538–558, 1968.
2. Cooper, Leon, and Drebes, Charles B.: An approximate solution method for the fixed charge problem, *Nav. Res. Logist. Quart.*, 16:101–113, 1967.

3. Cooper, Leon, and Steinberg, David: *Introduction to Methods of Optimization*, Philadelphia, W. B. Saunders Co., 1970.
4. Hadley, G.: *Nonlinear and Dynamic Programming*. Reading, Mass., Addison-Wesley, 1964.
5. Little, J. D. C., Murty, G., Sweeney, D. W., and Karel, C.: An algorithm for the traveling salesman problem. *Oper. Res.* 11:972–989, 1963.
6. Manne, Alan S.: On the job-shop scheduling problem. *Oper. Res.* 8:219–223, 1960.
7. Shapiro, Donald M.: Algorithms for the solution of the optimal cost and bottleneck traveling salesman problems. Doctoral Dissertation, Washington University, 1967.
8. Steinberg, David I.: The fixed charge problem. *Nav. Res. Logist. Quart.*, 17:217–235, 1970.
9. Weingartner, H. Martin: *Mathematical Programming and the Analysis of Capital Budgeting Problems*. Chicago, Markham Publishing Co., 1967.

16

Integer Linear Programming: Algorithms

16–1 AN INTRODUCTION AND OVERVIEW

In Chapter 15, we examined some of the many types of models which can be formulated as integer programming problems. In this chapter, we shall consider the question of how to solve such problems. Unlike the situation with linear programming problems, in which there is basically one algorithm—with variations—which is almost universally used to solve these problems, there are a large number of substantially different algorithms available for solving integer programming problems. Unfortunately, these algorithms have been, for the most part, rather disappointing in terms of their computational performance. Until recently, no one algorithm had proven itself superior to all the others. Furthermore, no algorithm has shown itself capable of being consistently able to solve large integer programming problems. The primary limitation on integer programming algorithms is the number of integer variables in the particular problem. Generally speaking, problems with more than several hundred integer variables cannot be solved, except in some special cases.

Because the current knowledge regarding integer programming is so unsatisfactory, research activity in this field is very intense, and there is some reason to hope that within the next decade substantial progress will be made toward developing really effective integer programming algorithms.

It is beyond the scope of this text to present in detail the many algorithms which have been developed. Instead, our aim is to provide the reader with an overview of the subject and to present several of the more important algorithms. For a more detailed and thorough treatment of the subject of integer programming, the reader may consult Garfinkel and Nemhauser,[10] Greenberg,[20] Hu,[22] and Salkin.[29]

Although there are a great many different integer programming algorithms, there are essentially three basically different methods of approach which are employed and which we may use to classify the algorithms: (1) Cutting plane methods; (2) Enumeration methods; and (3) Partitioning.

We shall discuss each of these approaches in general terms, and provide references to some of the algorithms which are representative of each category.

The basic idea of a cutting plane method is to derive new constraints—called "cutting planes" or "cuts"—for the problem so that the resulting feasible region for the related linear programming problem (the same problem without the integer restrictions) has an optimal extreme point with the desired integer properties.

Within the general cutting plane framework, there are several different approaches. In some methods, simplex tableaux are constructed which contain only integer elements, and this property is maintained at each iteration. Algorithms of this type have been developed by Glover,[12] Gomory,[15] and Young.[34] Such algorithms are only applicable to pure integer programming problems. Gomory's algorithm[15] begins with a solution which is feasible for the related dual linear programming problem, but not feasible for the primal, and works towards optimality by applying the dual simplex algorithm. In the algorithms of Glover[12] and Young,[34] the initial solution is a feasible integer solution, and the simplex method is used to find the optimal integer solution (by adding appropriate cutting plane constraints derived during the course of the computations). These latter two algorithms have the advantage that if the computations become excessive and the algorithm must be prematurely terminated, one at least has a feasible integer solution, whereas with Gomory's and other "dual feasible" algorithms this is not the case. There has been only a limited amount of computational experience with these algorithms, and probably no judgment should be made on their potential effectiveness.

There are also cutting plane methods in which the elements of the simplex tableau are not required to be integers. Gomory has developed several such methods, one of which is applicable only to the pure integer programming problem,[14] and the other of which may be used for either pure or mixed integer problems.[16] These two methods are perhaps the most widely used of the cutting plane methods and are discussed in detail in Sections 16–2 and 16–3. Some research has also been conducted toward applying modern group theory to developing cutting plane methods. See, for example, Gomory[17,18,19] and Shapiro.[30,31]

Our second category of integer programming methods, enumeration methods, may be subdivided into two types, frequently referred to as *branch and bound methods* and *search* or *implicit enumeration methods.* Among the search methods are those of Balas,[1,2] Cook,[4] Cooper,[5] Cooper and Echols,[8] Geoffrion,[11] Glover,[12] Jambekar and Steinberg,[23,24,33] Krolak,[25,26] and Lemke and Spielberg.[28]

Among the branch and bound methods are those of Dakin,[6] Driebeck,[7] and Land and Doig.[27] The Land and Doig algorithm is generally credited

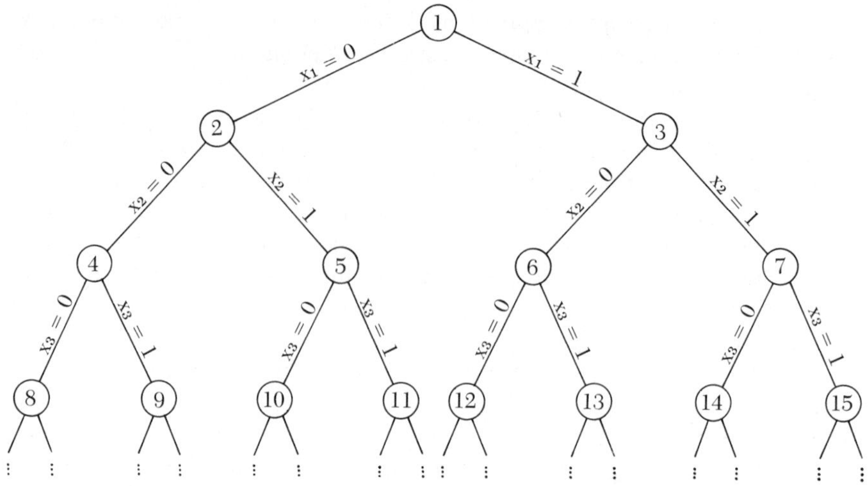

Figure 16–1 Tree of solutions for 0–1 programming problem.

with popularizing the technique. Dakins' method is a modification of Land and Doig's original algorithm, and refinements of Dakin's method appear to offer the most promising approach to solving integer programming problems. Most of the more recently developed commercial computer programs for solving integer programming problems are based upon Dakin's method. This method is presented in Section 16–4.

The basic approach in both branch and bound and search methods is to enumerate all possible integer solutions, either explicitly or implicitly. Throughout this enumeration process feasible integer solutions are found, and the best of these is saved for comparison with future integer feasible solutions generated. A set of solutions is said to have been implicitly enumerated by an algorithm if the entire set has been shown to contain no solutions which are better than the best currently known solution. Hence, solutions in this set need not be evaluated individually (explicitly). This implicit enumeration is accomplished by showing that all solutions in a given set violate certain (constraint derived) relations which are necessary for the existence of an improved solution. The effectiveness of an enumeration method depends to a great extent on its ability to eliminate in this way a large percentage of solutions without having to explicitly evaluate them.

In order to describe the enumeration method approach more fully, let us illustrate how such methods would be applied to solve an integer programming problem in which all the variables are required to be either 0 or 1. (Such a problem is sometimes called a 0–1 integer programming problem, or just a 0–1 programming problem.) The set of all solutions to such a problem may be described schematically by a "tree of solutions," as shown in Figure 16–1. In this figure, the numbered circles are called *nodes*, and the lines connecting the nodes are called *branches*. Node 1 represents the set of all solutions.

A sequence of nodes and branches from node 1 to any other node k is called a *path* to node k. Along each branch one constraint is imposed on the variables, so that node k represents the set of all solutions which satisfy the original constraints plus the constraints imposed by each of the branches in the path from node 1 to node k. For example, node 10 of Figure 16–1 represents the set of all solutions for which $x_1 = 0$, $x_2 = 1$, $x_3 = 0$, in addition to the original constraints. A node further down a given path thus represents a subset of solutions of any node above it on the same path.

To totally enumerate each possible solution for a 0–1 programming problem with n variables would require generating 2^n paths, the final node of each path corresponding to exactly one solution (possibly infeasible). The object of an enumeration method is to eliminate as many of these paths as possible. To see how this is accomplished, assume our problem to be a maximization problem, with optimal solution z^*, and suppose we have found a feasible integer solution with objective function value z_L. Since $z_L \leq z^*$, z_L is a lower bound on the optimal solution (although of course the value of z^* is unknown before the problem is solved). Now, suppose we have somehow determined that an upper bound on the set of all solutions represented by Node k is z_U^k. One such upper bound is the optimal solution to the related linear programming problem, for example. If $z_U^k < z_L$, then no path which includes node k can yield an improved solution; hence, we can eliminate all such paths without explicitly evaluating all the corresponding solutions. Node k is then called a *terminal node.*

Terminal nodes are obtained in one of three ways; node k is a terminal node if:

1. $z_U^k < z_L$, as discussed above.
2. There are no feasible solutions satisfying the constraints imposed† at Node k.
3. Node k represents a single feasible solution (i.e., values for all the variables have been uniquely determined).

Both branch and bound methods and search methods generate the nodes of the tree of solutions until all paths end in terminal nodes. In this way, all solutions will have been either explicitly or implicitly considered, and the best solution found is therefore optimal.

The difference between search and branch and bound methods is in the manner in which the nodes are generated; search methods generally follow a path until it ends at a terminal node before proceeding to another path. Branch and bound methods, on the other hand, examine the nodes in a less straightforward manner, and may generate, more or less simultaneously, many different paths. Branch and bound methods generally require much

† Frequently, the constraint $\bar{c}'\bar{x} \geq z_L$ is also imposed, where $\bar{c}'\bar{x} = z$ is the objective function, since only solutions better than the current best feasible solution are of interest.

more computer storage space than search methods and, in fact, computer storage may be exceeded in the course of attempting to solve a problem. However, the greater flexibility in the order in which the nodes are examined gives branch and bound methods a greater potential for efficiency in terms of time required for solution. Some algorithms combine the various features of both techniques.

We now turn to a consideration of our third category of integer programming algorithm, that of partitioning. The principal method of this type was developed by Benders,[3] and it is an algorithm for solving mixed integer programming problems. An outline of Benders' method is given below.

Consider the mixed integer programming problem

$$\text{Maximize } z = \bar{c}_1'\bar{y} + \bar{c}_2'\bar{x}$$

Subject to:

$$A_1\bar{y} + A_2\bar{x} \leq \bar{b}$$

$$\bar{x} \geq 0$$

$$\text{Components of } \bar{y} = 0, 1$$

Thus, the variables have been partitioned into two sets, the vector of continuous variables $\bar{x}$ and the vector of 0–1 variables $\bar{y}$. For any fixed vector $\bar{y}$, the above problem becomes the ordinary linear programming problem

$$\text{Maximize } \hat{z} = z - \bar{c}_1'\bar{y} = \bar{c}_2'\bar{x}$$

Subject to

$$A_2\bar{x} \leq \bar{b} - A_1\bar{y} \equiv \hat{b}$$

$$\bar{x} \geq \bar{0}$$

The dual of the above linear programming problem is

$$\text{Min } Z = \overline{w}'\hat{b} = \overline{w}'(\bar{b} - A_1\bar{y}) = \overline{w}'\bar{b} - \overline{w}'A_1\bar{y}$$

Subject to

$$A_2'\overline{w} \geq \bar{c}_1$$

$$\overline{w} \geq \bar{0}$$

At optimality, $Z^* = \hat{z}^*$. Hence, if $\bar{w}^*$ is the optimal dual solution, and $\bar{y}$ is fixed as above,

$$\overline{w}^{*\prime}\bar{b} - \overline{w}^{*\prime}A_1\bar{y} = \hat{z}^*$$

$$= z - \bar{c}_1'\bar{y}$$

Or, the corresponding value of the original objective function is

$$z = \overline{w}^{*\prime}\bar{b} + \bar{c}_1'\bar{y} - \overline{w}^{*\prime}A_1\bar{y}$$
$$= w^{*\prime}\bar{b} + [\bar{c}_1' - \overline{w}^{*\prime}A_1]\bar{y}$$

Thus, an improved solution for the original problem must satisfy

$$\overline{w}^{*\prime}\bar{b} + [c_1' - \overline{w}^{*\prime}A_1]\bar{y} > z$$

or,

$$[\bar{c}_1' - \overline{w}^{*\prime}A_1]\bar{y} > z - \overline{w}^{*\prime}\bar{b}$$

Observe that the above constraint involves only the integer variables $\bar{y}$. If we find a vector $\bar{y}^*$ which satisfies this constraint, then we can find a corresponding improved solution for the original problem by re-solving the linear programming problem with $\bar{y} = \bar{y}^*$. In turn, the optimal solution to this linear programming problem yields a new constraint for the integer variables $\bar{y}$. We then find a new $\bar{y}^*$ which satisfies the two constraints thus far generated, and repeat the procedure.

If we denote by S the set of constraints generated for the integer variables, Benders' method can be summarized as follows:

1. Find an integer vector $\bar{y}$ which is feasible for S (initially, S may be the null set). Call this vector $\bar{y}^*$.
2. With $\bar{y} = \bar{y}^*$, solve the linear programming problem

$$\text{Maximize } \hat{z} = \bar{c}_2'\bar{x}$$

Subject to

$$A_2\bar{x} \leq \hat{\bar{b}}$$
$$\bar{x} \geq \bar{0}$$

or its dual,

$$\text{Minimize } Z = \overline{w}'\hat{\bar{b}}$$
$$A_2\overline{w} \geq \bar{c}_2$$
$$\overline{w} \geq \bar{0}$$

If $\overline{w}^*$ denotes the optimal dual solution, and $z^* = \overline{w}^{*\prime}\hat{\bar{b}} + \bar{c}_1'\bar{y}^*$, then an improved solution for the original problem must satisfy

$$[\bar{c}_1' - \overline{w}^{*\prime}A_1]\bar{y} > z^* - \overline{w}^{*\prime}\hat{\bar{b}}$$

Add the above constraint to the set S and return to Step 1.

The above process is repeated until no feasible solution can be found in Step 1, at which point we terminate the algorithm. The best solution found is the optimal solution.

Step 1 requires the solution of a set of linear inequalities in integer variables, and this may be accomplished by any method, e.g., by an enumeration method. However, observe that S will generally contain fewer constraints than the original problem, and only a feasible integer solution is required, not an optimal solution.

Constraints of the type generated in Benders' method can also be used in conjunction with an enumeration method, to help cut down on the number of nodes examined. See, for example, Jambekar and Steinberg.[23,24,33]

16–2 GOMORY'S CUTTING PLANE ALGORITHMS

In this section and the next we shall investigate in some detail two of the cutting plane algorithms developed by Gomory.[13,14,16] We have singled out these particular algorithms for presentation for several reasons. First, these methods were among the first algorithms to be devised for solving integer programming problems, and hence are of some historical interest. Second, in many instances Gomory's algorithms are still the most effective algorithms available for solving large integer programming problems. Unfortunately, as noted in the previous section, the performance of these algorithms is completely unpredictable.

We shall begin our discussion by considering the pure integer programming problem (ILP):

$$\text{Maximize } z = \bar{c}'\bar{x} \tag{16–1}$$

Subject to

$$A\bar{x} = \bar{b} \tag{16–2}$$

$$\bar{x} \geq \bar{0}$$

$$x_j \text{ integer}, \quad j = 1, 2, \ldots, n \tag{16–3}$$

The problem with the integer constraints (16–3) deleted shall be called the related LP problem. The basic idea in a cutting plane algorithm is the following: Given an integer programming problem, solve the related LP problem. If the optimal solution contains only integer-valued variables, clearly, it must be the optimal ILP solution. If some variables in the optimal LP solution are not integers, a new constraint (the exact nature of which is discussed below) is added to the problem, and the new related LP problem is solved. Again, we check to see if the optimal solution to this new LP problem satisfies the integer requirements; if so, it is the optimal solution for the *original* ILP problem. If, on the other hand, some of the variables are again not integers, then another new constraint is added, and the process is repeated until one of the related LP problems is found to have an all integer optimal solution. The first such all integer solution obtained is the optimal solution for the original problem.

In order for the above process to be valid, the new constraint which is added at each step must possess the following properties:

(1) Every feasible *integer* solution to the original problem must also be a feasible solution to the new problem resulting from the addition of the constraint.
(2) The optimal solution to the LP problem solved at each step must become infeasible after the new constraint is added.

Property (1) states that the addition of the new constraint does not eliminate any of the integer solutions from the set of feasible solutions, thus ensuring that the optimal integer solution will not be made infeasible.

Property (2) is necessary because if the addition of the constraint did not eliminate the optimal LP solution, the latter would also be optimal for the new LP problem. Since we already know the former optimal LP solution is not an all-integer solution, nothing will have been gained by adding a constraint in which this solution is not excluded.

Each of the constraints added in the above manner are called cutting plane constraints, because they are hyperplanes which cut off a portion of the convex set of feasible solutions for the related LP problem.

Let us now investigate the nature of the particular cutting plane constraints devised by Gomory for solving the pure integer programming problem (16–1), (16–2), (16–3). For simplicity, we shall assume that the optimal solution to the related LP problem contains the first m columns of A in the basis. Hence, A can be partitioned into two submatrices, $A = [B, R]$, where B contains the first m columns of A (the basis matrix) and R contains the remaining $(n - m)$ columns of A. The corresponding optimal solution is $\bar{x}^* = [\bar{x}_B^*, \bar{x}_R^*]$, where $\bar{x}_R^* = \bar{0}$. Now, since $A\bar{x} = \bar{b}$, we have

$$[B, R]\begin{bmatrix} \bar{x}_B^* \\ \bar{x}_R^* \end{bmatrix} = \bar{b}$$
$$B\bar{x}_B^* + R\bar{x}_R^* = \bar{b} \tag{16–4}$$

Premultiplying equation (16–4) by B^{-1} yields

$$\bar{x}_B^* + (B^{-1}R)\bar{x}_R^* = B^{-1}\bar{b}$$

or

$$\bar{x}_B^* = B^{-1}\bar{b} - (B^{-1}R)\bar{x}_R^* \tag{16–5}$$

Since R consists of the columns of A, we can express equation (16–5) in terms of the columns of the simplex tableau; namely, letting $\bar{y}_j = B^{-1}\bar{a}_j$ and $\bar{y}_0 = B^{-1}\bar{b}$, we have

$$\bar{x}_B^* = \bar{y}_0 - \sum_{j \in N} \bar{y}_j x_j \tag{16–6}$$

where

$$N = \{j \mid \bar{a}_j \text{ is a column of } R\}$$
$$= \{j \mid x_j \text{ is nonbasic}\}$$

Suppose now that the rth basic variable of $\bar{x}_B^*$ is not an integer. From (16–6), we see that

$$x_{Br} = y_{r0} - \sum_{j \in N} y_{rj} x_j \tag{16–7}$$

In the current solution, all $x_j, j \in N$, are zero, and so y_{ro} is not an integer. Let us express y_{ro} as the sum of its integer part and its fractional part. Let w_{ro} be the integer (or "whole") part of y_{ro} (i.e., the greatest integer less than or equal to y_{ro}) and let f_{ro} be the remaining fractional part. Thus,

$$y_{ro} = w_{ro} + f_{ro} \tag{16–8}$$

$$w_{ro} \geq 0 \tag{16–9}$$

$$0 < f_{ro} < 1 \tag{16–10}$$

Let us similarly represent the $y_{rj}, j \in N$:

$$y_{rj} = w_{rj} + f_{rj} \tag{16–11}$$

$$0 \leq f_{rj} < 1 \tag{16–12}$$

If we substitute equations (16–8) and (16–11) into equation (16–7) we obtain

$$x_{Br} = w_{ro} + f_{ro} - \sum_{j \in N} (w_{rj} + f_{rj}) x_j$$
$$= \left(w_{ro} - \sum_{j \in N} w_{rj} x_j\right) + \left(f_{ro} - \sum_{j \in N} f_{rj} x_j\right) \tag{16–13}$$

Now, if we wish to modify the current solution so that the new solution is all-integer, then at least one of the currently nonbasic x_j ($j \in N$) must become positive (since there is a unique solution with all these $x_j = 0$, and it is $\bar{x}_B^*$). In such an all integer solution the quantity contained in the first pair of parentheses in equation (16–13) will be an integer, since each x_j must be an integer, and each w_{rj} is an integer by definition. Thus, the new value of x_{Br} can be expressed by

$$x_{Br} = \text{integer} + \left(f_{ro} - \sum_{j \in N} f_{rj} x_j\right) \tag{16–14}$$

Moreover, we also desire x_{Br} to be an integer; if this is to be attained, then the quantity $\left(f_{ro} - \sum_{j \in N} f_{rj} x_j\right)$ must be an integer. Since each $f_{rj} \geq 0$ and each $x_j \geq 0$, it is clear that

$$\sum_{j \in N} f_{rj} x_j \geq 0$$

Hence, the quantity $\left(f_{ro} - \sum_{j \in N} f_{rj} x_j\right)$ consists of one positive quantity, $\sum_{j \in N} f_{rj} x_j$, subtracted from a positive quantity f_{ro}, which is less than one (by definition; see equation (16–10)). The entire quantity $\left(f_{ro} - \sum_{j \in N} f_{rj} x_j\right)$, therefore, cannot be a positive integer. If it is to be an integer, then, it must be a negative integer (or zero):

$$f_{ro} - \sum_{j \in N} f_{rj} x_j \leq 0 \tag{16–15}$$

The inequality (16–15) is the cutting plane constraint derived by Gomory.[13,14] Observe that it does indeed satisfy the two properties described earlier in this section; namely, the current solution violates (16–15), since all $x_j = 0, j \in N$ and $f_{ro} > 0$. Furthermore, by our derivation of (16–15), every integer feasible solution of the original ILP problem will satisfy (16–15).

In our discussion thus far, we have not considered how to determine the cutting plane constraint if more than one basis variable is not an integer. In the most frequently employed technique, the x_{Br} whose fractional part f_{ro} is the largest is used to obtain the constraint (16–15). However, there are many other possible ways to select the cutting plane constraint. Gomory[14] has proven that with the proper choice for the cutting plane constraint at each step, the algorithm converges to the optimal integer solution in a finite number of iterations. A proof of convergence is also given by Hadley.[21]

It can also be shown that it is never necessary to have more than $(n + 1)$ constraints at one time in the related LP problem; that is, under certain conditions, it is possible to drop constraints previously added, so that the total number of constraints never exceeds $(n + 1)$ (see Hadley[21]). This fact makes sense intuitively, since the original problem only had n variables, all of which can be positive in a basic feasible solution for a set of $(n + 1)$ constraints. As we observed in Example 1 of Chapter 15, an optimal integer solution may consist of all positive variables.

When the constraint (16–15) is added to the optimal tableau for the related LP problem, the resulting problem is solved by the dual simplex algorithm, since the current solution is infeasible but still "optimal" (i.e., all $z_j - c_j \geq 0$). Addition of constraints of the type (16–15) is discussed in Section 12–5.

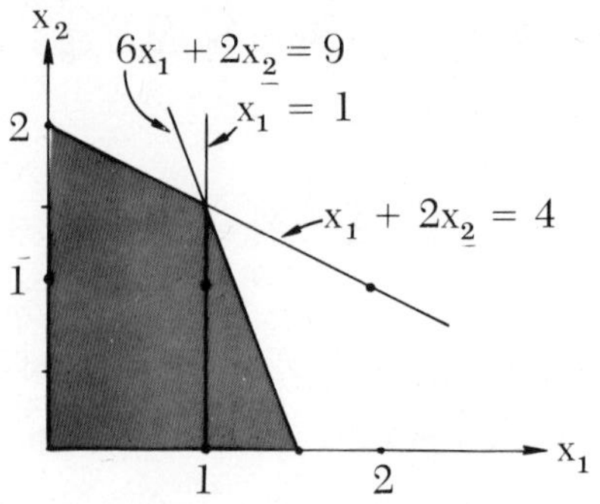

Figure 16–2 Feasible region for problem of Example 1.

EXAMPLE 1

Consider the following problem, illustrated in Figure 16–2:

$$\text{Maximize } z = x_1 - x_2$$

Subject to

$$x_1 + 2x_2 \leq 4 \tag{16–16}$$

$$6x_1 + 2x_2 \leq 9 \tag{16–17}$$

$$x_1, x_2 \geq 0$$

$$x_1, x_2 \text{ integer}$$

The optimal solution to the related LP problem occurs at $(x_1, x_2) = (\frac{3}{2}, 0)$. The optimal simplex tableau is given in Table 16–1 (where x_3 and x_4 are the slack variables added to constraints (16–16) and (16–17), respectively).

Since both x_{B1} and x_{B2} are noninteger, we may use either in the calculation of the cutting plane (16–15). We shall choose x_{B1} arbitrarily. From

TABLE 16–1 OPTIMAL TABLEAU–RELATED LP PROBLEM

Variables in basis	$\bar{b}$	$\bar{y}_1$	$\bar{y}_2$	$\bar{y}_3$	$\bar{y}_4$
$x_{B1} = x_1$	$\frac{3}{2}$	1	$\frac{2}{6}$	0	$\frac{1}{6}$
$x_{B2} = x_3$	$\frac{5}{2}$	0	$\frac{10}{6}$	1	$-\frac{1}{6}$
z	$\frac{3}{2}$	0	$\frac{8}{6}$	0	$\frac{1}{6}$

Table 16–1, we see that $y_{12} = \frac{2}{6}$, $y_{14} = \frac{1}{6}$, and $y_{10} = \frac{3}{2} = 1 + \frac{1}{2}$. Thus, (16–15) becomes

$$\tfrac{1}{2} - [\tfrac{2}{6}x_2 + \tfrac{1}{6}x_4] \leq 0$$

or

$$2x_2 + x_4 \geq 3 \tag{16–18}$$

However, since x_4 is the slack variable for (16–17), we have

$$x_4 = 9 - 6x_1 - 2x_2 \tag{16–19}$$

Substitution of (16–19) into (16–18) yields

$$2x_2 + 9 - 6x_1 - 2x_2 \geq 3$$

which reduces to

$$x_1 \leq 1 \tag{16–20}$$

We wish to emphasize that the constraints (16–18) and (16–20) are completely equivalent; we have calculated (16–20) only so that we may graph this example in terms of x_1 and x_2. The cutting plane constraint (16–20) is shown in Figure 16–2, and the optimal solution to this new related LP problem is easily seen to occur at $(x_1, x_2) = (1, 0)$, which is the desired integer solution.

However, suppose instead that we had chosen to calculate the cutting plane constraint based on x_{B2}. From Table 16–1, we see that $y_{22} = \frac{10}{6}$, $y_{24} = -\frac{1}{6}$, and $y_{20} = \frac{5}{2}$. Hence, $f_{22} = \frac{4}{6}$, $f_{24} = \frac{5}{6}$ (since f_{rj} must be non-negative, and $y_{24} = -\frac{1}{6} = -1 + \frac{5}{6}$). The new constraint would then have been

$$\tfrac{1}{2} - [\tfrac{4}{6}x_2 + \tfrac{5}{6}x_4] \geq 0$$

or

$$4x_2 + 5x_4 \geq 3 \tag{16–21}$$

To help us visualize this constraint, we again eliminate x_4 by substituting equation (16–19) into (16–21), yielding

$$4x_2 + 5(9 - 6x_1 - 2x_2) \geq 3$$

or

$$5x_1 + x_2 \leq 7 \tag{16–22}$$

If (16–22) is added to the original problem, instead of (16–20), the resulting LP problem has an optimal solution of $(x_1, x_2) = (\frac{7}{5}, 0)$, which is not all integer. Thus, the choice of constraints to be added affects the efficiency of the algorithm. Unfortunately, there is no *a priori* method of determining the best choice of a cutting plane constraint.

Computational experience with Gomory's method indicates that it is capable of solving some reasonably large ILP problems; however, it has also been known (see Hadley[21]) to run for over 2000 iterations without converging to the optimal ILP solution on problems which require less than twenty iterations to obtain the optimal solution for the related LP problem.

Gomory has developed another cutting plane algorithm for solving the pure integer programming problem.[15] In this algorithm, it is assumed that an initial tableau is available with all integer coefficients. Then, at each iteration a cutting plane is constructed in such a way that the pivot element is always unity, so that the resulting tableau again has all integer coefficients. However, this algorithm has not proved to be computationally promising.

16-3 GOMORY'S MIXED INTEGER PROGRAMMING ALGORITHM

Gomory[16] has developed a modified version of the algorithm described in the previous section, for solving mixed integer-continuous variable problems. The modification consists of a new formula for determining the cutting plane constraint to be added after each related LP problem has been solved.

Consider then the problem

$$\left.\begin{aligned} &\text{Maximize } z = \bar{c}'\bar{x} \\ \text{Subject to} \quad & A\bar{x} = \bar{b} \\ &\bar{x} \geq \bar{0} \\ &x_j \text{ integer}, j \in I \end{aligned}\right\} \tag{16–23}$$

Thus, the set I consists of the subscripts corresponding to those variables required to be integer. If $I = \{1, 2, \ldots, n\}$, the above problem is the pure integer programming problem discussed in the previous section.

Suppose, now, that we have solved the related LP problem, and the rth basic variable is not an integer, but is required to be an integer. Then, as in the previous section (see equation (16–7)), we can write

$$x_{Br} = y_{ro} - \sum_{j \in N} y_{rj} x_j \tag{16–24}$$

where $N = \{j \mid x_j \text{ is nonbasic}\}$. Again letting $y_{ro} = w_{ro} + f_{ro}$, as defined by equations (16–8) through (16–10), and substituting this relation into equation (16–24) we obtain

$$x_{Br} = w_{ro} + f_{ro} - \sum_{j \in N} y_{rj} x_j$$

Or, if x_{Br} is to be an integer,

$$f_{ro} - \sum_{j \in N} y_{rj} x_j = x_{Br} - w_{ro} = \text{integer} \tag{16–25}$$

Let us now partition N into two sets, as follows:

$$N^+ = \{j \mid y_{rj} \geq 0, \qquad j \in N\}$$
$$N^- = \{j \mid y_{rj} < 0\}$$

Then (16–25) becomes

$$f_{ro} - \sum_{j \in N^+} y_{rj} x_j - \sum_{j \in N^-} y_{rj} x_j = \text{integer} \tag{16–26}$$

Consider the following two cases:

1. $$\sum_{j \in N} y_{rj} x_j \geq 0$$

2. $$\sum_{j \in N} y_{rj} x_j < 0$$

CASE 1. $$\sum_{j \in N} y_{rj} x_j \geq 0$$

Then, by (16–26),

$$\sum_{j \in N} y_{rj} x_j = f_{ro} + P$$

where $P = 0, 1, 2, \ldots$ (i.e., P is a non-negative integer). Hence, any feasible solution to (16–23) must satisfy

$$\sum_{j \in N} y_{rj} x_j \geq f_{ro}$$

Moreover, since $\sum_{j \in N^-} y_{rj} x_j \leq 0$, for any feasible solution, it is clear that any feasible solution to (16–23) also must satisfy

$$\sum_{j \in N^+} y_{rj} x_j \geq f_{ro} \tag{16–27}$$

CASE 2. $$\sum_{j \in N} y_{rj} x_j < 0$$

Then, from (16–25) it must be true that

$$\sum_{j \in N} y_{rj} x_j = f_{ro} - P$$

where $P = 1, 2, 3, \ldots$ (i.e., P is a strictly positive integer). Therefore, for any feasible solution to (16–23),

$$\sum_{j \in N} y_{rj} x_j \leq f_{ro} - 1$$

$$\sum_{j \in N^+} y_{rj} x_j + \sum_{j \in N^-} y_{rj} x_j \leq f_{ro} - 1$$

Moreover, since $\sum_{j \in N^+} y_{rj} x_j \geq 0$, it must also be true that

$$\sum_{j \in N^-} y_{rj} x_j \leq f_{ro} - 1 \tag{16–28}$$

Now, suppose we multiply (16–28) by the strictly negative number $f_{ro}/(f_{ro} - 1)$, yielding

$$\sum_{j \in N^-} \left(\frac{f_{ro}}{1 - f_{ro}}\right) |y_{rj}| \geq f_{ro} \tag{16–29}$$

We can combine the two inequalities (16–27) and (16–29) obtained above for Cases 1 and 2, respectively, as follows:

$$\sum_{j \in N^+} y_{rj} x_j + \sum_{j \in N^-} \left(\frac{f_{ro}}{1 - f_{ro}}\right) |y_{rj}|\, x_j \geq f_{ro} \tag{16–30}$$

Observe that, regardless of which case holds, the inequality (16–30) must be satisfied by any feasible solution to (16–23), since the left-hand sides of (16–27) and (16–29) are both positive, and at least one must be greater than or equal to f_{ro}. Observe also that the current optimal solution to the related LP problem does not satisfy (16–30), since all x_j, $j \in N^+$ or $j \in N^-$, are currently zero, and $f_{ro} > 0$. Therefore, (16–30) can be used as the cutting plane constraint.

However, in deriving (16–30), we did not make use of the fact that some of these nonbasic variables may also be required to be integers. (Recall that in deriving the all integer cut (16–15) we did use this fact.) If we were to incorporate this additional information, the resulting cutting plane constraint might even be "better" than (16–30). By "better" we mean that the constraint will further restrict the feasible values of the nonbasic variables. This does not necessarily mean that such a constraint will actually be more effective in terms of helping to solve the problem. Hence, the smaller the coefficient of a variable x_j, in a particular constraint, the more the range for that variable is restricted, and the better the constraint is, with respect to that particular variable.

With the above comments in mind, we shall attempt to develop a cutting plane constraint in which the coefficient of each variable is as small as possible.

We proceed by first defining

$$\begin{aligned}
N_I &= N \cap I = \{j \mid x_j \text{ is nonbasic and required to be integer}\} \\
N_C &= \{j \mid x_j \text{ is nonbasic } (j \in N) \text{ and not required to be integer}\} \\
N_C^+ &= N^+ \cap N_C \\
&= \{j \mid y_{rj} \geq 0 \quad \text{and} \quad j \in N_C\} \\
N_C^- &= N^- \cap N_C \\
&= \{j \mid y_{rj} < 0 \quad \text{and} \quad j \in N_C\}
\end{aligned}$$

Also, as in Section 16–2, we will let $y_{rj} = w_{rj} + f_{rj}$, where w_{rj} is an integer and $0 \leq f_{rj} < 1$. Then, equation (16–25) becomes

$$f_{ro} - \sum_{j \in N_I} f_{rj} x_j - \sum_{j \in N_I} w_{rj} x_j - \sum_{j \in N_C} y_{rj} x_j = \text{integer}$$

Moreover, since $\sum_{j \in N_I} w_{rj} x_j$ must be an integer in any feasible solution to (16–23),

$$f_{ro} - \sum_{j \in N_I} f_{rj} x_j - \sum_{j \in N_C} y_{rj} x_j = \text{integer} \tag{16–31}$$

Now, if we apply the same arguments to (16–31) which we employed above to obtain (16–30), we will obtain the following cutting plane constraint:

$$\sum_{j \in N_I} f_{rj} x_j + \sum_{j \in N_C^-} y_{rj} x_j + \sum_{j \in N_C^+} \left(\frac{f_{ro}}{1 - f_{ro}}\right) |y_{rj}| \, x_j \geq f_{ro} \tag{16–32}$$

We can obtain still another cutting plane constraint by letting w_{rj}^* be the smallest integer greater than y_{rj}, so that $y_{rj} - f_{rj} + 1 = w_{rj}^*$. Then, equation (16–25) becomes

$$f_{ro} - \sum_{j \in N} y_{rj} x_j = \text{integer} \tag{16–25}$$

$$f_{ro} - \sum_{j \in N_I} (w_{rj}^* + f_{rj} - 1) x_j - \sum_{j \in N_C} y_{rj} x_j = \text{integer}$$

$$f_{ro} - \sum_{j \in N_I} (f_{rj} - 1) x_j - \sum_{j \in N_C} y_{rj} x_j = \text{integer} \tag{16–33}$$

since $\sum_{j \in N_I} w^*_{rj} x_j$ must be an integer in any feasible solution to (16–23). Observe that $\sum_{j \in N_I} (f_{rj} - 1)x_j \leq 0$. We can again apply the same arguments to (16–33) that we applied to (16–31) to obtain the cutting plane constraint

$$\sum_{j \in N_I} \left(\frac{f_{ro}}{1 - f_{ro}}\right)(1 - f_{rj})\, x_j + \sum_{j \in N_C^+} y_{rj}\, x_j + \sum_{j \in N_C^-} \left(\frac{f_{ro}}{1 - f_{ro}}\right) |y_{rj}|\, x_j \geq f_{ro} \tag{16–34}$$

Upon comparing coefficients in (16–32) and (16–34) we see that, for $j \in N_I$, we can choose either f_{rj} or $f_{ro}(1 - f_{rj})/(1 - f_{ro})$ as the coefficient of x_j, $j \in N_I$. Since our goal is to make the coefficient of each x_j as small as possible, we can choose the smallest of these two quantities.

In particular, note that if

$$\frac{f_{ro}(1 - f_{rj})}{1 - f_{ro}} > f_{rj}$$

then

$$f_{ro} - f_{ro} f_{rj} < f_{rj} - f_{rj} f_{ro}$$
$$f_{ro} < f_{rj}$$

Hence, f_{rj} is the smaller of the two if $f_{rj} > f_{ro}$. Therefore, we can obtain an even better cutting plane constraint by combining the above, and choosing the following constraint:

$$\sum_{j \in N} d_{rj} x_j \geq f_{ro} \tag{16–35}$$

where

$$d_{rj} = \begin{cases} y_{rj}, & j \in N_C^+ \\ \left(\dfrac{f_{ro}}{1 - f_{ro}}\right) |y_{rj}|, & j \in N_C^- \\ f_{rj}, & j \in N_I \text{ if } f_{rj} \leq f_{ro} \\ \left(\dfrac{f_{ro}}{1 - f_{ro}}\right)(1 - f_{rj}), & j \in N_I \text{ if } f_{rj} > f_{ro} \end{cases} \tag{16–36}$$

The above cutting plane constraint could also be used for a pure integer programming problem, in which case the sets N_C^+ and N_C^- would be empty.

16–4 A BRANCH AND BOUND ALGORITHM

In this section we shall study Dakin's method,[6] which is a modification of the original branch and bound algorithm proposed by Land and Doig.[27] As mentioned in Section 16–1, this method appears to have the potential to become one of the most effective techniques for solving integer programming problems, and is, in fact, the basic approach of most of the commercially written computer programs for solving integer programming problems. The method can be used for either pure or mixed integer programming problems. In general, suppose we are solving the problem

$$\text{Maximize } z = \bar{c}'\bar{x}$$

Subject to

$$A\bar{x} = \bar{b}$$

$$\bar{x} \geq \bar{0}$$

$$x_j \text{ integer}, \qquad j \in I$$

As in the previous section, I is the set of subscripts for those variables required to be integer; if $I = \{1, 2, \ldots, n\}$, the problem is a pure integer programming problem.

An outline of the algorithm follows:

1. Solve the related LP problem. If all $x_j, j \in I$, are integers, we have obtained the desired optimal solution. If at least one $x_j, j \in I$, is non-integer, proceed to Step 2.
2. Select one of the non-integer $x_j, j \in I$. Suppose

$$x_j = w_j + f_j \tag{16-37}$$

where w_j is an integer and

$$0 < f_j < 1 \tag{16-38}$$

Any feasible integer solution must therefore satisfy either $x_j \leq w_j$ or $x_j \geq w_j + 1$. We add each of these constraints, individually, to the constraint set, and solve the resulting two related LP problems.
3. For each of the two solutions found in Step 2:
(a) If the solution is a feasible integer solution, compare its objective function value with the best feasible integer solution found thus far. Save the better of the two. Select another non-integer $x_j, j \in I$, if any exists, and return to Step 2.
(b) If the resulting LP problem has no feasible solution, select another non-integer $x_j, j \in I$, if any exists, and return to Step 2.
(c) If the solution is not a feasible integer solution, repeat the process of Step 2 on the constraint set for this related LP problem.
4. The algorithm terminates when all possibilities have been exhausted in Step 3.

The algorithm generates a tree of solutions similar to that in Figure 16–1. Each non-terminal node has two branches emanating from it, corresponding to the two constraints $x_j \leq w_j$ and $x_j \geq w_j + 1$. However, unlike Figure 16–1, the same variable x_j may be constrained by more than one pair of branches.

Before describing the algorithm in more detail, we shall illustrate it with an example.

EXAMPLE 2

Consider the following problem, illustrated in Figure 16–3:

$$\text{Maximize } z = x_1 + x_2$$

Subject to

$$8x_1 + 63x_2 \leq 273$$

$$4x_1 - 3x_2 \leq 10$$

$$x_1, x_2 \geq 0$$

$$x_1, x_2 \text{ integer}$$

The tree of solutions generated by the procedure below is given in Figure 16–4.

We begin by solving the related LP problem (Step 1 of our outline), which yields $z = 8\frac{11}{12}$, $x_1 = 5\frac{1}{4}$, $x_2 = 3\frac{2}{3}$ (labeled Node #1 in Figure 16–4). Since neither x_1 nor x_2 is integer, we can choose either variable to begin implementation of Step 2. We arbitrarily select x_1, yielding the two branches $x_1 \leq 5$ and $x_1 \geq 6$, to Nodes 2*A* and 2*B*, respectively. The constraints thus imposed on Node 2*B* have no feasible solution, hence, no further branches emanating from this node are necessary. At Node 2*A*, we find the optimal solution for the related LP problem to be $z = 8\frac{44}{63}$, $x_1 = 5$, $x_2 = 3\frac{44}{63}$. Since x_2 is the only non-integer valued variable, we form the two branches $x_2 \leq 3$ and $x_2 \geq 4$, emanating from Node 2*A*, to Nodes 3*A* and 3*B*, respectively.

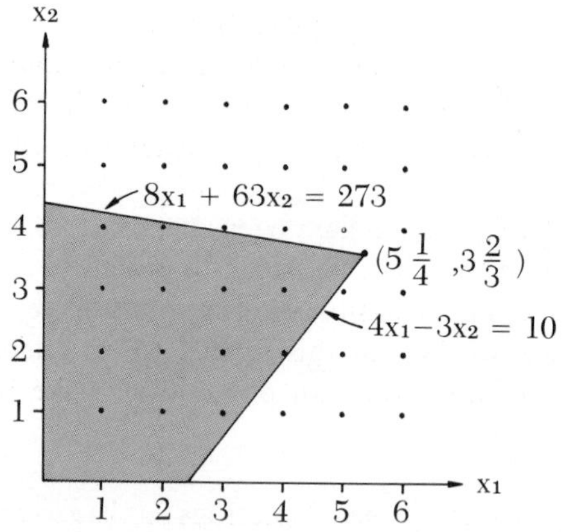

Figure 16–3 Feasible region for problem of Example 2

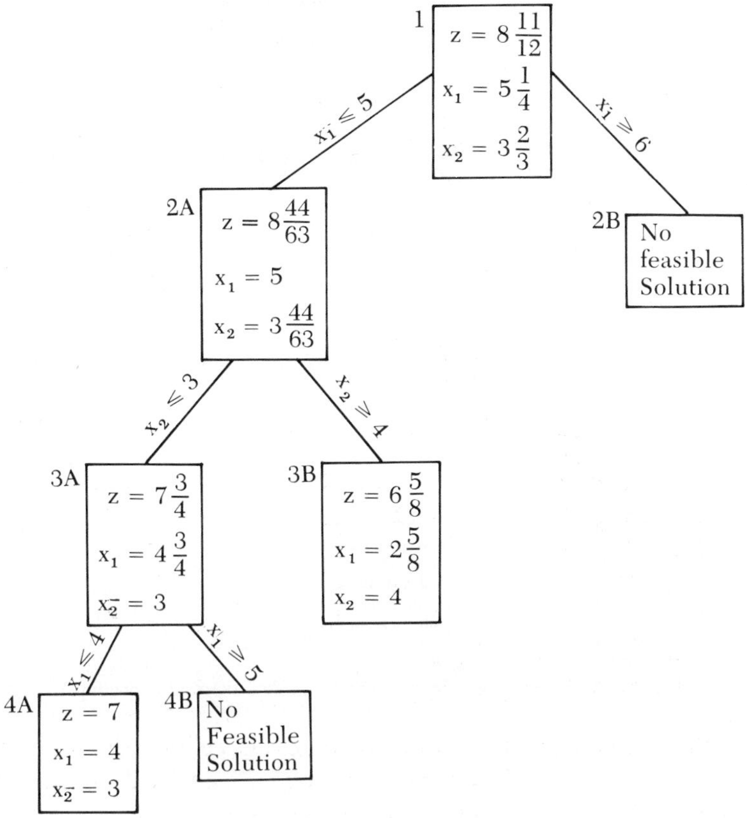

Figure 16–4 Tree of solutions for Example 2

Upon solving the related LP problems corresponding to Nodes 3*A* and 3*B*, we find that neither yields an integer solution and that Node 3*A* yields a larger value of z. Therefore, we shall proceed to branch on Node 3*A*, since it has a better potential for yielding the ultimate optimal integer solution. At Node 3*A*, only x_1 is not integer valued, so that the two branches emanating from this node have the constraints $x_1 \leq 4$ and $x_1 \geq 5$ imposed, respectively, on the nodes 4*A* and 4*B*. Upon solving the related LP problems corresponding to these nodes, we find that there is no feasible solution at Node 4*B*, and that Node 4*A* yields a feasible integer solution, $z = 7$, $x_1 = 4$, $x_2 = 3$. Hence, no further branching is necessary from either of these nodes, as they are both terminal nodes. We now must search the tree to see if there are any other nodes which might lead to a better integer solution than the one just obtained. The only possible such node is Node 3*B*, but for this node, $z = 6\frac{5}{8}$. This value is an upper bound on any possible feasible integer solutions for the constraints imposed at this node, and since we have already found a better integer solution, we now know that it must be the optimal integer solution. Hence, the algorithm terminates.

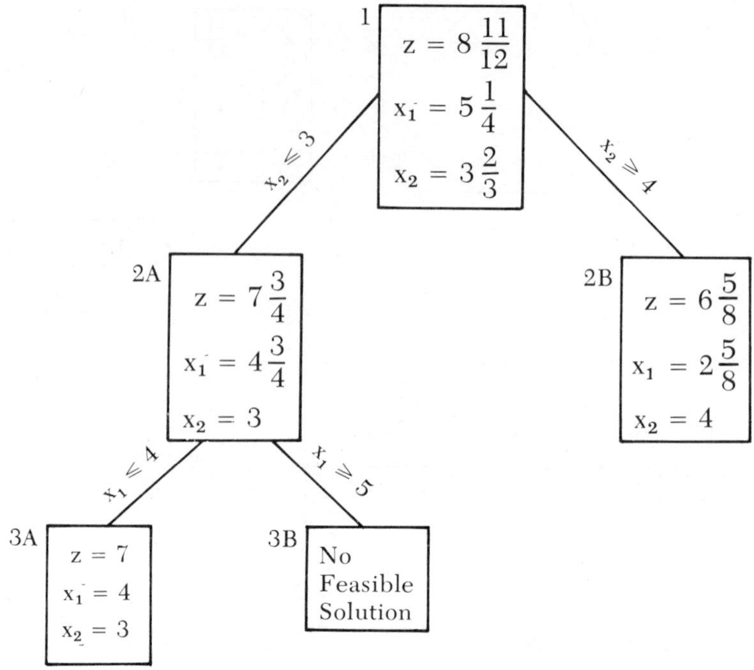

Figure 16–5 Tree of solutions for Example 2 (second procedure).

In the second step of the above procedure, we arbitrarily chose to branch on x_1 (i.e., the two branch constraints were $x_1 \leq 5$ and $x_1 \geq 6$). If, instead, we had branched on x_2, we would have generated the tree of solutions given in Figure 16–5. The nodes are again labeled in the order generated.

In Land and Doig's original algorithm, the constraint imposed along a branch required a variable to be equal to some fixed integer, and the number of branches emanating from any given node could be equal to the total number of possible integer values that particular variable could assume. This feature made the algorithm difficult to implement on a computer. In the version presented here, each nonterminal node has exactly two branches emanating from it, making computer implementation far easier.

In Example 2, we made one other somewhat arbitrary choice in the order in which the nodes were generated; this occurred when we chose to branch from Node 3*A* rather than Node 3*B*, because the objective function is larger at Node 3*A* than at Node 3*B*. This procedure is typical of a branch and bound method: At each stage, the most promising node is chosen to proceed from next, the idea being that the sooner a good (or, more important, the optimal) integer solution is found, the fewer the number of nodes that will have to be subsequently examined; we will be able to eliminate many nodes whose upper bound is less than the currently best integer solution.

However, the above algorithm can also be implemented as a search method. In this case, we always search along, say, the leftmost path (or the

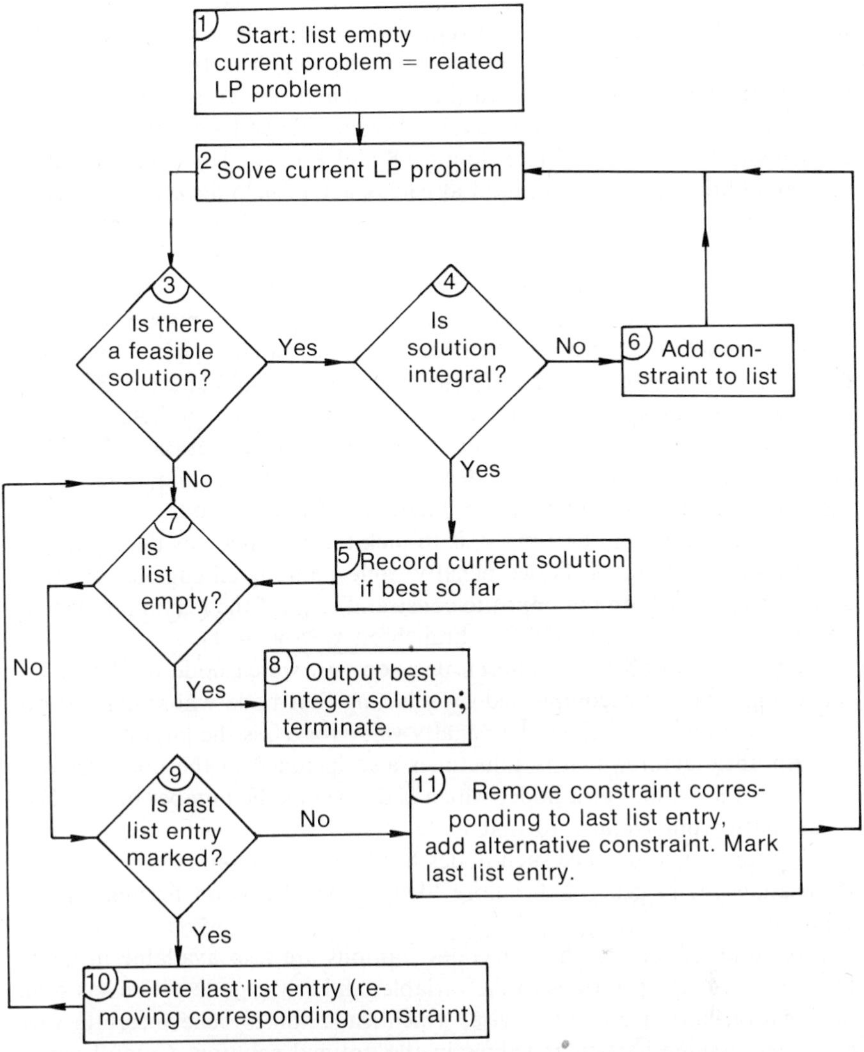

Figure 16–6 Flow diagram for Dakin's search method.

rightmost). In Example 2, the resulting tree of solutions would have coincided with that of Figure 16–4.

Dakin[6] provides a flow diagram of the algorithm as a search method, which is given in Figure 16–6. The "list" referred to in this diagram contains the additional constraints imposed by the branches; the "marking" is done to indicate whether we are currently on the first or second branch emanating from the particular node.

As we noted in Section 16–1, the search method will generally require considerably less storage space than the branch and bound method, because in the latter there may have been a large number of nonterminal nodes generated which have to be stored for possible future examination, whereas in the search method only the current node, current best solution, and the "list" need be stored. On the other hand, the search method may have to enumerate many more nodes than the branch and bound method.

In some of the commercially available computer codes which employ this branch and bound method (see Sommer[32] and Forrest, Hirst, and Tomlin[9]) several additional factors are used at each stage to help decide (a) which of the currently non-integer valued variables $x_j, j \in I$, to impose the additional constraints on; and (b) which node to choose for branching.

For (a), one approach is to compute a pair of "penalty" values for each variable. The penalty values for variable x_j are the amounts by which the objective function decreases when the two constraints $x_j \le w_j$ and $x_j \ge (w_j + 1)$ are imposed, respectively (see equations (16–37) and (16–38)). The actual penalty values computed may be only lower bounds on these amounts, and may be computed by performing one dual simplex iteration, by incorporating additional information derived from the integer requirements on the nonbasic variables, or by a combination of these methods.

A typical rule for choosing x_j is to pick the x_j which has the maximum penalty value (as one of its two penalty values) among all eligible variables, and then to branch on the opposite constraint (e.g., if the maximum penalty occurs for $x_j \le w_j$, we would first branch on $x_j \ge w_j + 1$).

Once the variable x_j is chosen decision (b), which node to choose for branching, may be accomplished by selecting that node whose current objective function value minus the penalty value for x_j, is the largest.

Another factor which is sometimes also included in the node decision process is the "rank" of a node, which is defined as the number of branches in the path from Node 1 to that node. Nodes with higher rank are given priority for selection, since such nodes have more constraints imposed upon them and will, in general, be more likely to yield a good feasible integer solution.

In some of these computer codes, options are also available in which the user can assign priorities to the variables and to the nodes, if he has some information, intuitive or otherwise, which leads him to believe certain variables will assume particular values in the optimal solution. Careful use of these features can greatly diminish the time it takes to solve the problem.

PROBLEMS

1. Solve by a cutting plane method:

$$\begin{aligned} \text{Maximize } z &= x_2 - x_1 \\ 2x_1 + x_2 &\le 9 \\ 2x_1 + 6x_2 &\le 21 \\ x_1, x_2 &\ge 0 \\ x_1, x_2 &\text{ integer} \end{aligned}$$

Illustrate graphically.

2. Solve the integer programming problem of Problem 1 by a branch and bound method.

3. Solve by a cutting plane method:

$$\text{Maximize } z = 32x_1 + 15x_2$$
$$2x_1 - 3x_2 \geq -9$$
$$16x_1 + 7x_2 \leq 52$$
$$x_1, x_2 \geq 0$$
$$x_1, x_2 \text{ integer}$$

Illustrate graphically.

4. Solve Problem 3 by a branch and bound method.

5. Solve by a cutting plane method:

$$\text{Maximize } z = x_1$$
$$x_1 + 2x_2 \leq 12$$
$$5x_1 + 2x_2 \leq 30$$
$$2x_1 - x_2 \leq 9$$
$$x_1, x_2 \geq 0$$
$$x_1, x_2 \text{ integer}$$

6. Solve Problem 5 by a branch and bound method.

7. Solve by a branch and bound method:

$$\text{Maximize } z = x_1 + x_2$$
$$x_1 + 2x_2 \leq 12$$
$$5x_1 + 2x_2 \leq 30$$
$$2x_1 - x_2 \leq 9$$
$$x_1, x_2 \geq 0$$
$$x_1, x_2 \text{ integer}$$

8. Solve the following mixed integer-continuous variable problem by a cutting plane method:

$$\text{Maximize } z = 5x_1 + 6x_2 + 9x_3$$
$$9x_1 + 10x_2 + 6x_3 \leq 50$$
$$6x_1 + 3x_2 + 19x_3 \leq 35$$
$$x_1, x_2, x_3 \geq 0$$
$$x_1, x_2 \text{ integer}$$

9. Solve problem 8 by a branch and bound method.

10. Consider the cutting plane constraint

$$\sum_{j\in N} x_j \geq 1 \tag{16–39}$$

for solving an all integer programming problem.

(a) Is (16–39) a valid cutting plane constraint, in the sense discussed in Section 16–2?

(b) Is (16–39) a "better" cutting plane constraint than those defined by (16–15) or by (16–35) and (16–36), in the sense discussed in Section 16–3?

REFERENCES

1. Balas, Egon: An additive algorithm for solving linear programs with zero-one variables. *Oper. Res.*, 13(4):517–526, 1965.
2. Balas, Egon: Discrete programming by the filter method. *Oper. Res.*, 15(5):915–957, 1967.
3. Benders, J. F.: Partitioning procedures for solving mixed-variables programming problems. *Numer. Math.*, 4:238–252, 1962.
4. Cook, Robert A.: An algorithm for integer linear programming. Doctoral Dissertation, Washington University, St. Louis, 1967.
5. Cooper, Leon: Hyperplane search algorithms for the solution of integer programming problems. *IEEE Transactions on Systems, Man & Cybernetics*, 3(3):234–240, 1973.
6. Dakin, R. J.: A tree-search algorithm for mixed integer programming problems. *Computer J.*, 9:250–255, 1966.
7. Driebeek, N. J.: An algorithm for the solution of mixed integer programming problems. *Manage. Sci.*, 12(7):576–587, 1966.
8. Echols, Robert E. and Cooper, Leon: Solution of integer linear programming problems by direct search. *J.A.C.M.*, 15 (No. 1):75–84, 1968.
9. Forrest, J. J. H., Hirst, J. P. H., and Tomlin, J. A.: Practical solution of large and complex integer programming problems with UMPIRE. Manage. Sci. 20(5) 736–773, 1974.
10. Garfinkel, Robert S., and Nemhauser, George L.: *Integer Programming*. New York, John Wiley and Sons, 1972.
11. Geoffrion, A. M.: An improved implicit enumeration approach for integer programming. *Oper. Res.*, 17(3):437–454, 1969.
12. Glover, Fred: A new foundation for a simplified primal integer programming algorithm. *Oper. Res.*, 16(4):727–740, 1968.
13. Gomory, R. E.: Outline of an algorithm for integer solutions to linear programs. *Bull. Amer. Math. Soc.*, 64:275–278, 1958.
14. Gomory, R. E.: An algorithm for integer solutions to linear programs. In: R. Graves and P. Wolfe (eds.), *Recent Advances in Mathematical Programming*. New York, McGraw-Hill Book Co., 1963.
15. Gomory, R. E.: All integer programming algorithm. IBM Research Center, Research Report RC-189 (1960); also in *Industrial Scheduling*. J. F. Muth and G. L. Thompson, (eds.) Englewood Cliffs, N.J., Prentice-Hall, 1963.
16. Gomory, R. E.: An algorithm for the mixed integer problem. The RAND Corp. P-1885, June 1960.
17. Gomory, R. E.: On the relation between integer and non-integer solutions to linear programs. *Proc. Nat. Acad. Sci.*, U.S.A., 53:(2):260–265, 1965.
18. Gomory, R. E.: Integer faces of a polyhedron. *Proc. Nat. Acad. Sci.*, U.S.A., 57(1): 16–18, 1957.
19. Gomory, R. E.: Some polyhedra related to combinatorial problems. *J. Lin. Alg. and Its Appl.* 2(4), 1969.

20. Greenberg, Harold: *Integer Programming*. New York, Academic Press, 1972.
21. Hadley, G.: *Nonlinear and Dynamic Programming*. Reading, Mass., Addison-Wesley, 1964.
22. Hu, T. C.: *Integer Programming and Network Flows*. Reading, Mass., Addison-Wesley, 1969.
23. Jambekar, Anil B.: An approach to integer programming. Doctoral Dissertation, Washington University, St. Louis, 1972.
24. Jambekar, Anil B., and Steinberg, David I.: Computational experience with a new algorithm for 0–1 integer programming. 43rd National O.R.S.A. Meeting, Milwaukee, Wis., May 1973.
25. Krolak, Patrick D.: The bounded variable algorithm for solving integer programming problems. Doctoral Dissertation, Washington University, St. Louis, Mo. (1968).
26. Krolak, Patrick D.: Computational results of an integer programming algorithm. *Oper. Res.*, 17(4):743–749, 1969.
27. Land, A. H., and Doig, A.: An automatic method of solving discrete programming problems. *Econometrica*, 28(3):497–520, 1960.
28. Lemke, C., and Spielberg, K.: Direct search algorithms for zero-one and mixed integer programming. *Oper. Res.*, 15(5):892–914, 1967.
29. Salkin, H. M.: *Integer Programming*. Reading, Mass., Addison-Wesley, 1974.
30. Shapiro, J. F.: Dynamic programming algorithms for the integer programming problem, I: The integer programming problem viewed as a knapsack problem. *Oper. Res.*, 16(1):103–121, 1968.
31. Shapiro, J. F.: Group theoretic algorithms for the integer programming problem. II: Extension to a general algorithm. *Oper. Res.*, 16(5):928–947, 1968.
32. Sommer, David C.: Computational experience with the OPHELIE mixed integer code. Control Data Corp., 1972.
33. Steinberg, David I. and Jambekar, Anil B.: An implicit enumeration algorithm for the general integer programming problems. 43rd National O.R.S.A. Meeting, Milwaukee, Wis., May 1973.
34. Young, R.: A simplified primal (all-integer) integer programming algorithm. *Oper. Res.*, 16(4):750–782, 1968.

Index